Solomon Deutsch

**Medical German**

A Manual Designed to Pid physicians in their Intercourse with German Patients

Solomon Deutsch

**Medical German**

*A Manual Designed to Pid physicians in their Intercourse with German Patients*

ISBN/EAN: 9783744606554

Printed in Europe, USA, Canada, Australia, Japan

Cover: Foto ©berggeist007 / pixelio.de

More available books at **www.hansebooks.com**

# MEDICAL GERMAN

## A MANUAL

Designed to Aid Physicians in their Intercourse with German Patients and in Reading Medical Works and Publications in the German Language

BY

## SOLOMON DEUTSCH, A. M., Ph. D.

*Author of Letters for Self-Instruction in the German Language, Practical Hebrew Grammar, etc.*

NEW YORK
1884

PRESS OF ISAAC FRIEDENWALD,
BALTIMORE, MD.

# PREFACE.

In the following pages, which have been prepared at the instance of a number of prominent New York and Brooklyn Physicians, the author has aimed not merely to accomplish the general object in view: to prepare a Manual which would be a welcome and reliable aid to the American Physician in his intercourse with German patients, and in reading German medical works and publications; but also to so arrange it that it might prove adequate to the needs of those who possess a merely rudimentary knowledge of German, as well as those who are more advanced in their German studies.

To this end the following general plan, elaborated by numerous minor, yet important features, has been adopted, and, the author trusts, carried out in a manner that will be found of practical use in accomplishing the purpose in view.

The Text is separated into two marked divisions:

THE FIRST, with its numerous subdivisions, gives an extensive and diversified collection of words and phrases that pertain directly or indirectly to the science and practice of medicine, such as:

*Parts of the Body* (limbs, bones, muscles, nerves, organs, senses, etc.);

*Diseases, their Names and various Symptoms;*

*Medical, Surgical and Hygienic Appliances* (Means of Transportation; Bandages; Surgical Instruments; Baths; Medical Officials

and Establishments: Articles of Food; Remedies and Expedients; Articles of Clothing and Furniture).

And for the benefit of those who are beginners, or not far advanced in the study of German, these words and terms are accompanied by brief grammatical annotations or hints, whenever deemed necessary. Thus the nouns are generally followed by *m.*, *f.*, or *n.*, to indicate their gender, and by the endings of the genitive singular and nominative plural. *Bold-faced type* is used whenever irregularities in nouns or verbs demand special attention. *Compound words*, with the exception of those that are self-explanatory, are preceded by their single components. *Phrases and Sentences* are prefixed by the words contained in them not before introduced. *Families of words* are often grouped together to show their derivation and natural connection, and thus by developing their meaning to aid the memory of the student.

THE SECOND DIVISION consists of *Conversations* supposed to be held at the sick-bed, and with patients in general, which have been so arranged as to cover as far as possible the *Questions* and *Answers* that are likely to occur when the diseases under which they are grouped are under consideration. And here the author begs leave to say that nothing, of course, was more remote from his thoughts than to suggest methods of treatment, and that his sole purpose was to offer a wide range of words and phrases *in this section*, as well as in that *under the head of Symptoms*, that might be applicable to the particular case. And in order that every German patient may readily understand the expressions used, the author has aimed to make them simple and direct. In the English translation, however, intended exclusively for the Physician, the more familiar scientific terms have generally been employed.

*The German and English Indexes*, which contain about 14,000 words, will be found of great value and very convenient for reference,

and will practically place at the disposal of the Physician almost every term needed by him to convey or understand the meaning of an expression.

In arranging and preparing the work the author availed himself of several German medical works of acknowledged merit, and of a few English authorities on Diagnosis and Treatment of Diseases. He also thought it expedient to consult a number of Physicians of high standing, who by their kind advice and suggestions did much to aid him in the completion of his work.

In this connection, he desires to acknowledge his indebtedness and express his sincere thanks to Dr. Frank W. Rockwell, of Brooklyn, who from the inception of the work manifested great interest in the undertaking, and not only aided him by encouragement and valuable advice, but also kindly assisted in revising the manuscript and correcting the proofs. He also takes pleasure in mentioning with gratitude Dr. R. E. Swinburn, the well-known oculist, through whose kindness he was enabled to furnish the Conversations on the Eye and Ear Diseases.

In conclusion, the author begs to express the earnest hope that this work, which is the result of much care and labor, may prove useful and acceptable to the Medical Profession.

# CONTENTS.

## PART I.
### Medical Terms Systematically Arranged.

#### I. THE HUMAN BODY.

| | PAGE. | | PAGE. |
|---|---|---|---|
| 1. General Expressions, | 3 | 8. The Organs of Digestion, | 26 |
| 2. The Bones, Cartilages and Ligaments, | 4 | 9. The Urinary Organs, | 28 |
| 3. The Muscles, | 14 | 10. The Sexual Organs and Propagation, | 29 |
| 4. The Vascular System, | 17 | 11. The five Senses, | 32 |
| 5. The Nervous System, | 20 | Phrases referring to all the Senses, | 36 |
| 6. The Skin, | 23 | | |
| 7. The Respiratory Organs, | 24 | | |

#### II. THE DISEASES.

##### A. Symptoms and Phenomena.

| | PAGE. | | PAGE. |
|---|---|---|---|
| 1. General Expressions, | 37 | 10. Defæcation, | 62 |
| 2. Pain and other Sensations, | 41 | 11. Excretion of the Urine, | 65 |
| 3. Conditions observable in the Skin, | 49 | 12. Attitude and General Condition of the Body, | 67 |
| 4. Arterial Pulsation, | 52 | 13. General Aspect of the Patient, | 70 |
| 5. Respiration, | 53 | 14. Derangement of General Sensation, | 76 |
| 6. Temperature, | 57 | | |
| 7. The Tongue, | 58 | 15. Alteration in the Mental Condition, | 77 |
| 8. Appetite and Thirst, | 59 | | |
| 9. Vomiting and Nausea, | 60 | | |

B. Names of Diseases, . . . . . . . . . . . 81

### III. MEDICAL, SURGICAL AND HYGIENIC APPLIANCES.

1. Means of Transportation, . . . . . 107
2. Bandages, Bands, . . 108
3. Surgical Instruments, . 109
4. Baths, . . . . . . 112
5. Medical Officials and Establishments, . . . 112
6. Articles of Food and Drink, . . . . . 114
7. Methods of Treatment, . . 118
8. Remedies and Expedients, . . . . . 120
9. Poisons, . . . . . . 126
10. Articles of Furniture and of Clothing, . . . 128

# PART II.

## Conversations.

1. VOCABULARY CONTAINING THE NEW WORDS OCCURRING IN THE CONVERSATIONS, . . . . . . . . . . . . . 135

## CONVERSATIONS.

1. At a Private Call, . . 158
2. " " " . . 160
3. " " " . . 163
4. " " " . . 164
5. At a Hospital, . . . 166
6. Headache, Vertigo, . . 170
7. Paralysis, . . . . . 173
8. Coma, . . . . . . 174
9. Neuralgia, . . . . . 174
10. Hemicrania, . . . . 176
11. Glossitis, . . . . . 176
12. Sore Throat, . . . . 177
13. Tonsilitis, . . . . . 178
14. Diseases of the Mouth, . 179
15. Diphtheria, . . . . 180
16. Laryngitis, . . . . . 181
17. Stricture of the Œsophagus, . . . . . 182
18. Cough, . . . . . . 183
19. Hæmoptysis, . . . . 183
20. Hæmatemesis, . . . . 184
21. Bronchitis, . . . . . 185
22. Asthma, . . . . . 186
23. Disease of the Chest, . 188
24. Phthisis, . . . . . 189
25. Pneumonia, . . . . 191
26. Peritonitis, . . . . . 193
27. Diseases of the Heart, . 195
28. Diseases of the Stomach, 196
29. Nausea, Vomiting, . . 197
30. Stomach-ache, . . . . 198
31. Acute Gastritis, . . . 199

## CONTENTS.

| | PAGE. | | PAGE. |
|---|---|---|---|
| 32. Chronic Gastritis, | 200 | 53. Rheumatism, | 227 |
| 33. Gastric Ulcer, | 201 | 54. Gout, | 228 |
| 34. Gastric Cancer, | 203 | 55. Fever, | 228 |
| 35. Colic, | 205 | 56. Typhoid Fever, | 230 |
| 36. Bilious Colic, | 206 | 57. Typhus, | 231 |
| 37. Enteritis, | 206 | 58. Scarlet Fever, | 234 |
| 38. Ileus Volvulus, | 208 | 59. Diseases of the Eye, | 236 |
| 39. Constipation, | 209 | 60. Ear Diseases, | 243 |
| 40. Diarrhœa, | 210 | 61. Menorrhagia, | 245 |
| 41. Dysentery, | 211 | 62. Amenorrhœa, | 246 |
| 42. Hemorrhoids, Piles, | 213 | 63. Leucorrhœa, | 248 |
| 43. The Fissure of the Anus, Prolapsus Recti, | 215 | 64. Prolapsus Uteri, | 249 |
| | | 65. Pregnancy, Childbed, | 250 |
| 44. Diseases of the Urinary Organs, | 216 | 66. Diseases of Children, | 252 |
| | | 67. Hydrocephalus, | 255 |
| 45. Hæmaturia, | 218 | 68. Thrush, | 255 |
| 46. Cystitis, | 219 | 69. False Croup, | 256 |
| 47. Nephritis, | 220 | 70. True Croup, | 257 |
| 48. Nephralgia, | 221 | 71. Measles, | 258 |
| 49. Acute Bright's Disease, | 221 | 72. Worms, | 259 |
| 50. Diabetes, | 224 | 73. Growth above the Palate, | 255 |
| 51. Syphilis, | 224 | 74. Toothache, | 260 |
| 52. Anæmia, | 226 | 75. At a Druggist, | 264 |

INDEX OF GERMAN WORDS AND TERMS, . . . . . . . . . 268
" ENGLISH " " " . . . . . . . . . 300

# PART I.

# MEDICAL TERMS SYSTEMATICALLY ARRANGED.

# MEDICAL TERMS SYSTEMATICALLY ARRANGED.

## I. THE HUMAN BODY.

## DER MENSCHLICHE KÖRPER.

### 1. *General Expressions.*

1 Mensch, *m.* (*plur.* -en), *man, human being.*
2 Geschlecht, *n.* (*plur.* -er), *species, race, sex.*
3 Menschengeschlecht, *n.*, *mankind, humankind.*
4 Race, *f.* (*also* Rasse), *race, breed.*
5 Menschenrace, *f.*, *race of men.*
6 Menschheit, *f.*, *humanity.*
7 menschlich, *a.*, *human.*
8 Körper, *m.*, *body.*
9 der menschliche Körper, *human body.*
10 bauen, *to build, construct.*
11 Bau, *m.*, *structure, construction.*
12 Körperbau, *m.*, *structure or frame of the body.*
13 schaffen, *to create, produce, make.*
14 beschaffen, *a.*, *made up, constituted.*
15 Beschaffenheit, *f.*, *natural condition, quality.*
16 Körperbeschaffenheit, *f.*, *the bodily constitution, diathesis.*
17 die Kraft (*plur.* **Kräfte**), *strength, vigor.*
18 Körperkraft, *f.*, *strength of body.*
19 körperlich, *a.* } *bodily, corporeal.*
20 leiblich, *a.*
21 Leib, *m.* (*plur.* **-er**), *body.*
22 Leibesbeschaffenheit, *f.*, *bodily constitution.*
23 Leibeskraft, *f.*, *bodily strength.*

24 gestalten, *to form, give shape or figure to.*
25 **die** Gestalt, *form, figure, frame.*
26 Leibesgestalt, *f., bodily form, figure.*
27 bilden, *to shape, form, make.*
28 Bildung, *f., formation, shape, structure.*
29 Körperbildung, *f., structure of body.*
30 legen, *to lay, put, place.*
31 anlegen, *to lay on, to found, establish.*
32 Anlage, *f., natural disposition, capacity, ability.*
33 Körperanlage, *f.* ⎫
34 **das** Temperament ⎬ *temperament.*
   (*plur.* -e), ⎭
35 formen, *to form, shape.*
36 **die** Form, *form.*
37 Figur, *f.* (ú), *figure.*
38 **gehen, ging, gegangen** (*with* **sein**), *to go.*
39 Gang, *m., gait.*
40 **wachsen, wuchs, gewachsen** (*w.* **sein**), *to grow.*
41 Wuchs, *m., growth, size.*
42 leben, *to live.*
43 Leben, *n., life.*
44 fähig, *a., able, fit for, capable of.*
45 lebensfähig, *a., vital.*
46 lebendig (bĕn'), *a., living, alive.*
47 Lebenskraft, *f., vital strength.*
48 dauern, *to last, endure, continue.*
49 **die** Dauer, *duration, lasting.*
50 Lebensdauer, *f., duration of life.*
51 warm, *a., warm.*
52 Wärme, *f., warmth.*
53 Lebenswärme, *f., vital warmth or heat.*
54 hauchen, *to breathe, exhale.*
55 Hauch, *m., breath.*
56 Lebenshauch, *m., breath of life.*
57 Geist, *m.* (*plur.* -**er**), *spirit, soul.*
58 Lebensgeist, *m., principle or spirit of life.*
59 **der** Funke, *spark.*
60 Lebensfunke, *m., spark of life.*

## 2. Die Knochen, Knorpel und Bänder.
### *The Bones, Cartilages, and Ligaments.*

61 Knochen, *m., bone* (related to *knuckle*).
62 Knöchlein or Knöchelchen, *n.* (*dim.*), *small bone, ossicle.*
63 Knöchel, *m., knuckle.*
64 verknöchern, *to ossify.*
65 knöchern, *a., made of bone, osseous.*
66 knochig, *a., bony.*
67 knochicht, *a., like bone.*

68 ähnlich, *a.*, *resembling, like, similar.*
69 knochenähnlich, *a.*, *osteoid, osseous.*
70 **die** Art, *kind, sort, peculiar nature.*
71 -artig (in comp'ds), *of the nature or kind of.*
72 knochenartig, *a.*, *bone-like, osteoid.*
73 Gebäude, *n.* (bauen, 10), *building, structure.*
74 Knochengebäude, *n.*, *bone structure, system of bones.*
75 rüsten, *to prepare, to fit out.*
76 Gerüste, *n.*, *scaffold, framework.*
77 Knochengerüste, *n.*, *skeleton.*
78 **das** System (tam'), *system.*
79 Knochensystem, *n.*, *system of bones.*
80 **die** Substanz (ă'), *substance.*
81 Knochensubstanz, *f.*, *bony substance.*
82 Masse, *f.*, *mass, bulk, quantity.*
83 Knochenmasse, *f.*, *bone or osseous substance.*
84 **weben, wob, gewoben,** *to weave.*
85 Gewebe, *n.*, *tissue, texture (weaving).*
86 Knochengewebe, *n.*, *bone tissue.*
87 Rippe, *f.*, *rib.*
88 Gerippe, *n.*, *skeleton.*
89 Knochengerippe, *bony framework.*
90 Knochenbildung, *f.*, *ossification, osteo-genesis.*
91 **das** Bein (*plur.* -e), *bone, leg.*
92 Gebeine (*pl.*), *bones.*
93 beinern, *a.*, *made of bone.*
beinicht, *a.*, *bony, hard as bone.*
94 beinähnlich, *a.* } *bone-like.*
95 beinartig, *a.*
96 Röhre, *f.*, *tube, pipe.*
97 Röhrenknochen, *m.*, *tubular, hollow or cylindrical bone.*
98 **binden, band, gebunden,** *to bind.*
99 verbinden, *to bind together, connect.*
100 Verbindung, *f.*, *connection.*
101 Knochenverbindung, *f.*, *articulation of the bones.*
102 fügen, *to fit together, join, connect.*
103 Fuge, *f.*, *joint, fitting together*
104 Knochenfuge, *f.*, *juncture or articulation of bones.*
105 **das** Band (*fr.* binden), *band, ligament, bandage, ligature.*
106 Bandfuge, *f.*, *ligamental juncture or suture.*
107 nähen, *to sew.*
108 die Naht (*pl.* **Nähte**), *commissure, suture.*
109 Knochennaht, *f.*, *suture (of the skull).*
110 **die** Haut (*pl.* **Häute**, *dim.* Häutchen), *skin, hide.*
111 Knochenhaut, *f.*, *periosteum.*
112 Pfanne, *f.*, *pan.*

113 Knochenpfanne, *f.*, *cotyle, glene.*
114 das Mark, *marrow.*
115 Knochenmark, *n.*, *bone-marrow.*
116 die Kapsel, *capsula.*
117 Knochenkapsel, *f.*, *bone capsula*
118 hohl, *a.*, *hollow.*
119 höhlen, } *to hollow out,*
120 aushöhlen. } *excavate.*
121 Höhle, *f.*, *hole, cavern, cavity.*
122 Knochenhöhle, *f.*, *bone cavity.*
123 Kern, *m.*, *kernel, central part, best part.*
124 Knochenkern, *m.*, *bone nucleus.*
125 Kopf, *m.*, *head.*
126 Knochenkopf, *m.*, *head of a bone.*
127 die Wand (*pl.* **Wände**), *wall.*
128 Knochenwand, *f.*, *osseous wall.*
129 Schale, *f.*, *shell.*
130 Knochenschale, *f.*, *osseous shell or envelope.*
131 Rand, *m.* (*pl.* **Ränder**), *edge, rim, margin.*
132 Knochenrand, *m.*, *edge, rim of the bone, acetabulum.*
133 Zacke, *f.*, *prong, serration.*
134 zackig, *a.*, *serrated.*
135 zackenförmig, *a.*, *jag-shaped or formed.*
136 Knochenrand mit Zacken, *the bone edge with serrations.*
137 vereinen, } *to make one,*
or } *unite,*
vereinigen, } *join together.*
138 Vereinigung, *f.*, *joining, union, combination.*
139 bewegen, *to move, agitate.*
140 beweglich, *a.*, *movable.*
141 unbeweglich, *immovable, motionless.*
142 die bewegliche Knochenvereinigung or -verbindung, *the movable articulation of the bones.*
143 die unbewegliche ——, *the immovable connection.*
144 Keil, *m.*, *wedge.*
145 einkeilen, *to wedge in, fasten with wedges.*
146 Einkeilung, *f.*, *incunneation, inclovation.*
147 Knorpel, *m.*, *cartilage.*
148 knorpelig, *a.*, *cartilaginous.*
149 knorpelicht, *a.* } *like*
150 knorpelartig, *a.* } *cartilage.*
151 knorpelähnlich, *a.* } *lage.*
152 überknorpeln, *to cover w. cartilage.*
153 Knorpelmasse, *f.*, *cartilage-substance.*
154 Knochenknorpel, *m.*, *osseous cartilage.*
155 wahr, *a.*, *true.*
156 falsch, *a.*, *false.*
157 wahrer Knorpel, *true cartilage.*
158 falscher Knorpel, *false cartilage.*

159 Knorpelband, *n.*, *fibro-cartilage*.
160 Knorpelhaut, *f.*, *cartilaginous membrane, perichondrium*.
161 zwischen, *between*.
162 Zwischenknorpel, *interarticular cartilage*.
163 das Polster, *cushion, pledget*.
164 Knorpelpolster, *n.*, *cartilaginous cushion*.
165 Knorpelring, *m.*, *annular cartilage*.
166 Knorpelfuge, *f.*, *synchondrosis*.
167 Gelenk, *n.*, *articulation, joint*.
168 Gelenkknorpel, *m.*, *articular cartilage*.
169 Gelenkband, *n.*, *ligament of joint*.
170 Gelenkring, *m.*, *joint-ring*.
171 Gelenkfügung, *f.*, *joint articulation, juncture*.
172 Gelenkhöhle, *f.*, *cotyle, cotyla*.
173 Fugengelenk, *n.*, *synarthrosis*.
174 **ziehen, zog, gezogen,** *to draw, pull*.
175 überziehen, *to draw over, cover, line*.
176 Überzug, *m.*, *crust, covering, lining, coating*.
177 Knorpelüberzug, *m.*, *cartilaginous covering or coating*.
178 flach, *a.*, *flat, plain, level*.
179 Fläche, *f.*, *flatness, plain surface, level*.
180 glatt, *a.*, *smooth, even*.
181 die glatte Fläche, *the smooth, even surface*.
182 eine überknorpelte (152) Fläche, *a surface covered with cartilage*.
183 Gelenkfläche, *f.*, *articular surface*.
184 enden, *to end*.
185 das Ende (*gen.* -s, *pl.* -n), *end, extremity*.
186 Gelenkende, *n.*, *articular extremity*.
187 die Kugel, *ball, globule, sphere*.
188 frei, *a.*, *free*.
189 das freie Gelenk, ⎫
   or                    ⎬ *arthrodia*.
190 Kugelgelenk,          ⎭
191 das Scharnier (French charnière), *hinge, corner, angle*.
192 Winkel, *m.*, *corner, angle, nook*.
193 Scharniergelenk, *n.* ⎫
    or                    ⎬ *gynglimus*.
194 Winkelgelenk, *n.*    ⎭
195 rollen, *to roll, revolve, rotate*.
196 drehen, *to turn, cause to revolve*.
197 Rollgelenk, *n.* ⎫ *trochoid ar-*
    or                ⎬ *ticulation,*
198 Drehgelenk, *n.* ⎭ *rotary joint*.
199 Gelenkkapsel, *f.*, *articular or synovial capsule*.
200 Kapselband, *n.*, *capsular ligament*.
201 die Faser, *fibre, filament*.

202 Stoff, *m.*, *substance, material, matter.*
203 Faserstoff, *m.*, *fibrin.*
204 faserig, *a.* } *fibrous, fibrinous.*
    fasericht, *a.* }
205 Faserband, *n.*, *fibrous ligament.*
206 **das** Bündel, *the bundle, fasciculus.*
207 Faserbündel, *n.*, *bundle of fibres.*
208 Sehne, *f.*, *sinew, tendon.*
209 Sehnenfaser, *f.*, *tendinous fibre.*
210 **fliessen, floss, geflossen,** *to flow, run.*
211 flüssig, *a.*, *fluid.*
212 Flüssigkeit, *f.*, *fluid, liquid, humor.*
213 schmieren, *to smear, grease, oil.*
214 Schmiere, *f.*, *grease, ointment.*
215 Gelenkschmiere, *f.*, *synovial fluid, axungia.*
216 Synovia, *synovia.*
217 Gelenkschmier- } *synovial*
    kapsel, or } *membrane,*
218 Gelenkkapsel, *f.* } *articular capsule.*
219 straff, *a.*, *tight, rigid.*
220 das straffe Gelenk, *the tight or rigid joint.*
221 vorder, *a.*, *anterior, forward, fore-, front-.*
222 hinter, *a.*, *hinder, back, hind-, back-.*
223 Vorderkopf (125), *m.*, *forepart of the head.*
224 Hinterkopf, *m.*, *back, or hindpart of the head.*
225 Wirbel, *m.* } *the crown, top*
    or } *of the head,*
226 Scheitel, *m.* } *vertex, parietal bone.*
227 **das** Haupt, *head,* (fig. *chief*), *principal, main.*
228 Hinterhaupt, *hind part of the head, occiput.*
229 Hinterhauptbein, *n.*, *occipital bone.*
230 **die** Stirn, *the front, brow, forehead.*
231 Stirnbein, *n.*, *frontal bone.*
232 Keilbein, *n.* (144), *sphenoid bone.*
233 Schläfe, *f.*, *the temple.*
234 Schläfenbein, *n.*, *temporal bone.*
235 Felsen, *m.* (or Fels, *gen.* -ens), *rock.*
236 Felsenbein, *n.*, *rocky or petrous bone.*
237 **das** Sieb (*pl.* -e), *sieve.*
238 Siebbein, *n.* } *ethmoid bone,*
    or } *cribri-form.*
239 Riechbein, *n.* }
240 **riechen, roch, gerochen,** *to smell.*
241 Gesicht, *n.* (*pl.* -er), *face, visage; sight, sense of seeing.*
242 Gesichtsknochen, *pl.*, *facial bones.*
243 Farbe, *f.*, *color.*
244 Gesichtsfarbe, *f.* } *complexion*
245 Hautfarbe, *f.* } *or skin-color.*

246 halb, *a.*, *half.*
247 Hälfte, *f.*, *half, moiety.*
248 Gesichtshälfte, *f.*, *half of countenance.*
249 Zug, *m.* (174), *line, lineament, feature.*
250 Gesichtszüge, *lineaments, features.*
251 Wange, *f.*, *cheek.*
252 Wangenbein, *n.*, *malar bone, zygoma.*
253 Kiefer, *m.*, *the mandible, jaw, maxilla.*
254 Kiefergelenk, *n.*, *maxillary joint.*
255 Kieferknochen, *m.*, *maxillary bone.*
256 Oberkiefer, *m.*, *upper jaw, superior maxilla.*
257 Oberkieferknochen (or -beine), *pl.*, *superior maxillary bones.*
258 Unterkiefer, *m.* or
259 Unterkieferbein, } *the lower jaw, inferior maxilla.*
260 das Kinn, *chin.*
261 Kinnbacken, *m.*, *jaw, maxilla*
262 Kinnlade, *f.*, *maxilla, gnathus.*
263 Kinnbackenbein, *n.* or -knochen, *m.* } *maxillary bone.*
264 Gaumen, *m.*, *the palate, the roof of the mouth.*
265 Gaumenbeine, *pl.*, *palatal bones.*
266 das Joch (*pl.* -e), *yoke; yoke-like structure, zygoma.*
267 Backe, *f.*, or
Backen, *m.* } *cheek.*
268 Jochbeine, or
269 Backenbeine, } *zygomatic bones, cheek-bones, malar-bones.*
270 Thräne, *f.*, *tear.*
271 Thränenbeine, *lachrymal bones.*
272 Nase, *f.*, *nose.*
273 Nasenbein, *n.*, *nasal bone.*
274 Spitze, *f.*, *point, peak* (Engl. spit = *a pointed wood*).
275 Nasenspitze, *f.*, *the tip of the nose.*
276 das Dach, *roof.*
277 Nasendach, *n.*, *nasal roof.*
278 das Loch, *hole, foramen.*
279 Nasenloch, *nostril.*
Nasenlöcher, *pl.*, *nares.*
280 die Muschel, *shell, concha.*
281 Nasenmuschel, *f.*
282 Nasenmuschelbein, *n.* } *concha inferior, inferior turbinated bone.*
283 **scheiden, schied, geschieden,** *to separate, divide.*
284 **die** Scheidewand (127), *partition wall, septum.*
285 Nasenscheidewand, *f.*, *septum narium.*
286 **scheren, schor, geschoren,** *to shear.*
287 **die** Shar, *share of a plow.*
288 Pflugschar, *f.*, *ploweshare.*
289 Pflugscharbein, *m.*, *vomer, bridge of the nose.*

290 Nasenknorpel, *m.*, *nasal cartilage.*
291 **beissen, biss, gebissen,** *to bite.*
292 Gebiss, *n.*, *a set of teeth.*
293 ein falsches Gebiss, *a set of false teeth.*
294 Zahn, *m.*, *tooth.*
295 Backenzahn, *m.* (267), *jaw tooth, molar tooth.*
296 **das** Auge (*pl.* **-en**), *eye.*
297 Augenzahn, *canine, corner, eye tooth.*
298 **schneiden, schnitt, geschnitten,** *to cut.*
299 Schneidezahn, *the incisive tooth, incisor.*
300 **die** Milch, *milk.*
301 Milchzahn, *m.*, *milk-tooth, deciduous tooth.*
302 weise, *a.*, *wise.*
303 Weisheit, *f.*, *wisdom.*
304 Weisheitszahn, *m.*, *the wisdom tooth.*
305 Vorderzahn, *m.* (221), *the front tooth, incisor.*
306 Hinterzahn, *m.* (222), *the back tooth, molar.*
307 Zahnhöhle, *f.* (wherein the tooth is inserted), *the socket of a tooth.*
308 Krone, *f.*, *crown.*
309 Zahnkrone, *crown of the tooth.*
310 Hals, *m.*, *the neck, throat.*
311 Zahnhals, *m.* (part between the root and the crown), *the neck.*
312 **die** Wurzel, *root.*
Zahnwurzel, *f.*, *tooth-root.*
313 Lücke, *f.*, *gap, chasm.*
314 Zahnlücke, *f.*, *the tooth-gap.*
315 Schmelz, *m.*, *enamel.*
316 Glasur, *f.*, *glazing, varnish.*
317 Zahnschmelz, *m.* ⎫
or            ⎬ *enamel of the teeth.*
318 Zahnglasur, *f.* ⎭
319 **das** Fleisch, *flesh, meat.*
320 Zahnfleisch, *n.*, *gum, gums.*
321 zahnen, *to cut one's teeth.*
322 **bekommen, bekam, bekommen,** *to receive.*
323 das Kind bekommt Zähne. *the child is cutting its teeth.*
324 **brechen, brach, gebrochen,** *to break.*
durchbrechen, *to break through.*
325 Zahndurchbruch, *m.*, *eruption of a tooth.*
326 Zahnen, *n.*, *dentition.*
327 Zahnung, *f.*, *teething, cutting of the teeth.*
328 Zunge, *f.*, *tongue.*
329 Zungenband, *n.*, *ligament of the tongue, frænulum.*
330 Zungenbein, *hyoid bone.*
331 Zungenwurzel, *f.*, *the base of the tongue.*
332 Zungenspitze, *f.* (274), *the end of the tongue.*
333 Zungenrücken, *m.*, *back of the tongue.*
334 Zungenfleisch, *n.*, *lingual parenchyma.*

335 Mund, *m.* (*pl.* Munde), mouth.
336 Mundhöhle, *f.*, *oral cavity.*
337 **graben, grub, gegraben,** *to dig.*
338 Grube, *f.*, *pit, ditch.*
339 Schläfengrube, *temporal fossa*
340 Rumpf, *m.*, *trunk, abdomen.*
341 Wirbel, *m.* (225), *vertebra.*
342 Wirbelgelenk, *n.*, *vertebral articulation.*
343 Säule, *f.*, *the column.*
344 Wirbelsäule, *f.*, *spinal or vertebral column.*
345 Rücken, *m.*, *back, dorsum.*
346 Rückenbein, *n.*, *spine, vertebra.*
347 Grat, *m.* (*pl.* Grate), *edge, ridge* (from which die Gräte, *fish-bone*).
348 Rückgrat, *m.*, *backbone, vertebral column, spine.*
349 Rückenwirbel, *m.* ⎫
350 Rückgratswirbel, *m.* ⎬ *vertebra, dorsal vertebra.*
351 Rückgratsgelenk, *n.*, *vertebral joint, turning joint.*
352 Rückenmark, *n.*, *spinal marrow.*
353 Kanal, *m.* (a'hl), *canal.* Rückenmarkskanal, *m.*, *canal of the spinal marrow.*
354 Halswirbel, *m.*, *vertebrae of the neck, cervical vertebrae.*
355 Halsbein, *n.*, *collar bone, clavicle.*
356 Halsgelenk, *n.*, *cervical articulation.*
357 Halshöhle, *f.*, *cervical cavity.*
358 Nacken, *m.* ⎫ *nape, the nape*
359 Genick, *n.* ⎭ *of the neck.*
360 **tragen, trug, getragen,** *to bear, carry.*
361 Atlas, *m.*, *the atlas.*
or
362 Träger, *m.*, *the bearer, carrier.*
363 umdrehen, *to turn about, twirl;* sich ——, *to rotate.*
364 Umdreher, *m.*, *rotator, trochanter.*
365 **die** Brust, (*pl.* **Brüste**), *breast, thorax, chest.*
366 **das** Blatt, *leaf.*
367 Brustbein, *n.* ⎫ *breast*
368 Brustblatt, *n.* ⎬ *bone,*
369 Brustknochen, *m.* ⎭ *sternum.*
370 Brustknorpel, *m.*, *costal cartilage.*
371 Brustwirbel, *m.*, *thoracic vertebra.*
372 Brusthöhle, *f.*, *thoracic cavity.*
373 Kasten, *m.*, *chest.*
374 Brustkasten, *m.*, *chest, thorax.*
375 männlich, *a.*, *male.*
376 weiblich, *a.*, *female.*
377 die weibliche Brust, *the teat.*
378 Bauch, *m.*, *abdomen, belly.*
379 Lende, *f.*, *loin, rein.*
380 Bauchwirbel, *m.*, *abdominal vertebra.*
381 Lendenwirbel, *m.*, *lumbar vertebra.*
382 Hüfte, *f.*, *hip, haunch, ischium.*

383 Hüftbein, *n.* } *ischium,*
384 Hüftknochen, *m.* } *os coxæ.*
385 weich, *a.*, *soft, tender.*
386 Weichen (*pl.*), *groin.*
387 Weichenband, *n.*, *Fallopian ligament.*
388 **die** Gegend, *region.*
389 Weichengegend, *f.*, *inguinal region.*
390 die wahren Rippen (87), *true ribs.*
391 die falschen Rippen, *false ribs*
392 **das** Glied, *limb.*
393 Gliedmassen (*pl.*), *the members, limbs.*
394 ober, *a.*, *over, above, upper, situated above.*
395 unter, *a.*, *below, under; lower, under.*
396 die oberen or vorderen Gliedmassen, *the upper or anterior (fore) members.*
397 die unteren or hintern ——, *the lower or hinder ——.*
398 Extremitäten (*pl.*), *the extremities.*
399 **die** Schulter, } *shoulder.*
400 **die** Achsel, }
401 Achselbein, *n.* } *scapula, shoulder-blade.*
402 Schulterblatt, *n.* }
403 Schulterbein, *n.* } *humerus.*
404 Schulterknochen, *m.* }
405 **schliessen, schloss, geschlossen,** *to close, shut, lock up.*
406 Schlüssel, *m.*, *key.*

407 Schlüsselbein, *n.*, *collar bone.*
408 Achselhöhle, *f.* } *arm-pit,*
409 Achselgrube, *f.* } *axilla.*
410 Arm, *m.*, *arm.*
411 Oberarm, *m.*, *upper part of the arm.*
412 Oberarmkopf, *m.*, *head of humerus.*
413 Knorren, *m.*, *knotty or bony excrescence, protuberance; knuckle.*
414 Gelenkknorren, *m.*, *articular protuberance.*
415 Gelenkrolle, *f.*, *articular roller, articular trochlea.*
416 Unterarm, *m.* } *the forearm.*
    Vorderarm, *m.* }
417 Elle, *f.*, *ell, cubit.*
418 Speiche, *f.*, *spoke.*
419 **biegen, bog, gebogen,** *to bow, bend.*
420 Bogen, *m.*, *bow.*
421 Ellenbogen, *m.*, *the elbow.*
422 Höcker, *m.*, *hump, knob, bunch.*
423 Ellbogenhöcker, *m.*, *olecranon.*
424 Ellbogengelenk, *n.*, *bend of the elbow.*
425 **die** Hand (*pl.* **Hände**), *hand.*
426 recht, *a.*, *right.*
427 link., *a.*, *left.*
428 die rechte, linke Hand, *the right, left hand.*
429 Handrücken, *m.*, *the back of the hand.*

430 Hohlhand, *f.* ⎫ *the hollow*
or                ⎬ *or palm*
431 Handfläche, *f.* ⎭ *of the hand.*
432 Handgelenk, *n.* ⎫ *the wrist.*
433 Handwurzel, *f.* ⎭
434 mittel, *a., middle.*
435 Mittelhand, *f., the metacarpus, palm.*
436 Mittelhandknochen, *m., metacarpal bone.*
437 Finger, *m., finger.*
438 Fingerband, *n., digital ligament.*
439 Fingergelenk, *n., knuckle.*
440 Fingerglied, *n.* (392), *finger-articulation.*
441 Fingerbein, *n., phalanx.*
442 Daumen, *m., the thumb.*
443 zeigen, *to show, indicate.*
444 Zeigefinger, *m., fore-finger, index.*
445 Mittelfinger, *m., middle finger.*
446 Ringfinger, *m., ring finger.*
447 der kleine Finger, ⎫ *little finger or ear finger.*
or                    ⎬
448 Ohrfinger, *m.* ⎭
449 **das** Becken, *the pelvis, basin.*
450 Beckenknochen, *m., pelvic bone.*
451 Fuss, *m., foot.*
452 Fussknochen, *m., bone of the foot.*
453 Hüftblatt, *n.* (382), *ischium, os coxae.*
454 Darm, *m., gut, intestine.*
455 Darmbein, *n., ilium.*

456 **sitzen, sass, gesessen,** *to sit.*
457 Sitzbein, *n., ischium.*
458 Gefäss, *n., the breech, buttock.*
459 sich schämen, *to be ashamed.*
460 **die** Scham, *shame; privy parts, pudenda.*
461 Schambein, *n., os pubis.*
462 Schambeinfuge, *f., pubic symphysis.*
463 Gelenkpfanne, *f.* (112). *articular cavity, acetabulum.*
464 **das** Kreuz (*pl.* -e). *cross; loins, reins, lumbi.*
465 heilig, *a., holy, sacred.*
466 Kreuzbein, *n.* ⎫ *the sacrum,*
467 heiliges Bein, ⎬ *os sacrum,*
468 Heilbein, *n.* ⎭ *chine bone.*
469 Kreuzwirbel, *m., sacral vertebra.*
470 Schwanz, *m., tail, cauda.*
471 Schwanzbein, *n., coccyx.*
472 Steiss, *m.,* ⎫
473 **der** Hintere (222), ⎬
    *buttocks, rump, coccygeal region.*
474 Kuckuk, *m., cuckoo.*
475 Steissbein, *n.* ⎫ *coccyx.*
476 Kuckuksbein, *n.* ⎭
477 Steissbeinwirbel, *m.* ⎫
    Steisswirbel, *m.* ⎭
    *coccygeal vertebra.*
478 Schenkel, *m., the thigh, femur crus.*
479 Schenkelbein, *n., thigh-bone, femur.*

480 Schenkelgelenk, *n.*, *hip-joint.*
481 Schenkelband, *n.*, *crural ligament.*
482 Schenkelbogen, *m.*, *arcus cruralis.*
483 Oberschenkel, *m.*, *upper part of the thigh.*
484 dick, *a.*, *thick.*
485 Dickbein, *n.*, *thigh.*
486 Oberschenkelbein, *n.*, *femur.*
487 Gelenkknopf, *m.* ⎫ *condyle.*
 Gelenkkopf, *m.* ⎭
488 Hügel, *m.*, *hill ; knob, projection.*
489 der grosse Rollhügel (1316), *the great trochanter.*
490 der kleine Rollhügel, *the lesser trochanter.*
491 **das** Knie (*pl.* -e), *knee.*
492 Scheibe, *disc, discus.*
493 Kniescheibe, *f.*, *the knee-pan.*
494 Kniekehle, *f.*, *the hough or ham.*
495 Schiene, *f.*, *shin, tibia.*
496 Schienbein, *n.*, *shin-bone, tibia.*

497 Unterschenkel, *m.*, *lower part of the thigh, leg.*
498 Spann, *m.*, *the instep.*
499 Wade, *f.*, *the calf of the leg.*
500 Wadenbein, *n.*, *the perone, fibula.*
501 Vorderfuss, *m.*, *fore-foot.*
502 Fusswurzel, *f.*, *tarsus.*
503 **springen, sprang, gesprungen,** *to spring.*
504 Sprungbein, *n.*, *astragalus.*
505 Hacke, *f.* ⎫ *the heel.*
506 Ferse, *f.* ⎭
507 Fersen- or Hackenbein, *n.*, *heel-bone.*
508 Knöchel (63), *m.*, *the ankle, malleolus.*
509 Mittelfuss, *m.*, *the metatarsus*
510 Zehe, *f.*, *the toe.*
511 die grosse Zehe, *big toe, great toe.*
512 die kleine Zehe, *little toe.*
513 Sohle or Fusssohle, *f.*, *the sole (of the foot).*
514 Fussrücken, *m.*, *the upper part of the foot.*

## 3. Die Muskeln, *the Muscles.*

515 Muskel, *m.*, *muscle.*
516 muskelig, *a.* ⎫ *muscular.*
517 musculös, *a.* ⎭
518 Muskelgewebe, *n.*, *muscular tissue.*
519 Muskelfaser, *f.*, *muscular fibre.*
520 Muskelhaut, *f.*, *muscular membrane.*

521 Muskelband, *n.*, *muscular band or membrane.*
522 Muskelbinde, *f.*, *muscular ligament.*
523 Bewegung, *f.* (139), *motion, movement.*
524 Muskelbewegung, *f.*, *muscular motion.*

525 schichten, *to dispose in layers or strata.*
526 die Schicht, *layer, stratum.*
527 Muskelschicht, *f., muscular layer, stratum.*
528 hüllen, *to envelop, wrap up.*
529 Hülle, *f., envelope, covering, sheath.*
530 Muskelhülle, *muscular envelope, tunic, sheath.*
531 Muskelkraft, *f.* (17), *muscular strength or power.*
532 stark, *a., strong.*
533 muskelstark, *a., muscular, having strong muscles.*
534 Muskelstärke, *f., muscular strength.*
535 Muskulatur, *f.* (tū'r), *the muscular structure.*
536 fleischig (319), *a., fleshy.*
537 fleischlos, *a., fleshless.*
538 Fleischmasse, *f., fleshy substance.*
539 Fleischfaser, *f., sarcous fibre.*
540 Bindegewebe (98), *n., connective tissue.*
541 Flechse, *f., sinew, tendon.*
542 Sehnenhäute (110), *f., tendinous membranes.*
543 Schleim, *m., slime, mucus, phlegm.*
544 Beutel, *m., sac, follicle.*
545 Schleimbeutel, *m., mucous follicle.*
546 Scheide, *f., sheath, vagina.*
547 Schleimscheide, *f., bursa mucosa.*
548 Faserknorpel, *m., fibro-cartilage.*
549 Sesambein, *n.* (or beinchen), *sesamoid bone.*
550 Muskelfaserstoff, *m.* ⎱ *muscu-*
   or ⎬ *lar*
551 Syntoninstoff, ⎰ *fibrin.*
552 willkührlich, *a., voluntary.*
553 unwillkührlich, *a., involuntary.*
554 die willkührlichen Muskeln, *the voluntary muscles.*
555 die unwillkührlichen Muskeln, *the involuntary muscles.*
556 Ringmuskeln, *circular muscles.*
557 Schliessmuskeln (405), *sphincters.*
558 Hohlmuskeln (118), *hollow muscles.*
559 beugen, *to cause to bend, bend, curve.*
560 strecken, *to stretch.*
561 abziehen (174), *to draw off or away.*
562 anziehen, *to draw on (towards one's self).*
563 Beuger, *m., flexor muscle.*
564 Strecker, *m., extensor muscle.*
565 Abzieher, *m., abductor muscle.*
566 Anzieher, *m., adductor muscle.*
567 wirken, *to act, operate, work.*
568 zusammenwirken, *to cooperate, act together.*
569 gegen, *against* (in compds.: *counter.*)

570 gegenwirken, *to counteract, react.*
571 einwärts, *inward, inwards.*
572 auswärts, *outwards.*
573 Einwärtsdreher, m. ⎫
574 Einwärtswender, m. ⎬ *pronator.*
575 Einwärtsroller, m. ⎭
576 Auswärtsdreher, m. ⎫
577 Auswärtswender, m. ⎬ *supinator.*
578 Auswärtsroller, m. ⎭
579 zusammenwirkende Muskeln, *cooperative muscles.*
580 gegenwirkende Muskeln, *counteracting muscles.*
581 zusammenziehende (174) Muskeln, *constrictors.*
582 Genossen (*pl.*), *associates, companions.*
583 Gegner, or ⎱ *opponents,*
  Antagonisten, ⎰ *antagonists.*
584 sehnig, a. (208), ⎫ *sinewy,*
  or ⎬ *tendinous,*
  sehnicht, ⎭ *nervous.*
585 sehnige Muskelbinden, *sinewy muscular ligaments.*
586 Schädel, *skull, pan, cranium.*
587 Schädelmuskeln, *cranial muscles.*
588 Gesichtsmuskeln (241), *facial muscles.*
589 Augenmuskeln (296), *muscles of the eye.*
590 Nasenmuskeln (272), *nasal muscles.*
591 Backenmuskeln (267), *buccinator muscles.*
592 Mundmuskeln (335), *muscles of the mouth.*
593 kauen, *to chew, masticate.*
594 Kaumuskeln, *masseters.*
595 Rumpfmuskeln (34), *trunk muscles.*
596 Halsmuskeln, *cervical muscles.*
597 nicken, *to nod.*
598 Kopfnicker, m., *sterno cleido mastoideus, head-nodder.*
599 Nackenmuskeln (358), *cervical muscles.*
600 Brustmuskeln, *thoracic muscles.*
601 **das** Fell (*pl. -e*), *skin, tunic, hide.*
602 zwerch (in compds.), *athwart, crosswise.*
603 Zwerchfell, n. ⎱ *diaphragm.*
604 Diaphragma, n. ⎰
605 Rückenmuskeln, *dorsal muscles.*
606 Bauchmuskeln, *abdominal muscles.*
607 Bauchmuskelwand, f., *muscles of the abdominal walls.*
608 pressen, *to press, squeeze.*
609 Bauchpresse, *abdominal pressure, nixus, straining.*
610 Beckenmuskeln (449), *pelvic muscles.*
611 Gefässmuskeln (622), *vascular muscles.*
612 Schulterblattmuskeln, *scapular muscles.*
613 Deltamuskel, *deltoid muscle.*

614 Oberarmmuskeln, *brachial muscles.*
615 Vorderarmmuskeln, *forearm muscles.*
616 Handmuskeln, *muscles of the hand.*
617 Oberschenkelmuskeln, *femoral muscles.*
618 Unterschenkelmuskeln, *muscles of the leg.*
619 Wadenbeinmuskeln, *peroneal muscles.*
Fussmuskeln, *muscles of the foot.*
620 Achillessehne, *f.* ⎫ *tendon*
    or        ⎬ *of*
621 Achillesflechse, *f.* ⎭ *Achilles.*

## 4. Das Gefässsystem, *the Vascular System.*

622 Gefäss, *n., vessel, tube, canal.*
623 Gefässgewebe, *n., vascular tissue or texture.*
624 das Herz (*gen.* **-ens**) (*pl.* **-en**), *heart.*
625 quer, *a., athwart, crosswise, oblique.*
626 Querscheidewand, *f.* (283), *the transverse partition wall*
627 die rechte Hälfte, *the right half.*
628 die linke Hälfte, *the left half.*
629 die Kammer, *chamber, ventricle.*
630 Herzkammer, *f., chamber of the heart, ventricle.*
631 Vorkammer, *f.* ⎫ *auricle,*
    or        ⎬ *atrium,*
632 Vorhof, *m.* ⎭ *vestibule.*
633 Mündung, *f., mouth, orifice.*
634 Herzklappe, *f., valve of the heart.*
635 Vorkammerklappe, *f., auricular valve.*
636 Vorhofkammerklappe, *auriculo-ventricular valve.*
637 Herzbein, *n., sternum.*
638 Herzbeutel, *m.* ⎫
639 Herzsack, *m.*   ⎬ *peri-*
640 Herzfell, *n.*     ⎬ *cardium.*
641 Herzbündel, *n.* ⎮
642 Herzhaut, *f.* ⎭
643 Herzdrüse, *f., cardiac gland.*
644 Herzgrube, *f., pit of the stomach.*
645 Herzgegend, *f., cardiac region.*
646 Herzhöhle, *f., heart-cavity.*
647 Herzgekröse, *n., mesocardium.*
648 Herzbewegung, *f.* (139), *heart motion.*
649 Herzspitze, *f., point or apex of the heart.*
650 Herzohr, *n.* ⎫ *auricle,*
651 Herzöhrchen, *n.* ⎭ *dim.*
652 das Muskelbündel in den Herzkammern, *the muscles of the ventricles of the heart.*
653 rauschen, *to rustle, murmur.*
654 schallen, *to sound, ring forth.*
655 Geräusch, *n., rustling, noise.*

656 Schall, *m.*, *sound, noise.*
657 Herzgeräusch, *n.* ⎫ *cardiac*
658 Herzschall, *m.* ⎭ *sound.*
659 **rinnen, rann, geronnen,** *to flow, drop.*
660 gerinnen, *to coagulate.*
661 Herzgerinnsel, *n.*, *heart-clot.*
662 **die** Ader, *vein.*
663 Puls, *m.*, *pulse.*
664 pulsieren, *to pulsate.*
665 **schlagen, schlug, geschlagen,** *to beat.*
666 Pulsschlag, *m.*, *pulsation, pulse-beat.*
667 Pulsader, *f.* ⎫
668 Schlagader, *f.* ⎬ *Artery.*
669 Arterie, *f.* (ri-e) ⎭
670 Herzröhre, *f.* ⎫
671 die grosse Pulsader, ⎪
672 die grosse Schlagader ⎬ *Aorta*
673 die Aorta, ⎪
674 Körperpulsader, *f.* ⎭
675 **steigen, stieg, gestiegen,** *to step, mount, ascend or descend* (according to the prefix or adjuncts).
676 absteigen, *to descend.*
677 aufsteigen, *to ascend.*
678 die absteigende Aorta, *descending aorta.*
679 die aufsteigende Aorta, *ascending aorta.*
680 die bogenförmige (420) Aorta, *the arch of the aorta.*
681 **flechten, flocht, geflochten,** *to twine, twist.*

682 Adergeflecht. *n.*, *plexus of veins.*
683 **das** Blut, *blood.*
684 bluten, *to bleed.*
685 blutig, *a.*, *bloody, gory.*
686 blutlos ⎫ *bloodless, exsan-*
687 blutleer ⎭ *guine.*
688 leer, *a.*, *empty.*
689 farbig, *a.* (243), *colored.*
690 blutfarbig, *a.*, *blood-colored.*
691 blutroth. *a.*, *blood-red, hematine.*
692 Blutader, *f.* ⎫ *vein.*
693 Vene, *f.* ⎭
694 Blutgefässe (622), *blood-vessels.*
695 Blutmasse, *f.*, *mass of blood.*
696 Blutgerinnsel, *n.*, *blood-coagulum.*
697 Klumpen. *m.*, *clump, lump, mass.*
698 Kuchen, *m.*, *cake.*
699 Blutklumpen, *m.* ⎫ *coagulum,*
 or ⎬ *crassa-*
700 Blutkuchen, *m.* ⎭ *mentum.*
701 Blutkügelchen, *n.* ⎫
702 Blutkörperchen, *n.* ⎬
703 Bluttheilchen, *n.* ⎭
 *blood globule, blood corpuscle.*
704 Theil, *m.* (das Theilchen), *part, deal.*
705 **das** Wasser, *water.*
706 Blutwasser, *n.* ⎫ *serum, blood-*
707 Serum, *n.* ⎭ *water.*
708 wässerig, *a.* ⎫ *serous.*
709 serös, *a.* ⎭
710 dunkel, *a.*, *dark.*

711 venös, *a.*, *venous.*
712 roth, *a.*, *red.*
713 hellroth, *a.*, *light-red.*
714 arteriell, *a.*, *arterial.*
715 dunkles ⎫
716 venöses ⎬ Blut
717 arterielles ⎪
718 hellrothes ⎭
    *dark,* ⎫
    *venous,* ⎬ *blood.*
    *arterial,* ⎪
    *light-red* ⎭
719 kreisen, *to circle about, revolve.*
720 Kreis, *m.*, *circle.*
721 **laufen, lief, gelaufen,** *to run, move rapidly.*
722 umlaufen, *to move round rapidly.*
723 Umlauf, *m.*, *moving round, circulation.*
724 **die** Bahn, *path, road, course.*
725 Blutkreislauf, *m.* ⎫
726 Blutumlauf, *m.* ⎬ *circulation of the blood.*
727 Blutbahn, *f.* ⎭
728 der grosse Kreislauf, *greater (systemic) circulation.*
729 der kleine Kreislauf, *lesser (pulmonary) circulation.*
730 die grosse Blutbahn, *greater blood-passage.*
731 die kleine Blutbahn, *lesser blood-passage.*
732 Lunge, *f.*, *lung, pulmo.*
733 Lungenblutbahn, *f.* ⎫
734 Lungenkreislauf, *m.* ⎭
    *pulmonary circulation.*

735 Körperblutbahn, *f.*, *circulation of the body.*
736 **stossen, stiess, gestossen,** *to thrust, strike, push.*
737 Stoss, *m.*, *shock, impulse.*
738 Herzpuls, *m.*, *the pulse of the heart.*
739 Herzschlag, *m.*, *heart-beat.*
740 Herzstoss, *m.*, *heart's impulse.*
741 pochen, *to palpitate.*
742 Herzpochen, *n.*, *palpitation of the heart.*
743 ausdehnen, *to extend.*
744 die Ausdehnung des Herzens, ⎫
745 die Diastole, ⎬
    *expansion of the heart, diastole.*
746 Zusammenziehung (174) des Herzens, ⎫
747 Systole, ⎭
    *contraction, systole of the heart.*
748 ernähren, *to nourish, feed.*
749 Ernährung, *nutrition.*
750 Ernährungsflüssigkeit, *nutritious fluid.*
751 **saugen, sog, gesogen,** *to suck, absorb.*
752 Lymphe, *f.*, *lymph.*
753 Lymphgefässe, ⎫ *lymphatic vessels,*
    or ⎬ *or lymphatic*
754 Saugadern, ⎭ *absorbents.*
755 Knoten, *m.*, *knot, node; ganglion, tuberculum.*

756 Lymphdrüse, *f.* } *lympha-*
757 Saugdrüse, *f.* } *tic gland*
758 Lymphknoten, *m.*
759 Lymphweg, *m.* } *lymph*
760 Lymphbahn, *f.* } *channel.*
761 Lymphkörperchen, *n.* }
762 Lymphplasma, *n.*
    *lymph corpuscle.*
763 Lymphcoagulum, *a.*, *coagulated lymph.*
764 speisen, *to feed, eat.*
765 Speise, *f.*, *food, nutriment.*
766 Speisesaft, *m.* } *chyle.*
767 Chylus, *m.*

768 Speisesaftgefässe, } *chyle*
769 Speisegefässe, } *ducts.*
770 das Haar ( *pl.* Haare), *hair.*
771 Haargefässe, } *capillary, ca-*
772 Capillaren, } *pillary vessels.*
773 Lungenblutader, *f.*, *pulmonary vein.*
774 Hohlader, *f.* (118), *vena cava.*
775 Pforte, *f.*, *gate, door, porta.*
776 Pfortader, *f.*, *portal vein.*
777 Pfortaderblut, *n.*, *portal blood*
778 Pfortaderblutumlauf, *m.*, *portal circulation.*

## 5. Das Nervensystem, *the Nervous System.*

779 Nerv, *m.* (*pl.* -en), *nerve.*
780 Nervenmasse, *f.* } *nerve*
    or } *sub-*
781 Nervensubstanz, *f.* } *stance.*
782 Nervengewebe, *n.*, *nerve-tissue.*
783 Nervengeflecht, *n.*, *nervous plexus.*
784 das Netz (*pl.* -e), *net.*
785 Nervennetz, *nerve plexus.*
786 Nervenfaser, *f.*, *nerve-fibre.*
787 Nervenbündel, *n.*, *nerve-bundle*
788 Nervenschicht (526), *f.*, *nerve layer.*
789 Nervenmark, *n.*, *nerve medulla, nerve pulp.*
790 Nervensaft, *m.*, *nervous fluid.*
791 Nervenhaut, *f.*, *retina.*
792 Nervenhülle, *f.*, *nerve envelope, coat, sheath.*

793 Nervenstamm, *m.*, *nerve trunk*
794 Ast, *m.*, *branch, bough.*
    Nervenast, *m.*, *nerve branch.*
795 verbreiten, *distribute, spread, diffuse.*
796 Nervenverbreitung, *f.*, *nervous distribution.*
797 Nervenendigung, *f.*, *nerve termination.*
798 Nervenwurzel, *f.*, *nerve root.*
799 Nervenwarze, *f.* } *nervous*
800 Nervenwärzchen, *n.* } *papilla*
801 Nervenkern, *m.*, *ganglion.*
802 Nervenkörperchen, *n.*, *nerve corpuscle.*
803 nervig, *a.* } *nervous, fig.*
    nervicht, *a.* } *also for sinewy*
804 nervös, *a.*, *nervous.*
805 nervenlos, *a.*, *nerveless, weak.*
806 nervenartig, *a.*, *nerve-like.*

807 das Hirn, } *brain, brains,*
808 das Gehirn, } *cerebrum.*
809 Hirnnerv, *m.* } *cerebral*
810 Gehirnnerv, *m.* } *nerve.*
811 Gehirnnervensystem, *n.*, *the system of the cerebral nerves*
812 Rückenmarksnerv, *m.* (352), *spinal nerve.*
813 Rückenmarksnervensystem, *n.*, *the spinal nervous system.*
814 Nervenknoten, *m.* or } *nerve tubercle,*
815 Ganglien (*plur.*) } *ganglion.*
816 Ganglien System, *n.*, *ganglionic system.*
817 **halten, hielt, gehalten,** *to hold, keep, contain.*
818 markhaltig, *a.*, *medullated.*
819 marklos, *a.*, *non-medullated.*
820 die centrale Faser, } *axis*
821 die Achsenfaser, } *fibre.*
822 Markscheide, *f.*, *medullary sheath.*
823 Nervenzelle, *f.*, *nerve cell.*
824 Ganglienkugeln (*pl.*), *ganglion globules.*
825 reizen, *to irritate.*
826 Nervenreiz, *m.*, *nerve irritation.*
827 Reizbarkeit, *f.*, *irritability.*
828 **empfinden, empfand, empfunden,** *to be sensible of, feel.*
829 empfindlich, *a.*, *sensible, sensitive.*
830 Empfindlichkeit, *f.* } *sensi-*
831 Sensibilität, *f.* } *bility.*
832 Empfindungsfaser, *f.*, *sensitive fibre.*
833 Sinn, *m.*, *sense, organ of perception.*
834 Sinnesnerv, *m.*, *nerve of sense.*
835 sensitive or Gefühlsnerven, *sensitive nerves or nerves of sensation.*
836 Mitempfindung, *f.*, *sympathy, irradiation.*
837 überstrahlen, *to irradiate.*
838 Mitbewegung, *f.* } *reflex, ir-*
839 Überstrahlung, *f.* } *radiation*
840 Reflexempfindungen, *reflex sensations.*
841 Hirnhöhle, *f.*, *cavity of the brain.*
842 das grosse Gehirn, *cerebrum.*
843 das kleine Gehirn, *cerebellum.*
844 Mittelgehirn (434), *central brain.*
845 Brücke, *f.*, *bridge, pons.*
846 **winden, wand, gewunden,** *to wind, twist, coil.*
847 Windungen, *windings, coils, convolutions.*
848 Furche, *f.*, *furrow.*
849 spalten, *to split, cleave.*
850 Spalt, *m.*, *fissure.*
851 Längenspalt, *m.*, *longitudinal fissure.*
852 Halbkugel (503), *f.* } *hemis-*
853 Hemisphäre, *f.* } *phere.*
854 Lappen, *m.*, *flap, lobe.*
855 Gehirnlappen, *lobe of the brain.*

856 Querspalte, *f.* (625), *transverse fissure.*
857 die harte Hirnhaut, *dura mater.*
858 Spinngewebe, *n.*, *spider's web.*
859 Spinnwebenhaut, *f.*, *arachnoid membrane.*
860 die weiche Hirnhaut, *pia mater.*
861 Hirnfläche, *f.*, *surface of the brain.*
862 Hirnflüssigkeit, *f.*, *cerebral fluid.*
863 Hirnnervenfasern (*pl.*), *fibres of the cerebral nerves.*
864 Geruchsnerv, *m.* (239), *olfactory nerve.*
865 Sehnerv, *m.*, *optic nerve.*
866 gemeinschaftlich, *a.*, *common.*
867 der gemeinschaftliche Augenmuskelnerv, *nervus motor oculi.*
868 Rollmuskelnerv, *m.*, *trochlearis or pathetic nerve.*
869 der äussere Augenmuskelnerv, *nervus abducens.*
870 der dreigetheilte Hirnnerv, *nervus trigeminus.*
871 Gesichtsnerv, *m.* (241), *facial nerve.*
872 Hörnerv, *m.*, *auditory nerve.*
873 Zungenschlundkopfnerv, *m.*, *nervus glosso-pharyngeus.*
874 Lungen-Magennerv, *m.*, *nervus pneumogastricus.*
875 Beinerv, *m.*, *accessory nerve.*

876 Zungenfleischnerv, *m.*, *nervus hypoglossus.*
877 Rückgratskanal, *m.*, *vertebral canal.*
878 Rückenmarkszapfen, *spinal or rachidian cone (or uvula)*
879 Rückenmarksfaden, *m.*, *spinal or rachidian filament (string).*
880 Rückenmarksrinde, *f.*, *medullary cortex.*
881 Rückenmarkshaut, *f.*, *spinal or rachidian meninge or membrane.*
882 Rückenmarksliquor, *m.*, *spinal or rachidian liquor.*
883 Strang, *m.*, *string, cord, fasciculus.*
884 Seitenstrang, *m.*, *funiculus seu fasciculus lateralis.*
885 Bewegungsnervenfaser, *f.*, *fibre of the motory nerve.*
886 Spinalganglion, *spinal ganglion.*
887 sympathetische Nervenfaser, *sympathetic nerve fibre.*
888 Halsnerv, *m.*, *cervical nerve.*
889 Halsgeflecht, *n.*, *cervical plexus.*
890 Armgeflecht, *n.*, *brachial plexus.*
891 Rückennerven (*pl.*), *dorsal nerves.*
892 Brustnerven, *thoracic nerves.*
893 Zwischenrippennerven, *intercostal nerves.*
894 Lendennerv, *m.*, *lumbar nerve.*

895 Bauchwirbelnerv, *m.*, *nerve of the lumbar vertebra.*
896 Lendengeflecht, *n.*, *lumbar plexus.*
897 Schenkelnerv, *m.*, *crural nerve*
898 Kreuzbeinnerv, *m.*, *sacral nerve.*
899 Hüftnerv, *m.*, *sciatic nerve.*
900 Schamgeflechtsnerv, *m.*, *nerve of the pudic plexus.*
901 Steissbeinnerv, *m.*, *coccygeal nerve.*
902 vasomotorisches, or
903 gefässbewegendes- Nerven-system, } *vasomotor system.*
904 sympathetisches Nervensystem, *sympathetic nervous system.*
905 Grenze, *f.*, *boundary, border, limit.*
906 Grenzstrang, *m.*, *principal trunk, funiculus marginalis.*
907 Ganglienkette, *f.*, *ganglion chain.*
908 Sympatheticus, *m.*, *sympathetic nerve.*
909 Sonne, *f.*, *sun.*
910 Sonnengeflecht, *n.*, *solar plexus.*

## 6. Die Haut, *the Skin, Hide* (110).

911 Zelle, *f.*, *cell, cellula.*
912 Zellgewebe, *n.*, *cellular tissue.*
913 umhüllen (528), *to wrap up or around, envelop.*
914 Umhüllungszellgewebe, *n.*, *investing cellular tissue.*
915 Blase, *f.*, *bladder, vesicle; cyst.*
916 Wasserhaut, *f.*, *hyaloid membrane.*
917 seröse Haut, *serous membrane*
918 Sack, *m.*, *sack; pocket; cyst.*
919 die Eingeweide (*plur.*), *intestines, bowels, viscera.*
920 Schleimhaut, *f.*, *the mucous membrane.*
921 Zellschichte, *f.*, *cellular layer.*
922 äusser, *a.*, *outer, external.*
923 die äussere Haut, *external skin, integumentum commune.*
924 Schleimnetz, *n.*, *mucous reticulum.*
925 Oberhäutchen, *n.*, *epithelium, cuticula.*
926 fett, *a.*, *fat, adipose.*
927 das Fett, *fat, grease.*
928 Fetthaut, *f.*, *fat membrane.*
929 Fettgewebe, *n.*, *adipose tissue.*
930 ablagern, *to deposit.*
931 Ablagerung, *f.*, *deposit.*
932 Fettablagerung, *f.*, *fat deposit.*
933 das Leder, *leather.*
934 Lederhaut, *f.*, *corium, derma, cutis vera.*
935 der (or das) Talg, *tallow, sebum.*
936 Talgdrüse, *f.*, *sebaceous gland.*

937 Schweiss, *m.*, *sweat, perspiration*.
938 Schweissdrüse, *f.*, *sweat gland*
939 Schweisskanal, *m.*, *sudriparous canal*.
940 Schleimschicht, *f.*, *mucous layer*.
941 Farbestoff, *pigment*.
942 das Horn, *horn*.
943 hörnern, *a.*, *horny, of horn*.
944 hornig, *a.*, *horny*.
945 hornicht, *a.*, *horn-like*.
946 hornartig, *a.*, *corneous*.
947 Platte, *f.*, *plate*.
948 Hornplatte, *f.*, *corneous plate*.
949 Schwiele, *f.*, *hard skin, callosity*.
950 schwielig, *a.*, *callous*.
951 Nagel, *m.*, *nail*.
952 Nagelkörper, *m.*, *body of the nail*.
953 Nagelwurzel, *f.*, *root of the nail*.
954 keimen, *to germinate, bud*.
955 Keim, *m.*, *germ*.
956 Haarkeim, *hair-germ*.
957 haarig, *a.*, *hairy, villous*.
958 Zwiebel, *f.*, *bulb (esp. onion)*.
959 Haarzwiebel, *f.*, *hair-bulb*.
960 Haarwurzel, *f.*, *hair-root*.
961 Schaft, *m.*, *shaft*.

962 Haarschaft, *m.*, *hair shaft*.
963 Zopf, *m.*, *plica*.
964 Rinde, *f.*, *rind; crust; cortex*.
965 Rindenmasse, *f.*, *cortical substance*.
966 Markmasse, *f.* (114), *medullary substance*.
967 das Oel (*pl.* Oele), *oil*.
968 Kopfhaar, *n.*, *hair of the head*.
969 Bart, *m.*, *beard*.
970 Barthaar, *n.*, *beard-hair*.
971 Achselhaar, *n.*, *axillary hair*.
972 Schamhaar, *n.*, *hair of the pubes*.
973 Gefässdrüsen (*pl.*), *vascular glands*.
974 ausscheiden, *to secrete*.
975 Ausscheidungsdrüsen, *secretory glands*.
976 Saugaderdrüsen, *lymphatic glands*.
978 Blutdrüsen, } *vascular glands*.
979 Blutgefässknoten, }
980 Thränendrüsen, *lachrymal glands*.
981 Speichel, *m.*, *saliva*.
982 Speicheldrüsen, *salivary glands*.
983 Brustdrüsen, *mammary glands*.

## 7. Die Athmungswerkzeuge, *the Respiratory Organs*.

984 athmen, *to breathe*.
985 Athem, *m.*, *breath respiration*.
986 holen, *to fetch*.
987 schöpfen, *to scoop, draw*.

988 Athem holen, }
989 Athem schöpfen, } *to take, draw, or fetch breath, to breathe*.

990 tief athmen, *to fetch a long, deep breath.*
991 schwer athmen, *to breathe hard.*
992 Athemzug, *m., drawing of the breath, respiration, breath.*
993 Athmungsprozess, *m., respiratory process.*
994 ausser Athem, *out of breath, breathless.*
995 den Athem an sich halten, *to hold one's breath.*
996 einathmen, *to inhale, inspire.*
997 ausathmen, *to breathe out, expire.*
998 eng, *a., narrow, close.*
999 weit, *a., wide.*
1000 verengen, or verengern, *to make narrow, close, to contract.*
1001 erweitern, *to widen, expand, dilate.*
1002 Verengerung (*f.*) der Brusthöhle, *narrowing or contraction of the thoracic cavity.*
1003 Erweiterung der Brusthöhle, *expansion or dilatation of the thoracic cavity.*
1004 Athmungsmuskeln, *respiratory muscles.*
1005 die Luft (*pl.* Lüfte), *air, atmosphere.*
1006 die Luftwege (*pl.*), *air passages.*
1007 Mundrachenhöhle, *f., pharyngo-oral cavity, fauces.*
1008 Nasenhöhle, *f., nasal cavity.*
1009 Kehle, *f., glottis, throat.*
1010 Luftröhre, *f., trachea, bronchus.*
1011 Luftkanal, *m., trachea.*
1012 Luftröhrenkopf, *m., larynx.*
1013 Lungenblatt (732), *pulmonary lobe.*
1014 Zweig, *m.* 
1015 Ast, *m.* } *branch, bough.*
1016 sich verästeln, or
1017 sich verzweigen, } *ramify, branch out.*
1018 Verästelung, *f., ramification.*
1019 Luftröhrenast, *m., bronchus.*
1020 Luftröhrenstamm, *bronchial trunk, stem.*
1021 die Gabel, *fork, bifurcation.*
1022 Gabelung der Luftröhre, *bifurcation of the trachea.*
1023 einschneiden (298), *to cut in, incise.*
1024 Einschnitt, *incision.*
1025 Läppchen, *n.* (854), *lobule.*
1026 Hauptlappen, *m., principal lobe.*
1027 Flügel, *m.* (*wing, from* fliegen, *to fly*), *lobe.*
1028 Lungenlappen, *m.* or
1029 Lungenflügel, *m.* } *lobe of the lungs.*
1030 Lungenfell, *n.* (601), *pleura.*
1031 Mittelfell, *n., mediastinum.*
1032 Mittelfellhöhle, *f., cavum mediastini.*
1033 Schild, *m., shield.*

1034 Schilddrüse, *f.*, *thyroid or shield-shaped gland.*
1035 Bröse, *f.* ⎫ *thoracic*
    Brustdrüse, *f.* ⎭ *gland.*
1036 Lungengewebe, *n.*, *pulmonary tissue.*
1037 Lungengeflecht, *n.*, *pulmonary plexus.*
1038 Lungengefäss, *n.*, *pulmonary vessel.*
1039 Lungendrüse, *f.*, *pulmonary or bronchial gland.*
1040 Lungenkammer, *f.*, *pulmonary ventricle.*
1041 Lungenspitze, *f.*, *apex pulmonis.*
1042 Lungenbläschen, *n.* ⎫
1043 Luftbläschen, *n.* ⎭ *pulmonary vesicle; air vesicle.*
1044 Lungenzellen, ⎫ *pulmonary*
1045 Luftzellen, ⎭ *cells; air cells.*
1046 Lungenschall, *m.*, *respiratory sound.*

## 8. Die Verdauungswerkzeuge, *the Organs of Digestion.*

1047 verdauen, *to digest.*
1048 Verdauung, *f.*, *digestion.*
1049 verdaulich, *a.*, *digestible.*
1050 unverdaulich, *a.*, *indigestible.*
1051 Verdauungsapparat, *m.*, *digestive apparatus.*
1052 Verdauungsprozess, *m.*, *process of digestion.*
1053 wechseln, *to change.*
1054 Wechsel, *m.*, *change.*
1055 Stoffwechsel, *m.*, *change of matter.*
1056 nähren (748), *to nourish.*
1057 Nahrung, *f.*, *nourishment, food.*
1058 nahrhaft, *a.*, *nourishing.*
1059 Nahrungsmittel, *n.*, *food, nutriment.*
1060 Nährstoff, *m.*, *nutritive matter.*
1061 Rachen, *m.*, *pharynx; mouth; jaws.*
1062 Schlund, *m.*, *pharynx.*
1063 Rachenhöhle, *f.* ⎫
1064 Schlundhöhle, *f.* ⎭ *pharyngeal cavity; œsophagus, gullet.*
1065 Speiseröhre, *f.*, *œsophagus, gullet.*
1066 Schlundkopf, *m.*, (125), *pharynx.*
1067 Magen, *m.*, *stomach.*
1068 Magengegend, *f.* (388), *gastric region.*
1069 Magenwand, *f.*, *gastric wall.*
1070 Magengrund, *m.*, *fundus of the stomach.*
1071 Magengefäss, *gastric vessel.*
1072 Magengeflecht, *gastric plexus.*
1073 Magennerv, *m.*, *gastric nerve*

1074 Magendrüse, *f.*, *gastric gland*.
1075 Magenhaut, *f.*, *gastric membrane, coat of the stomach*.
1076 Magengekröse, *n.*, *meso-gastrium*.
1077 Pförtner, *m.* (775),
1078 Magenmund, *m.*
1079 Magenpförtner, *m.*
 *orifice of the stomach; pylorus*.
1080 Blindsack, *m.*, *cæcal pouch or sack*.
1081 Pförtnerklappe, *f.*, *pyloric valve*.
1082 Bauchspeichel, *m.* (981), *pancreatic juice*.
1083 Bauchspeicheldrüse, *f.*, *pancreas*.
1084 Einspeichelung, *f.*, *insalivation*.
1085 Brei, *m.*, *pap, pulp, broth*.
1086 Speisebrei, *m.* (765)
1087 Chymus, *m.* } *chyme*
1088 Magenbrei, *m.*
1089 Magensaft, *m.*, *gastric juice*.
1090 Speisebreibildung, *f.*
1091 Chymification, *f.* } *chymification*.
1092 Darmkanal, *m.*, *intestinal canal*.
1093 Darmsaft, *m.*, *intestinal juice or liquor*.
1094 Dünndarm, *m.*, *small intestine*.
1095 Dünndarmgekröse, *n.*, *mesentery*.
1096 Zwölffingerdarm, *m.*, *duodenum*.
1097 **die** Leber, *liver*.
1098 Leberlappen, *m.*, *hepatic lobe*.
1099 Galle, *f.*, *gall, bile; choler*.
1100 gallenbitter, *a.*, *as bitter as gall*.
1101 Gallsäure, *f.*, *colic acid*.
1102 Gallengang, *m.* } *biliary*
1103 Gallenkanal, *m.* } *duct*.
1104 Gallenblase, *f.*, *gall-bladder*.
1105 **die** Milz, *spleen*.
1106 Leerdarm, *m.* (688), *jejunum*.
1107 Krummdarm, *m.*, *ileum* (krumm, *crooked*).
1108 Gekröse, *n.*, *mesentery*.
1109 Gekrösendärme (*pl.*), *mesenteric intestines*.
1110 Schlange, *f.*, *snake, serpent*.
1111 schlangenförmig, *a.*, *snake-formed, serpentine*.
1112 Dickdarm, *m.*, *large intestine*.
1113 Gekrösdrüsen, *mesenteric glands*.
1114 Vorverdauung, *f.*, *fore-digestion*.
1115 Nachverdauung, *f.*, *after-digestion*.
1116 blind, *a.*, *blind*.
1117 Blinddarm, *m.*, *cæcum*.
1118 Regenwurm, *m.*, *earthworm, angleworm*.
1119 regenwurmähnlich, *a.*, *earthworm-like*.

1120 fortsetzen, *to continue, proceed.*
1121 Fortsatz, m., *process, apophysis.*
1122 Wurmfortsatz, m., *vermiform process.*
1123 Grimm, m., *fury, wrath, rage.*
1124 grimmen (*impers.*), *to gripe, to have the gripes.*
1125 Grimmdarm, m., *colon.*
1126 die Mast, *mast (for fattening), food.*
1127 mast, a., *fat, well-fed.*
1128 mästen, *to feed with mast, fatten.*
1129 Mastdarm, m., *rectum.*
1130 Mastdarmgeflecht, n., *rectal plexus.*
1131 Mastdarmgefäss, n., *hemorrhoidal vessel.*
1132 Mastdarmknoten (*pl.*), *hemorrhoids.*
1133 After, m., *anus, buttocks.*
1134 Aftergegend, f., *anal region.*
1135 Afterdarm, m., *rectum.*

## 9. Die Harnwerkzeuge, *the Urinary Organs.*

1136 Harn, m. ⎱ *urine.*
1137 Urin, m. ⎰
1138 Harn lassen, ⎫
1139 harnen, ⎬ *to urine, make water.*
1140 urinieren, ⎭
1141 Harnapparat, m., *urinary apparatus.*
1142 Harngefässe (*pl.*), *urinary vessels.*
1143 Harnabsonderung, f. ⎱
    Harnausscheidung, f. ⎰ *urinary secretion.*
1144 Harnstoff, m., *urea.*
1145 Harnsäure, f., *uric acid.*
1146 Niere, f., *kidney.*
1147 Nierenwurzel, f., *renal radix.*
1148 Nierenpyramide, f., *renal pyramid.*
1149 Kelch, m., *cup, chalice, calix.*
1150 Nierenkelch, *renal calix.*
1151 Nierenbecken, n., *pelvis of the kidney.*
1152 Nierenleiter, m., *renal conductor, ureter.*
1153 Harnwege (*pl.*) ⎱ *urinary*
1154 Harngänge (*pl.*) ⎰ *passages*
1155 Harnkanälchen (*pl.*), *uriniferous canalicules.*
1156 Rindensubstanz, f., *cortical substance.*
1157 Marksubstanz, f., *medullary substance or matter.*
1158 Harnblase (915) ⎱ *bladder.*
    or Blase, f. ⎰
1159 Blasenhals, m., *neck of the bladder.*
1160 Harnröhre, *urethra.*

## 10. Die Geschlechtsorgane und die Fortpflanzung, *The Sexual Organs and Propagation.*

1161 Geschlechtsorgan, *n.* (2), *sexual organ.*
1162 das männliche (375) Geschlecht, *the male sex.*
1163 das weibliche (376) Geschlecht, *the female sex.*
1164 Geschlechtstheile (704), *genital parts.*
1165 Trieb, *m., instinct: impetus.*
1166 Geschlechtstrieb, *sexual instinct.*
1167 reif, *a., ripe.*
1168 unreif, *unripe.*
1169 Geschlechtsreife, *f., puberty.*
1170 die geschlechtliche Unreife, *f., impuberty.*
1171 Mannheit, *f., virility.*
1172 zeugen, } *to beget, generate.*
1173 erzeugen, }
1174 Zeugung, *f., begetting, generation.*
1175 zeugungsfähig (44), *a., able to beget, potent.*
1176 zeugungsunfähig, *a., impotent.*
1177 Zeugungsunfähigkeit, *f., impotence.*
1178 Zeugungsglied (392), *n., genital member.*
1179 Zeugungskraft, *f.* (17), *procreative or prolific power.*
1180 begatten, *to copulate, couple.*
1181 Begattung, *f., copulation, coition.*
1182 Begattungsorgane (*pl.*), *copulative organs.*
1183 die männlichen Schamtheile, *the natural parts of men.*
1184 die weiblichen Schamtheile, *the natural parts of women*
1185 das männliche Glied, }
1186 die Ruthe (*rod, yard*) } *penis*
1187 der Penis, }
1188 die Eichel, *gland, glans penis.*
1189 Vorhaut, *f., prepuce, foreskin.*
1190 Vorhautsbändchen, *n., foreskin band, frænum.*
1191 Hoden, *m., testicle.*
1192 Hodensack, *m., scrotum.*
1193 Nebenhoden, *m., epididymis.*
1194 der (männliche) Samen, *semen.*
1195 das Thier (*pl.* -e) (*dim.* Thierchen), *animal (animalcule).*
1196 das Samenthierchen, *zoosperm, spermatozoon.*
1197 Samenstrang (883), *spermatic cord.*
1198 ausspritzen, *to inject, ejaculate.*
1199 Ausspritzung, *f., ejaculation.*
1200 Ausspritzungsgang, *m., ejaculatory duct.*

1201 Samenbläschen (s-chen), *n.*, seminal vesicle.
1202 die Prostata, *prostate gland*, prostata.
1203 die äusseren (weiblichen) Schamtheile, *vulva.*
1204 Schamgegend, *f.*, *pubic region.*
1205 Schamberg, *m.*, *mons veneris*
1206 Scheide, } *vagina.*
1207 Mutterscheide,
1208 Schamlippe, *f.*, *lip of the pudenda, wing.*
1209 Schamspalte, *f.*, *pudendal fissure, rima pudendi.*
1210 Lefze, *lip or margin.*
1211 das Ei (*pl.* Eier), *egg.*
der Eierstock, *ovary, ovarium.*
1212 Gebärmutter, *f.*, *uterus, womb.*
1213 Gebärmutterhals, *m.*, *neck of the uterus.*
1214 Gebärmutterhöhle, *f.*, *cavity of the uterus.*
1215 Muttermund, *m.*, *mouth of the womb.*
1216 Mutterband, *n.*, *uterine ligament.*
1217 Keimlager (955), *n.*, *germinal bed, stroma.*
1218 Eibläschen (s-chen), *n.*, *egg globule, vesicle.*
1219 die Frucht (*pl.* **Früchte**) *fruit.*
1220 Fruchthalter (817), *m.*, *womb, uterus.*
1221 Trompete, *f.*, *trumpet.*
1222 Muttertrompete, *f.* }
1223 Eileiter (1211), *m.* } *oviduct, Fallopian tube.*
1224 der Leiter, *guide, conductor.*
1225 Jungfrau, *f.* } *virgin.*
Jungfer, *f.*
1226 Jungfernhäutchen (110), *hymen.*
1227 Wasserlefzen (*pl.*), *nymphæ*
1228 kitzeln, *to tickle.*
1229 Kitzler, *m.*, *clitoris; tickler.*
1230 Milchdrüse (300), *f.*, *lactiferous gland.*
1231 die Brust (365), *mamma, breast.*
1232 Brustwarze, *nipple, teat.*
1233 Warzenhof, *m.*, *areola mammæ.*
1234 Busen, *m.*, *bosom, breast.*
1235 Milchleiter (1224) } *lactiferous duct.*
or
1236 Milchgang,
1237 Milchkanal, *m.*, *lactiferous canal.*
1238 Milchgefäss, *n.*, *lacteal vessel.*
1239 heirathen, *to marry.*
1240 Heirath, *f.*, *marriage.*
1241 heirathsfähig, *a.*, *marriageable, nubile.*
1242 das Alter, *age.*
1243 heirathsfähiges Alter, *age of puberty.*
1244 Heirathsfähigkeit, *f.*, *nubility.*
1245 sinnlich, *a.* (833), *sensual.*

1246 Sinnlichkeit, *f.*, *sensuality*.
1247 **die** Wollust, *lust, voluptuousness*.
1248 wollüstig, *a.*, *lustful, voluptuous*.
1249 schwanger, *a.*, *pregnant, big with child, in the family way*.
1250 Umstand, *m.*, *circumstance*.
 in andern Umständen sein, *to be in the family way*.
1251 schwängern, *to impregnate*.
1252 Schwangerschaft, *f.*, *pregnancy*.
1253 hochschwanger, *a.*, *far advanced; quick with child*.
1254 gelüsten, *to desire, itch for, hanker after or for*.
1255 Gelüst, *n.*, *the hankering, itching*.
1256 wunderlich, *a.*, *strange, odd, whimsical*.
1257 der wunderliche Appetit einer Schwangern, *pica, vitiated appetite*.
1258 entbinden (98), *to deliver (a woman), to put or bring to bed*.
1259 entbunden werden, *to lie in, to be delivered*.
1260 niederkommen, *to be brought to bed, to be delivered*.
1261 Niederkunft, *f.* ⎫ *delivery,*
 or       ⎬ *confine-*
1262 Entbindung, *f.* ⎭ *ment.*
1263 **gebären, gebar, geboren,** *to give birth to, to be delivered*.
1264 **die** Geburt, *birth*.
1265 schwere Geburt, *difficult labor*.
1266 **die** Zeit, *time*.
1267 Unzeit, *f.*, *wrong time*.
 zur Unzeit gebären, *to miscarry*.
1268 unzeitig, *a.*, *wrong-timed, untimely*.
1269 unzeitige Geburt, ⎫ *abortion,*
 or             ⎬ *miscar-*
1270 Fehlgeburt,    ⎭ *riage.*
1271 geboren werden, *to be born*.
1272 vorzeitig, *a. and adv.*, *premature, before the proper time*.
1273 vorzeitig niederkommen, *to be delivered prematurely*.
1274 Woche, *f.*, *week*.
1275 in Wochen kommen, *to be delivered*.
1276 in Wochen sein, *to be in childbed, lying in*.
1277 die Wochen halten, *to be confined, to lie in*.
1278 die Wöchnerin,   ⎫ *lying*
1279 die Sechswöchnerin ⎬ *in*
1280 die Kindbetterin,  ⎭ *woman*
1281 mit einem Sohne niederkommen, *to be brought to bed of a son*.
1282 Leibesfrucht, *f.* (20, 1219), *foetus, fetus*.
1283 neugeboren,   ⎫ *new-born.*
 ebengeboren, ⎭
1284 weh,  ⎫ *a. and adv., pain-*
 wehe, ⎭ *ful, sore.*
1285 **das** Weh, *woe, pain, pang*.

1286 die Wehen (*pl.*), *pains, throes.*
1287 Weh thun (*with dative*), *give pain to.*
1288 Wehen haben, *to be in labor.*
1289 **die** Noth (*pl.* **Nöthe**), *need, trouble, misery.*
1290 in Kindesnöthen sein, *to be in labor.*
1291 das Fruchtwasser, *amniotic fluid.*
1292 Nabelstrang, *m.*, *navel string, umbilical cord.*
1293 die Nachgeburt, *afterbirth, secundines.*
1294 an der Mutterbrust saugen (751), *to suck one's mother.*
1295 säugen, *to suckle, to give suck, nurse.*
1296 die Brust geben, ⎫
1297 stillen, ⎭ *to give suck or the breast to a child, to still.*
1298 Amme, *f.*, *nurse.*
1299 Säugamme, *f.*, *wet-nurse.*
1300 Flasche, *f.*, *the bottle.*
1301 Saugflasche, *f.* ⎫ *nursing*
1302 Saugglas, *n.* ⎭ *bottle.*
1303 aufziehen, *to bring up.*
1304 Ein Kind mit der Saugflasche (or dem Saugglase) aufziehen, *to bring up a child by hand.*
1305 entwöhnen, *to wean.*
1306 Entwöhnen, *n.*, *weaning.*

## 11. Die fünf Sinne, *the five Senses.*

1307 **sehen, sah, gesehen,** *to see.*
1308 Sehorgan, *n.* ⎫
1309 das Sehwerkzeug, ⎭ *organ of vision, visual organ.*
1310 das Organ des Gesichts (241), *organ of sight.*
1311 Augenhöhle, *f.* (296), *orbit (socket) of the eye.*
1312 heben, or aufheben, *to lift, raise, elevate.*
1313 Heber, or Aufheber, *m.*, *elevator.*
1314 niederziehen, *to draw down, depress.*
1315 Niederzieher, *m.*, *depressor (muscle).*
1316 rollen, *to roll, revolve.*
1317 Rollmuskel, *m.* ⎫ *rotator, tro-*
Roller, *m.* ⎭ *chlearis.*
1318 Augenbraue, *f.*, *eyebrow.*
1319 runzeln, *to knit the brows, to wrinkle, corrugate.*
1320 Augenbraunrunzler, *m.*, *corrugator.*
1321 **das** Augenlid, *eye-lid.*
1322 **die** Augenwimper, ⎫
Augenwimmer, ⎬ *eye-lash.*
1323 **die** Wimper, ⎭
1324 Augenlidknorpel, *m.*, *tarsal cartilage.*

1325 Augenlidbindehaut, *f.*, *palpebral conjunctiva*.
1326 Thränenapparat, *m.* (270), *lachrymal apparatus*.
1327 Thränengang, *m.*, *lachrymal duct*.
1328 Thränenkanal, *m.*, *lachrymal canal*.
1329 Thränenkarunkel, *m.* }
1330 Thränenhügel, *m.*   *caruncula lachrymalis*.
1331 Thränensack, *m.*, *lachrymal sack*.
1332 **der** Thränensee, *lacus lachrymalis*, *the lachrymal lake*.
1333 Thränenwärzchen, *n.*, *lachrymal papilla*.
1334 **die** Butter, *butter*.
1335 Augenbutter, *f.*, *the secretion from lids*.
1336 Augapfel, *m.*, *eyeball, globe, or apple of the eye*.
1337 Augenhaut, *f.*, *tunic, or coat of the eye*.
1338 das Weisse im Auge, *the white of the eye*.
1339 die harte, or weisse Augenhaut, } *sclerotic*.
1340 Sclerotica, *f.*
1341 Cornea, or } *cornea, horny or corneous tunic*.
1342 Hornhaut, *f.*
1343 Gefässhaut, *f.* ⎫
1344 Aderhaut des Auges, ⎬
1345 schwarze Augenhaut, ⎪
1346 Chorividea, *f.* ⎭
  *vascular membrane, choroid*

1347 Strahl, *m.* (*pl.* **-en**), *ray* (*of light*).
1348 Strahlenband, *n.*, *ciliary ligament, ligamentum ciliare*.
1349 Falte, *f.*, *fold*.
1350 Kranz, *m.*, *crown*.
1351 Faltenkranz, *m.*, *ring rays; plaited band*.
1352 Strahlenkörper, *m.*, *ciliary body*.
1353 blenden, *to make blind, to dazzle*.
1354 Blendehaut, *f.*, *the iris*.
1355 Regenbogen, *m.* (420), *rainbow*.
1356 Regenbogenhaut, *f.* } *iris*.
1357 Iris, *f.*
1358 Sehloch, *n.* (278), ⎫
1359 Pupille, *f.* ⎬ *pupil*.
1360 Sehe, *f.* ⎭
1361 Augenkammer, *f.* (629), *chamber of the eye*.
1362 die vordere (221) Augenkammer, *anterior chamber*.
1363 die hintere (222) Augenkammer, *the posterior chamber*.
1364 der Verengerer (1000) der Pupille, *constrictor*.
1365 der Erweiterer (1001) der Pupille, *dilator*.
1366 Traube, *f.*, *grape*.
1367 Traubenhaut, *f.*, *tunica uvea*.
1368 Nervenhaut, *f.* } *retina*.
  Netzhaut (784), *f.*

1369 Fleck, *m.*, *spot, stain. macula.*
1370 der gelbe Fleck, *macula lutea.*
1371 Strahlenblättchen, *n., zonula ciliaris, or Zinnii.*
1372 feucht, *a., moist.*
1373 Feuchtigkeit, *f., moisture, dampness; humor.*
1374 wässerige (708) Feuchtigkeit, *aqueous humor (of the eye).*
1375 Linse, *f., lens.*
1376 Linsenkapsel, *f., capsule of the lens.*
1377 Glaskörper, *m., vitreous body.*
1378 Glasfeuchtigkeit, *f.* ⎫ *vitre-*
 or                         ⎬ *ous*
1379 Glasflüssigkeit, *f.*   ⎭ *humor*
1380 Glashaut, *f., hyaloid, or vitreous membrane.*
1381 **das Gehörorgan**, *the auditory organ.*
1382 hören, *to hear.*
1383 Gehör, *n., hearing, audition.*
1384 Gehörwerkzeug, *n., organ of hearing.*
1385 **das** Ohr (*pl.* **-en**), *ear.*
 das äussere (922) Ohr, *external ear.*
1386 Ohrhöhle, *f.* ⎫ *cavity of the*
 Ohrloch, *n.*  ⎭ *ear.*
1387 Ohrblatt, *n.*    ⎫ *lobe of*
1388 Ohrläppchen, *n.* ⎭ *the ear.*
1389 der äussere Gehörgang, *external auditory duct.*

1390 **das** Schmalz, *grease, (melted) fat.*
1391 Ohrschmalz, *n.*     ⎫ *cerumen*
 or                       ⎬ *auris,*
 Ohrenschmalz,*n.*       ⎭ *ear-wax.*
1392 **das** Ohrwachs, ⎫ *id.*
 Ohrenwachs, *n.*   ⎭
1393 **die** Trommel, *drum.*
1394 Pauke, *f., drum, tympanum.*
1395 Trommelfell,*n.* ⎫ *drum of*
 or                  ⎬ *the ear,*
1396 Paukenfell, *n.* ⎭ *tympanum*
1397 das mittlere Ohr,*middle ear.*
1398 Trommelhöhle,*f.* ⎫ *tympan-*
1399 Paukenhöhle, *f.*  ⎭ *ic cavity.*
1400 Vorhof, *m., vestibule, auricle.*
1401 **das** Fenster, *window.*
1402 Vorhoffenster, *n., vestibular window,fenestra vestibuli.*
1403 Schnecke, *f., cochlea.*
1404 Schneckenfenster, *n., cochlear window.*
1405 Ohrtrompete, *f.* ⎫ *Eusta-*
1406 Eustachische      ⎬ *chian*
 Trompete, *f.*      ⎭ *tube.*
1407 Gehörknöchelchen, *auditory ossicle.*
1408 Hammer, *m., hammer.*
1409 Amboss, *m., anvil.*
1410 Linsenknöchelchen, *n., lenticulus (of the stapes).*
1411 Steigbügel, *m., stapes.*
1412 die Bogengänge des Labyrinths, *semi-circular canals.*

1413 Gehörsteinchen, *n.* ⎫
1414 Ohrkrystall, *m.*   ⎬ *otolith.*
1415 Ohrsand, *m.*       ⎭
1416 Gehörnerv, *m.*, *auditory nerve.*
1417 Schneckennerv, *m.*, *cochlear nerve.*
1418 Vorhofsnerv, *vestibular nerve.*
1419 **das Geruchsorgan,** *the organ of smell.*
1420 **der** Geruch, *smell, scent, odor.*
1421 Geruchssinn, *m.*, *sense of smelling.*
1422 Geruchswerkzeug, *n.*, *olfactory organ.*
1423 Nasenknochen, *m.* (272), *nasal bone.*
1424 Nasenwurzel, *f.*, *root of the nose.*
1425 Nasenrücken, *m.*, *bridge of the nose.*
1426 Nasenflügel, *m.* (1027), *wing of the nose, lateral cartilage.*
1427 Nasenschleim, *m.*, *nasal mucus.*
1428 Nasenschleimhaut, *f.*, *nasal mucous membrane.*
1429 Riechhaut, *f.*, *olfactory membrane.*
1430 Riechnerv, *m.*, *olfactory nerve.*
1431 Riechzelle, *f.*, *olfactory cell.*
1432 **das Geschmacksorgan,** *the organ of taste.*
1433 schmecken, *to taste.*
1434 **der** Geschmack, *taste.*
1435 Geschmackssinn, *m.*, *sense of taste.*
1436 Zungenmuskel, *m.* (328), *lingual muscle.*
1437 Zungennerv, *m.*, *lingual nerve.*
1438 Zungenrand, *m.* (131), *border, edge of the tongue.*
1439 Zungenhaut, *f.*, *lingual membrane.*
1440 Zungenwarze, *f.*, *papilla of the tongue.*
1441 Zungenwärzchen, *n.* ⎫
1442 Geschmackswärzchen, *n.* ⎬ *gustatory papilla.*
1443 Geschmacksnerv, *m.*, *gustatory nerve.*
1444 Kegel, *m.*, *cone.*
1445 kegelförmig, *a.*, *conical.*
1446 die kegelförmigen Wärzchen, *conical papillæ.*
1447 Pilz, *m.*, *fungus, favus.*
1448 pilzförmig, *a.*, *fungiform.*
1449 die pilzförmigen Wärzchen, *fungiform papillæ.*
1450 die kelchförmigen (1149) Wärzchen, *n.*, *papillæ maximæ or circumvallate papillæ.*
1451 Zungenbändchen, *n.*, *lingual frenulum.*
1452 Gaumensegel, *n.* (264), *palate sail, velum palatinum.*
1453 Gaumenbogen, *m.*, *palatine arch.*

1454 **die** Mandel, *almond.*
1455 Mandeln (*pl.*), *tonsils.*
1456 Zungendrüse, *f.*, *lingual gland.*
1457 **das Stimmorgan**, *the vocal organs.*
1458 Stimme, *f.*, *voice.*
1459 **sprechen, sprach, gesprochen,** *to speak.*
1460 Sprache, *f.*, *language, speech.*
1461 Sprachwerkzeuge (*pl.*), *organs of speech.*
1462 Kehlkopf, *m.* (1009), *larynx.*
1463 Adamsapfel, *m.*, *Adam's apple, pomum Adami.*
1464 Kehlkopfsknorpel, *m.*, *laryngeal cartilage.*
1465 Schildknorpel, *m.*, *thyroid cartilage.*
1466 Ringknorpel, *m.*, *cricoid cartilage.*
1467 Giesskanne, *f.*, *watering pot.*
1468 Giesskannenknorpel, *m.*, *arytenoid cartilage.*
1469 rund, *a.*, *round.*
1470 rundlich, *a.*, *roundish.*
1471 die rundlichen Knorpel (*pl.*), *roundish cartilages.*
1472 Kehlkopfhöhle, *f.*, *laryngeal cavity.*
1473 Ritze, *f.*, *rift, rima glottidis.*
1474 Stimmritze, *f.*, or Glottis, *f.*, *glottis.*
1475 Stimmritzband, *n.*, *ligamentum glottidis.*

1476 decken, *to cover.*
1477 Deckel, *m.*, *cover.*
1478 Kehldeckel, *m.* ⎫ *laryngeal*
 or ⎬ *cover, epi-*
1479 Epiglottis, ⎭ *glottis.*
1480 Tasche, *f.*, *pocket.*
1481 Kehlkopftasche, *f.*, *laryngeal pocket, ventriculus laryngis.*

**PHRASES REFERRING TO ALL THE SENSES:**

1482 die Augenbrauen in die Höhe ziehen, *to raise the eyebrows.*
1483 die Augenbrauen senken, *to lower the eye-brows.*
1484 die Ohren klingen mir, *my ears tingle (tinkle).*
1485 löchern, *a.* (1357), *to make holes in, to perforate.*
1486 Ohrlöcher stechen, *to pierce the ear.*
1487 durch die Nase sprechen, *to speak through the nose.*
1488 schnüffeln, ⎫ *to snuffle.*
1489 näseln, ⎭
1490 sich schnäuzen, *to blow one's nose.*
1491 sich in die Zunge beissen, *to bite one's tongue.*
1492 ein gutes (schlechtes) Gesicht haben, *to have a good (bad) sight, eye-sight.*
1493 kurzsichtig sein, *to be short sighted, near sighted.*

1494 weitsichtig sein, *to be far sighted, presbyopic.*
1495 ein gutes Gehör haben, *to be quick of hearing.*
1496 ein schlechtes Gehör haben, *to be dull of hearing.*
1497 einen guten, feinen Geschmack haben, *to have a good, a delicate taste.*
1498 den Geschmack verloren haben, *to be out of taste, to have lost one's taste.*
1499 er hat keinen Geruch, *he is destitute of the sense of smell.*
1500 Gefühl, *n., feeling, touch, sensation.*
1501 das erkennt man am Gefühl, *that is known by the touch.*

## II. THE DISEASES.

## DIE KRANKHEITEN.

### A. SYMPTOMS AND PHENOMENA.

#### 1. General Expressions.

1502 sich befinden, *to be, to find one's self (in a condition).*
1503 das Befinden, *the state of health.*
1504 Wie ist Ihr Befinden? *how is your health?*
1505 Wie befinden Sie sich? *how do you do? how are you?*
1506 sich wohl befinden, *to be well (in health).*
1507 gesund, *a., sound, healthy, hale.*
1508 kerngesund (123), *a., thoroughly sound.*
1509 frisch und gesund, *hale and sound.*
1510 gesund und munter, *safe and sound, safe and well.*
1511 Gesundheit, *f., health.*
1512 bei or in guter Gesundheit, *in good health.*
1513 Zustand, *m., the state, condition.*
1514 Gesundheitszustand, *m., state of health.*
1515 gesund sein, *to be well, to be in good health.*
1516 gesund werden, *to get well again.*
1517 **genesen, genas, genesen** (*with* **sein**), *to recover, to be restored to health.*
1518 gesund aussehen, *to look well.*
1519 krank, *a., ill, sick.*
1520 krank aussehen, *to look ill.*

1521 Krankheit, *f.*, *illness, sickness, disease.*
1522 Kinderkrankheit, *f.*, *disease peculiar to children.*
1523 Nachkrankheit, *f.*, *subsequent, secondary disease.*
1524 Rückfall, *m.*, *relapse.*
1525 kränklich, *a.*, *sickly, ailing, infirm.*
1526 Kränklichkeit, *f.*, *infirmity, sickliness.*
1527 krank sein, *to be ill.*
1528 sich krank fühlen, *to feel one's self ill, unwell.*
1529 krank werden, *to be taken ill, to fall ill or sick, to grow sick.*
1530 schwer ⎫
1531 gefährlich ⎬ krank sein,
     sehr ⎭
    *to be dangerously ill.*
1532 todtkrank, *a.*, *fatally sick.*
1533 ungesund, *a.*, *unsound, unhealthy, sickly.*
1534 unwohl, *a.*, *unwell, indisposed.*
1535 das Unwohlsein, *indisposition, illness.*
1536 unpässlich, *a.*, *indisposed, poorly.*
1537 Unpässlichkeit, *f.*, *indisposition, complaint.*
1538 das Übelbefinden, *indisposition, uneasiness, painful sensation.*
1539 sich sehr unwohl fühlen, *to feel very uneasy.*

1540 schwach, *a.*, *weak, infirm, feeble.*
1541 nervenschwach, *a.*, *neurasthenic, nervous.*
1542 Schwäche, *f.*, *weakness, infirmity.*
    der Zustand der Schwäche, *condition of weakness.*
1543 schwächen, *to weaken, debilitate.*
1544 schwächlich, *a.*, *feeble, weakly, sickly, delicate.*
1545 schwächliche Gesundheit, *weak health.*
1546 sich beschweren, *to make complaint, to complain.*
1547 Beschwerde, *f.*, *complaint, trouble, pain, malady.*
1548 **leiden, litt, gelitten,** *to suffer pain.*
    leiden **an**, *to suffer pain* **from** *or with.*
    **an** den Augen (Zähnen) leiden, *to have a pain in one's eyes (teeth).*
1549 **das** Leiden, *suffering, pain, ailment.*
1550 langes Leiden, *long ailment.*
1551 leidend, *a.*, *suffering.*
1552 fehlen, *to fail, to be wanting, missing or deficient.*
    Was fehlt Ihnen? *what ails you? what is the matter with you?*
    Was fehlt Ihnen am Auge? *what is the matter with your eye?*

1553 schmerzen, *to pain, ail, ache*
1554 kräftig (17), *a., vigorous, strong.*
1555 kräftigen, *to strengthen, invigorate.*
1556 entkräften, *to debilitate, enervate, enfeeble.*
1557 Entkräftigung, *f., enervation, enfeeblement.*
1558 stärken (532), *to strengthen.*
1559 Stärke, *f., strength.*
1560 der Kranke bessert sich, *the patient is getting better.*
1561 herstellen, *to restore (sick persons).*
1562 Herstellung, *f., recovery.*
1563 eine Krankheit vertreiben, *to expel or remove a disease.*
1564 Krankheitssitz, *m., seat of disease.*
1565 matt, *a., faint, languid, exhausted.*
1566 Mattigkeit, *f., faintness, lassitude.*
1567 siech, *a., sick, sickly.*
1568 siechen, *to be sickly, languish.*
1569 hindsiechen, *to pine away, to waste away under sickness.*
1570 Siechthum, *n., sickness.*
1571 Siechbett, *n.* ⎫ *sick-bed.*
1572 Krankenbett, *n.* ⎭
1573 auf dem Siechbett (Krankenbett) liegen, *to keep one's bed, to be bed-ridden.*
1574 bettlägerig, *a., bed-ridden.*
1575 das Bett *or* das Zimmer hüten, *to keep one's bed, one's room.*
1576 vom Krankenbette aufstehen, *to recover from illness.*
1577 gebrechlich, *a.* (324), *decrepit, frail.*
1578 Gebrechlichkeit, *f., frailty, feebleness.*
1579 Gebrechen, *n., defect, frailty*
1580 Leibesgebrechen, *n., bodily defect or infirmity.*
1581 Leibesfehler, *m., bodily deformity.*
1582 Leibesbeschwerde, *bodily complaint, malady.*
1583 **fallen, fiel, gefallen,** *to fall.*
1584 hinfällig, *ready to fall down, decrepit, deciduous.*
Hinfälligkeit, *f., decrepitude, prostration.*
1585 erschöpfen (987), *to exhaust, drain.*
1586 Erschöpfung, *f., exhaustion, dialysis.*
1587 alterschwach, (1242), *a., weak from old age.*
1588 Alterschwäche, *f., weakness of old age.*
1589 Aussehen, *n.* (1307), *aspect, expression.*
1590 krankhaftes ⎫ Aussehen, ungesundes ⎭ *sickly look, unhealthy look.*
1591 blass, *a., pale, wan.*

1592 leichenblass, *a.*, *pale as death, ghastly.*
1593 Blässe, *f.*, *paleness, pallor.*
1594 Leichenblässe, *f.*, *death-like paleness, ghastliness.*
1595 Saft, *m.*, *sap, juice.*
1596 schlechte ⎫
 böse ⎬ Säfte,
 ungesunde ⎭
 *bad or unhealthy humors.*
1597 schaden, *to harm, hurt, damage.*
1598 Schaden, *m.*, *injury, hurt.*
1599 Leibesschaden, *m.*, *bodily deformity.*
1600 behaglich, *a.*, *pleasant, comfortable.*
1601 unbehaglich, *a.*, *unpleasant, uneasy.*
1602 Unbehaglichkeit, *f.*, *discomfort, uneasiness.*
1603 quälen, *to torment.*
1604 die Qual, *torment, torture.*
1605 peinigen, *to cause pain, to afflict, torment.*
1606 die Pein, *torment, pain.*
1607 martern, *to torture.*
1608 die Marter, *torture, pang.*

**die Krankheit ist,** *the illness is:*
1609 leicht, *a.*, *light, slight.*
1610 schwer, *a.*, *hard, severe.*
1611 bedeutend, *important.*
1612 bedenklich, *critical, serious.*
1613 verzweifelt, *desperate, past remedy.*

**die Krankheit ist,** *the illness is:*
1614 gefährlich, *dangerous.*
1615 lebensgefährlich, *perilous, mortal.*
1616 ungefährlich, *not dangerous.*
1617 primär, *idiopathic, primary.*
1618 secundär, *symptomatic.*
1619 recidive, *relapsing.*
1620 (vorschützen, *to pretend, feign.*)
1621 vorgeschützt, *pretended, feigned.*
1622 simuliert, *feigned.*
1623 heilbar, *curable.*
1624 unheilbar, *incurable.*
1625 tödtlich, *fatal, deadly.*
1626 heftig, *violent.*
1627 hitzig, *acute, inflammatory.*
1628 akut (ū'), *acute, violent, rapid.*
1629 schnell, *rapid.*
1630 chronisch, *chronic, slow.*
1631 langwierig, *chronic, obstinate.*
1632 langsam, *slow.*
1633 fieberhaft, *feverish, aguish.*
1634 hartnäckig, *obstinate, stubborn.*
1635 eingewurzelt, *inveterate.*
1636 entzündlich, *inflammatory.*
163~ schleichend, *lingering, slow, insidious.*

die Krankheit ist, *the illness is:*

| | | | |
|---|---|---|---|
| 1638 | ansteckend, *infectious, catching.* | 1656 | neuralgisch, *neuralgic.* |
| 1639 | contagiös, *contagious.* | 1657 | sympathisch, *sympathetic* (*s. headache*). |
| 1640 | nicht ansteckend, *non-contagious.* | 1658 | momentan, *momentary* (*m. unconsciousness*). |
| 1641 | epidemisch, *epidemic.* | 1659 | central (tráhl), *central* (*c. paralysis*). |
| 1642 | endemisch, *endemic.* | | |
| 1643 | (die Seuche, *epidemic.*) | 1660 | functionell, *functional* (*f. palsies*). |
| 1644 | seuchenartig, *a., epidemical.* | 1661 | local, *local* (*l. palsy*). |
| 1645 | übertragbar, *transferable.* | 1662 | peripher, *peripheral* (*p. palsies*). |
| 1646 | gutartig, *a., benign.* | 1663 | reflex, *reflex.* |
| 1647 | bösartig, *a., malignant, malign.* | 1664 | intermittierend, *intermittent.* |
| 1648 | erblich, *a., hereditary.* | 1665 | hysterisch, *hysterical.* |
| 1649 | schrecklich, *formidable.* | 1666 | alternierend, *alternating.* |
| 1650 | gewöhnlich, *common.* | | |
| 1651 | selten, *rare.* | | serös (709). |
| 1652 | einfach, *simple.* | 1667 | traumatisch, *traumatic* (produced by a wound or *trauma*). |
| 1653 | allmählig, *gradual* (*gradual paraplegia*). | | |
| 1654 | still, *quiet* (*quiet delirium*). | 1668 | das Stadium, *stage.* die Stadien der Krankheit, *the stages of the disease.* |
| 1655 | furibund, *fierce* (*f. delirium*). | | |
| | nervös (804). | | |

## 2. *Pain and other Sensations.*

1669 schmerzen* (1553), *to pain, cause pain, to feel pain.*

1670 der Schmerz (-es, *pl.* -en), *pain, ache, smart.*

*Take notice! This verb is exclusively used in the 3d person, and of things only, as die Wunde (or *sie*) schmerzt, der Zahn (or *er*) schmerzt, das Geschwür (or *es*) schmerzt, but never der Mann schmerzt, or ich schmerze, etc. The person feeling the pain generally in the accusative, as die Wunde schmerzt mich, dich, ihn, uns.

1670 Schmerz empfinden (828), *to feel pain.*
1671 verursachen, *to cause.*
1672 machen, *to make.*
1673 lindern, *to mitigate, alleviate.*
  Schmerz verursachen, machen, lindern, *to cause, give, alleviate pain.*
  Ich empfinde grossen Schmerz, *I feel great pain.*
1674 die Wunde macht (verursacht) mir Schmerz, *the wound gives me pain.*
1675 weh or wehe (1284) thun, *to ache.*
  der Kopf thut mir weh (or wehe), *my head aches.*
  der Zahn thut mir weh, *I have an aching tooth.*
  es thut ihm kein Zahn mehr weh (proverbial phrase), *he is dead* (lit. *he has no aching tooth any more*).
1676 mir ist weh, *I feel pain.*
  mir wird weh,
1677 mir wird unwohl, } *I feel ill, faint, sick.*
  mir wird übel,
1678 mir wird weh ums Herz, *my heart aches.*
  ihm thut kein Finger weh (proverbial phrase), *no finger aches him* (i. e. *he is perfectly well*).
1679 schmerzlich, *a.* } *painful,*
1680 schmerzhaft, *a.* } *smarting.*
1681 schmerzvoll, *a., painful.*

1682 schmerzfrei, *a.*
  or } *without pain, painless.*
1683 schmerzlos, *a.*
1684 Schmerzhaftigkeit, *f., painfulness.*
1685 Schmerzlosigkeit, *f., absence of pain.*
1686 schmerzstillend (1297), *assuaging or allaying pain, anodyne.*
  **der Schmerz ist,** *the pain is:*
1687 gering, *slight.*
1688 dumpf, *dull.*
  heftig (1626).
1689 scharf, *sharp.*
1690 drückend, *pressing, oppressive.*
1691 brennend, *burning.*
1692 klopfend, *beating, pulsating.*
1693 hämmernd, *hammering.*
1694 stechend, *pungent.*
1695 schneidend, *cutting.*
1696 bohrend, *boring.*
1697 lancinierend, *lancinating.*
1698 spannend, *straining.*
1699 ziehend, *pulling, drawing.*
1700 zusammenschnürend, *constricting.*
1701 reissend, *tearing.*
1702 nagend, *gnawing.*
1703 fast unerträglich, *unendurable, almost intolerable.*

**der Schmerz ist,** *the pain is:*

1704 quälend, \) *excruciating,*
1705 qualvoll, \} *torturing.*
1706 kribbelnd, *itching.*
1707 reizend, *irritating.*
1708 fliegend, *flying, rushing.*
1709 blitzgleich, *lightning-like.*
  nervös (804).
  neuralgisch (1656).
1710 krampfhaft, \)
1711 krampfartig, \}
1712 krampfähnlich, \)
  *spasmodic, convulsive.*
1713 intensiv, *intense.*
1714 erschöpfend, *exhausting.*
1715 markiert, *marked.*
1716 marternd, *racking, torturing.*
1717 kneifend, *pinching, nipping.*
1718 kneipend, *griping.*
1719 zwickend, *twinging.*
1720 innerlich, *internal.*
1721 äusserlich, *external.*
1722 oberflächlich, *superficial*
1723 unregelmässig, *irregular.*
  intermittierend (1664).
1724 plötzlich, *sudden.* —
  hartnäckig (1634).
1725 constant, *constant.*
1726 dauernd, *continuing.*
1727 andauernd, *persistent.*
1728 localisiert, *localised.*

**der Schmerz ist,** *the pain is:*

1729 einseitig, *one-sided.*
1730 begrenzt auf einen Nerven, *limited to one nerve.*
1731 ausgebreitet, *diffuse.*
1732 ausstrahlend nach allen Seiten, *radiating on all sides.*
1733 reflectiert, *reflex.*
1734 weit vom Sitze der Krankheit entfernt, *far away from the seat of the disease.*
1735 den Sitz mehr oder weniger wechselnd, *more or less changing its seat.*
1736 auf- und abschiessend, *shooting up and down.*
1737 ähnlich electrischen Entladungen, *similar to electric shocks.*
1738 variirend an Intensität, *varying in intensity.*
1739 von Zeit zu Zeit verstärkt, *becoming from time to time exalted.*
1740 von Temperatur oder Witterungswechsel beeinflusst, *influenced by a change of temperature or weather.*
1741 vermehrt durch Bewegung, *increased by movement.*

**der Schmerz ist,** *the pain is:*
1742 vermehrt durch aufrechte Stellung, *augmented by the erect posture.*
1743 vermehrt durch irgend welches Geräusch (655), *increased by sound of any kind.*
1744 heftiger durch Berührung
1745 oder Druck, *intensified by touch or pressure.*
Berührung, *f., touch.*
Druck, *m., pressure.*
1746 Schmerzparoxysmen (*pl.*), *paroxysmal pains.*
heftige Paroxysmen von Schmerz, *violent paroxysms of pain.*
Empfindlichkeit (830) gegen Berührung, *sensitiveness to touch.*
1747 ein dumpfes (1688) Wehgefühl, *a heavy aching feeling.*
1748 Gefühl von Wundsein, *a feeling of soreness.*
1749 Druckempfindung, *f., sense or perception of pressure.*
1750 ein heftiger Schmerzanfall, *a violent attack of pain.*
1751 der Sitz des Schmerzes, *the seat of the pain.*
1752 eine einfache (1652) unbehagliche (1601) Empfindung, *a mere uneasy sensation.*
1753 mildern, *to soften, alleviate, relieve.*
1754 horizontal, *a.* (táhl), *horizontal, recumbent.*
1755 Lage, *f., position, posture, condition.*
1756 erhöhen, *to raise, increase.*
1757 gelind, *soft, smooth, gentle.*
1758 verschlimmern, *to make worse, aggravate.*
(ein Schmerz) gelindert durch horizontale Lage, *relieved by recumbent position.*
erhöht durch gelinden Druck, gemildert durch starken Druck, *increased by slight pressure, relieved by strong pressure.*
verschlimmert oder gelindert durch die Zunahme von Nahrung, *augmented or relieved by the taking of food.*
1759 Steifheit, *f., stiffness.*
1760 verbunden mit Steifheit der Muskeln und Gelenke, *combined with stiffness of the muscles and joints.*
1761 Anschwellung, *f., swelling, inflation, intumescence.*
1762 Härte, *f., hardness.*

1763 Oberfläche, *f.*, *surface, superficies, outside.*
1764 Röthe, *f.*, *redness.*
1765 Krampf, *m.*, *cramp, spasm, convulsion.*
  das Unbehagen (1602).
1766 Erstickung, *f.*, and das Ersticken, } *suffocation.*
  schmerzhafte Anschwellung und Härte um die Gelenke der Gliedmassen herum, *painful swelling and hardness about the joints of the extremities.*
  schmerzlose Anschwellung der Oberfläche ohne Röthe, *painless swelling of the surface, devoid of redness.*
  Krämpfe in den Beinen und in den Waden (499), *cramps in the legs and in the calves.*
  heftige krampfhafte (1710) Schmerzen im Leibe und den Extremitäten, *severe spasmodic pains in the abdomen and extremities.*
  Gefühl des Unbehagens und des Erstickens, *sensation of uneasiness and of suffocation.*
  ein Erstickungsgefühl im Halse, *a strangling sensation in the throat.*

1767 Ameise, *f.*, *ant, formica.*
1768 kriechen, *to creep.*
1769 ein Gefühl von Ameisenkriechen, *n.* }
1770 Ameisenlaufen (721), } *formication.*
1771 Knochenschmerz, *m.* }
1772 Knochenweh, *n.* }
1773 Beinweh, *n.* } *pain in the bones, osteocopus.*
1774 Muskelschmerz, *m.*, *myodynia.*
1775 Nervenschmerz, or } *nervous pain,*
1776 Neuralgie, } *neuralgia.*
  Schmerz in den Gliedern, *pain in the limbs.*
  Schmerz in den Waden, *pain in the calves of the leg.*
1777 zucken, *to feel convulsive twitches, quiver, jerk.*
  die Muskeln zucken, *the muscles jerk.*
1778 Kopfschmerz, *m.* or } *pain in the head,*
1779 Kopfweh, *n.* } *headache*
  Kopfweh haben, *to have headache.*
  an Kopfschmerzen leiden, *to suffer from headache.*
1780 den Kopf einnehmen, *to affect, benumb the head.*
  einen schweren, eingenommenen Kopf haben, *to have a heaviness in the head.*

1781 Eingenommenheit des Kopfes *heaviness, dizziness.*
einseitiger (1729) Kopfschmerz, *one-sided headache.*
hysterischer (1665) Kopfschmerz, *hysterical headache.*
1782 Seitenkopfweh, *n.*, *hemicrania.*
heftiger Schmerz an der Stirn, *severe pain fixed to the forehead.*
1783 Stirnkopfschmerz, *m.*, *frontal headache.*
1784 Gesichtsschmerz, *m.*, *neuralgia facialis.*
der nervöse Gesichtsschmerz, *the tic douloureux.*
1785 Mundschmerz, *m.*, *stomalgia.*
1786 Zungenschmerz, *m.*, *glossalgia.*
1787 Augenschmerz, *m.* }
1788 das Augenweh, }
*pain in the eyes, ophthalmalgia.*
1789 Ohrenschmerz, *m.* } *otalgia,*
1790 Ohrenweh, *n.* } *earache.*
1791 Zahnschmerz, *m.* }
1792 Zahnweh, *n.* }
*odontalgia, toothache.*
1793 Halsweh, *n.*, *pain in the neck or throat.*
1794 würgen, *to strangle, choke.*
1795 Würgen, *n.*, *strangling, choking.*
1795 Beschwerden und Würgen im Halse, *obstruction in the throat.*
1796 Nackenweh, *n.*, *trachelagra.*
1797 Genickschmerz, *m.*, *pain in the neck.*
1798 Schulterschmerz, *m.*, *omalgia.*
Schmerz in der rechten Schulter und Weiche, *pain in the right shoulder and loin.*
1799 Rückenschmerz, *m.* } *dorsal*
1800 Rückenweh, *n.* } *stitches*
1801 Rückgratsschmerz, *m.*, *rhachialgia.*
1802 Rückenmarksschmerz, *m.*, *myelalgia.*
1803 Seitenstiche haben, }
Seitenstechen, *n.* }
*to have stitches in one's side.*
1804 Rippenweh, *n.*, *pleurodynia.*
1805 Wirbelschmerz, *m.*, *spondylalgia.*
1806 Brustschmerz, *m.* } *pain in*
1807 Brustweh, *n.* } *the chest.*
1808 Brustbeinschmerz, *m.*, *sternodynia.*
1809 Herzweh, *n.*, *heartache, cardialgia.*
1810 Herzdrücken, *n.*, *oppression of the heart.*
1811 Leibschmerz, *m.* } *pain in*
1812 Leibweh, *n.* } *the bowels*
1813 Bauchschmerz, *m.* } *belly-*
1814 Bauchweh, *n.* } *ache.*

1815 Bauchgrimmen, *n.*, *gripes, belly-ache.*
1816 kolikartig, *a.*, *colic-like.*
kolikartiger Schmerz, localisiert (1728) im Unterleib, *a colic-like pain, localized in the abdomen (hypogastrium).*
1817 Schmerz im linken Hypochondrium, *pain in the left hypochondrium.*
1818 Unterbauchschmerz, *m.*, *hypogastralgia.*
1819 Magenschmerz, *m.* ⎫
1820 Magenkrampf, *m.* ⎬
*gastrodynia, stomachache, pain in the stomach.*
epigastrischer Schmerz, *epigastric pain.*
hypogastrischer Schmerz, *hypogastric pain.*
1821 Brennen in der Magengegend (1068), *a burning sensation in the epigastric regions.*
1822 Empfindlichkeit (830) in der Magengrube, *tenderness (or sensitiveness) in the pit of the stomach.*
1823 **verderben, verdarb, verdorben**, *to spoil, corrupt.*
1824 der verdorbene Magen, *a spoiled stomach, indigestion, a stomach out of order.*
sich den Magen verderben, *to get an indigestion.*
1825 sich den Magen überladen, *to overload one's stomach (one's self) with food.*
1826 Magenschwäche, *f.*, *weakness of stomach, frigidity of the stomach.*
1827 Verdauungsschwäche, *f.* ⎫
Dyspepsie, ⎬
*dyspepsia.*
den Magen stärken (1558), *to strengthen or refresh the stomach.*
1828 Darmweh, *n.* ⎫
1829 Darmstrenge, *f.* ⎬
*gripes, pain in the bowels, colic.*
1830 Lungenschmerz, *m.*, *pulmonalgia.*
1831 Leberschmerz, *m.*, *hepatalgia.*
1832 Schwere, *f.*, *weight, heaviness.*
1833 Spannung, *f.*, *tension.*
unbehagliches Gefühl von Schwere und Spannung in der Lebergegend, *unpleasant sensation of weight and tension in the hepatic region.*
dumpfer (1688) Schmerz, strahlend (1732) von der Lebergegend bis zum oberen Theile der Brust, *a dull pain, radiating from the seat of the liver to the upper portion of the thorax.*

1834 Milzweh, *n.*, *splenalgia.*
permanenter Schmerz in der linken Seite, in der Gegend der Milz, *persistent pain in the left side in the region of the spleen.*
1835 Kreuzschmerzen, *pain in the loins.*
1836 Lendenschmerz, *m.*, *lumbago.*
1837 Hüftschmerz, *m.*, *pain in the hip-joint.*
1838 Hüftgelenkschmerz, *m.*, *hip-gout.*
constanter Schmerz in der Weiche und in der Gegend des Harnleiters, *a constant pain in the loin and in the course of the ureter.*
1839 Nierenweh, *n.*, *nephralgia.*
1840 Nierenschmerz, *m.*, *nephritic colic.*
Schmerz in der Nierengegend *pain in the renal region.*
1841 Harnröhrenschmerz, *m.*, *urethralgia.*
scharfes (1689) Brennen und schneidender Schmerz in der Harnröhre, *acute burning and lancinating pain along the urinary passages.*
1842 Blasenschmerz, *m.*, *cystodynia.*
1843 Steinschmerzen, *stone-colic.*
1844 Scheidenschmerz, *m.*, *colpalgia.*
1845 Mutterschmerz, *m.*, *hysteralgia.*
1846 Geburtsschmerzen, } *labor pains.*
Geburtswehen,
1847 Steissweh, *n.*, *pain in the buttocks.*
1848 Afterschmerz, *m.*, *proctalgia.*
1849 gichtischer Afterschmerz, *proctagra.*
heftige Muskelschmerzen, namentlich in den Waden, *severe muscular pains, especially in the calves of the legs.*
1850 Schenkelschmerz, *m.*, *pain in the thigh.*
1851 taub, *deaf, numb, benumbed.*
1852 Taubsein, *n.*, *numbness.*
Gefühl von Taubsein in dem Schenkel der ergriffenen Seite, *numbness of the thigh of the affected side.*
1853 Knieschmerz, *m.*, *pain in the knee.*
1854 Gelenkschmerz, *m.*, *gout, arthalgia.*
1855 Gichtschmerz, *m.*, *pains of the gout.*
1856 Gliederschmerz, *m.*, *arthritic pains.*
Krämpfe in den Beinen, *cramps in the legs.*
Schmerzen in den Gliedmassen, *pain in the limbs or extremities.*
1857 Fussschmerz, *m.*, *pain of the foot.*

## 3. Conditions Observable in the Skin.

**Die Haut ist,** *the skin is:*

| | | | |
|---|---|---|---|
| 1858 | hart, *hard, gritty.* | | |
| | schwielig (950). | | |
| 1859 | zart, *delicate, tender.* | | |
| 1860 | sanft, *soft.* | | |
| | weich (385). | | |
| 1861 | locker, *slack, loose.* | | |
| 1862 | schlaff, *flabby, lax, flaccid.* | | |
| 1863 | gespannt, *tense, tight.* | | |
| | straff (219). | | |
| 1864 | runzlich, a. (1319), *wrinkled* | | |
| 1865 | voll, *full.* | | |
| 1866 | rauh, *harsh.* | | |
| 1867 | dünn, *thin.* | | |
| | dick (484). | | |
| 1868 | trocken, *dry.* | | |
| | feucht (1372). | | |
| | empfindlich (829). | | |
| 1869 | unempfindlich, *insensible, numb.* | | |
| 1870 | geschwollen, *swollen, tumefied.* | | |
| 1871 | aufgeblasen, *emphysematous.* | | |
| 1872 | aufgedunsen, *puffy.* | | |
| 1873 | ödematös, *œdematous.* | | |
| 1874 | warm, *warm.* | | |
| 1875 | heiss, *hot.* | | |
| 1876 | sehr heiss und juckend, *very hot and itchy.* | | |
| 1877 | kühl, *cool.* | | |
| 1878 | kalt, *cold.* | | |

**Die Haut ist,** *the skin is:*

| | |
|---|---|
| 1879 | eisigkalt, *icy.* |
| 1880 | weiss, *white.* |
| 1881 | bleich, *pale.* |
| | blass, *pallid* (1501). |
| 1882 | wachsähnlich, } *waxy.* |
| 1883 | wachsartig, } |
| | roth (712). |
| | hellroth (713). |
| 1884 | dunkelroth (710), *dark-red.* |
| 1885 | hektischroth, *hectic-red.* |
| 1886 | schmutzig, *dirty.* |
| 1887 | fahl, *fallow.* |
| 1888 | erdfahl, *earth-colored, earthy.* |
| 1889 | blau, *blue.* |
| 1890 | livid, *livid.* |
| 1891 | bläulichroth, *violet.* |
| 1892 | braun, *brown.* |
| 1893 | bräunlichroth, *brownish-red.* |
| 1894 | gelb, *yellow.* |
| 1895 | gelbgrün, *yellowish-green.* |
| 1896 | strohartig, *straw-colored.* |
| 1897 | klebrig, *clammy.* |
| 1898 | teigartig, *doughy.* |
| | runz(e)lig (1864). |
| 1899 | **die** Runzel, *wrinkle.* |
| 1900 | **der** Ausschlag, } |
| 1901 | der Hautausschlag, } *eruption, rash.* |

1902 Efflorescenz, *f.*  ⎫ *efflores-*
1903 **die** Hautblüthe, *f.* ⎬ *cence, ex-*
　　der flache (178) ⎭ *anthem.*
　　Hautausschlag,
　　Ausschlag in zackigen (134) Flecken (1369), *eruption in crescentic patches.*
1904 Knötchen, *n.* ⎫ *nodule,*
1905 **die** Papel, ⎬ *papule,*
1906 Stippchen, *n.* ⎭ *papula.*
1907 Knoten, *m.*, *node, tubercle.*
1908 Quaddel, *f.*, *pomphus, seu urtica.*
　　Blase (915).
1909 Bläschen, *n.*, *vesicle, small blister.*
1910 Blasen ziehen, *to produce blisters.*
1911 Blasen bekommen, *to blister, to get blisters.*
1912 Eiterblase, *f.* ⎫
　　Eiterbläschen, *n.* ⎬ *pustule.*
1913 Eiterblatter, *f.* ⎭
1914 **die** Pustel,
1915 bedecken, *to cover.*
　　mit Pusteln bedeckt, *pustular.*
1916 erhaben, *raised, elevated.*
1917 bestehen, *to consist of.*
1918 Hirse, *f.*, *millet.*
1919 hirsenförmig, *miliary.*
1920 erscheinen, *to appear.*
1921 sickern, *to trickle, ooze.*
1922 Eiter, *m.*, *pus, matter.*
　　ein Ausschlag aus leicht erhabenen rothen Flecken bestehend, welche zusammen fliessen, *an eruption consisting of slightly raised red spots which coalesce.*
　　der Ausschlag zeigt kleine hirsenförmige Bläschen, *the eruption shows small miliary vesicles.*
　　der Ausschlag ist rauh (1866) *the eruption is coarse.*
　　auf der Spitze (274) erscheint ein Bläschen, *at the top appears a vesicle.*
　　der Eiter sickert heraus, *the matter oozes.*
1923 Hitzbläschen, *n.*, *pimple, pustule.*
1924 Hitzblatter, *f.*, *blister, vesicle.*
1925 schürfen, *scratch, scrape (connected with scharf, sharp).*
1926 Schorf, *m.*, *scurf, scratch, crust.*
1927 Hautabschürfung, *excoriation.*
1928 Schramme, *f.*, *light scratch, scar, slight wound.*
1929 schrunden *and* sich schrunden, *to split, crack, chop.*
1930 Schrunde, *rhagades, fissure.*
1931 Schuppe, *squama, scale, pityriasis.*
1932 Schuppen im Gesichte, *tetters*
1933 Schuppen auf dem Kopfe, *scurf.*
1934 Abschuppung, *desquamation*

1935 sich abschälen, *to peel.*
1936 Fetzen, *m., patch.*
die Epidermis schält sich in grossen Fetzen ab, *large patches of cuticle fall from the surface.*
1937 Kruste, *f., crust, scab.*
1938 Grind, *m., scab.*
1939 Kopfgrind, *m., scurf, dandruff.*
1940 grindig, *a., scabbed, scurfy.*
grindicht, *a., like scurf.*
1941 räudig, *a., mangy.*
1942 kratzen, *to scrape, scratch, grate.*
1943 krätzig, *itchy, scabio*
1944 Krätze, *f., itch, scabies.*
1945 Flechte, *f., tetter, herpes.*
1946 nasse Flechte, *ulcerated eruption.*
1947 nässende Flechte, *eczema.*
1948 trockne Flechte, *dandruff.*
1949 nässender Grind, *impetigo.*
1950 Gänsehaut, *f., goose-skin, cutis anserina.*
1951 jucken, *to itch.*
1952 Hautjucken, *n., itching, itch, pruritus cutaneus.*
1953 Hautbeissen, *n., pruritus.*
1954 Kitzel, *m.* (1228), *tickling, pruriency.*
der ganze Körper juckt ihm, *his whole body itches.*
1955 der heftige Hautreiz, *the cutaneous irritation.*
1956 Kratzwunde, *f., scratch.*
1957 ritzen, *to scratch, graze.*

1958 Ritz, *m., scratch.*
1959 Strieme, *f., stripe, streak, scar.*
1960 Hautbrennen, *n.* }
1961 Hautbrand, *m.* }
*a burning or itching sensation of the skin.*
1962 Hautschwiele, *f.* (949), *callosity, hard swelling.*

1963 **schwitzen,** *to sweat, perspire.*
1964 das Schwitzen, *sweating, perspiring.*
1965 ausdünsten, *to evaporate, perspire.*
1966 Ausdünstung, *evaporation.*
1967 ausschwitzen, *to exude.*
1968 Ausschwitzung, *exudation.*
1969 Exsudat, *exudation.*
1970 Flüssigkeit, *f.* (211), *fluidity, humor.*
1971 die ausgeschwitzte Flüssigkeit, *exudate.*
der gelinde (1757) Schweiss, *moisture, humidity.*
1972 schweissfeucht, *a., moist, damp.*
der kalte Schweiss, *cold sweat*
1973 übermässiges Schwitzen, *excessive sweating.*
einseitiges (1729) Schwitzen, *one-sided sweating.*
locales Schwitzen, *local sweating.*

1974 allgemeines Schwitzen, *general sweating.*
1975 Blutschwitzen, *n.*, *hæmidrosis.*
klebrige (1897) Schweisse, *clammy sweats.*
1976 profuser Schweiss, *profuse perspiration.*
1977 Nachtschweiss, *m.*, *night-sweat.*
1978 Todesschweiss, *m.*, *death-sweat, sweat of agony.*
1979 schwitzend, *a.* } *perspiring,*
1980 schweissig, *a.* } *sweaty.*

1981 **triefen, troff, getroffen,** *to drip, trickle.*
vonSchweiss triefen, *to stream with perspiration.*
1982 sauer, *a.*, *sour, acid.*
die Haut brennend heiss, mit saurem Schweiss bedeckt, *skin of pungent heat, covered with acid perspiration*
1983 ausströmen, *to stream forth, to emit.*
einen Ammoniakgeruch ausströmen, *emitting an ammoniacal odor.*

## 4. *Arterial Pulsation.*

1984 fühlen, *to feel.*
sich fühlen, *to feel one's self.*
1985 befragen, *to consult.*
den Puls fühlen, befragen, *to feel, consult the pulse.*
1986 Minute, *f.*, *minute.*
sein Puls schlägt (665) so oft in der Minute, *his pulse beats so many times a minute.*
**der Puls ist,** *the pulse is:*
voll, *full* (1865).
1987 gross, *large.*
1988 klein, *small.*
erhaben, *high* (1916).
schnell, *quick* (1629).
1989 rasch, *rapid.*
1990 frequent, *frequent.*
1991 kurz, *short.*
1992 träg(e), langsam (1632). } *slow.*

**der Puls ist,** *the pulse is:*
1992 (wird langsamer, *slackens.*)
hart, *hard* (1858).
gespannt, *tense* (1863).
weich, *soft* (385).
1993 nachgebend, *yielding.*
stark, *strong* (532).
1994 mässig stark, *only moderately strong.*
schwach, *feeble* (1540).
immer schwächer werdend, *sinking.*
1995 doppelt schlägig, *double beating.*
1996 ungleich, *uneven, unequal.*
1997 kaum fühlbar, *scarcely perceptible.*
nicht fühlbar, *imperceptible.*

**der Puls ist,** *the pulse is:*

1997 nur an einem Handgelenk (432) fühlbar, *felt only at one wrist.*
1998 schnellend unter dem Finger, *bounding.*
1999 hüpfend, *jerking.*
2000 schwankend, *intercurrent.*
2001 schlagend mit Intercadenzen, *subsultory.*
2002 unterbrochen, } *inter-*
2003 aussetzend, } *mittent.*
von Zeit (1266) zu Zeit schneller schlagend, *intercurrent.*

**der Puls ist,** *the pulse is:*

2004 stillstehend, *standing still, stagnating.*
2005 stockend, *stopping, ceasing to circulate.*
2006 regelmässig, *regular.*
unregelmässig, *irregular* (1723).
2007 fadenförmig, *thready.*
2008 drahtförmig, *wiry.*
2009 schwirrend, *vibrating.*
2010 wallend, *undulating.*
2011 fiebernd, *feverish.*
die Schlagadern (668) am Halse pulsieren (664) heftig, *the carotids throb violently.*

## 5. *Respiration.*

2012 Einathmung, *f.* (996), *respiration.*
2013 Ausathmung, *f., exhalation.*
2014 kurzer Athem, *shortness of breath.*
2015 aufathmen, *to recover breath.*
2016 ausser Athem sein, *to be out of breath.*
2017 athemlos, *a., breathless.*
2018 schwere Athemzüge (992), *heavings, gasps.*
2019 Mundathmen, *n., oral respiration, r. through the mouth.*
2020 Nasenathmen, *n., nasal respiration, r. through the nose.*
2021 Röhrenathmen (96), *tubal respiration, r. through a tube.*
2022 Rippenathmen, *costal respiration.*
2023 Athemnoth, *f.* (1289), *dyspnoea, suffocation.*
2024 Engathmigkeit, } *shortness*
2025 Kurzathmigkeit, } *of breath.*
er leidet an Athemnoth, *he suffers for want of breath.*
2026 oberflächliches (1763) Athmen, *enfeebled breathing.*
2027 Häufigkeit, *f., frequency.*
die vermehrte (1741) Häufigkeit des Athmens, *the increased frequency of breathing.*

2028 beschleunigen, *to accelerate, hasten.*
ausserordentlich beschleunigtes Athmen, *extremely rapid breathing.*
2029 mühevoll, *a., irksome, distressed.*
2030 beschwert, *a., embarrassed.*
2031 beeinträchtigen, *to injure, impede.*
beeinträchtigtes Athmen, *impeded breathing.*
2032 Athmungsbeschwerde, *f., difficulty in respiration.*
unregelmässiges Athmen, *irregular breathing.*
eine lang gezogene (174) Einathmung, *a long-drawn inspiration.*
eine krampfhafte (1710) Exspiration, *a spasmodic expiration.*
2033 auftreten, *to come on or forth, appear.*
Athemnoth in Paroxysmen auftretend, *dyspnœa coming on in paroxysms.*
2034 erschwert, *impeded, laborious.*
2035 geräuschvoll (655), *a., noisy.*
2036 schnarchen, *to snore.*
2037 laut, *a., loud, audible.*
2038 zischen, *to hiss, whiz.*
2039 mühsam, *a., labored.*
2040 **pfeifen, pfiff, gepfiffen,** *to whistle.*
2041 röcheln, *to wheeze, rattle in the throat.*
sein Athem ist bedeutend erschwert und geräuschvoll, *his breathing is seriously impeded and noisy.*
eine schnarchende Inspiration, *a snoring inspiration.*
ein Athmen ist laut, zischend, und mühsam, *his breathing is audible, hissing and labored.*
eine Inspiration mit einem pfeifenden oder röchelnden Geräusch, *an inspiration with a wheezing or whistling noise.*
2042 übel, *a., evil, bad.*
2043 übler Geruch (1421), *bad smell.*
2044 aus dem Munde riechen, *to have a foul breath.*
2045 übelriechend, *smelling badly, offensively.*
sein Athem ist übelriechend, *his breath is offensive.*
2046 leise, *a., noiseless.*
sein Athem ist leicht und leise, *his breathing is gentle and noiseless.*
2047 schluchzen, *to hiccough, sob.*
eine Einathmung mit schluchzendem Tone, *an inspiration attended with a whoop.*
2048 Rasselgeräusch, *n., rale, rhonchus.*

2049 grossblasig, *a.* (915), *large bubbling.*
2050 sonor (nō′r), *a., sonorous.*
  sonore und grossblasige Geräusche, *sonorous and large bubbling sounds.*
2051 Schleimrasseln, *n., mucous rattle, râle, or rhonchus.*
  hochpfeifende (2040) Geräusche, *high pitched whistling sounds.*
2052 knistern, *to crackle, crepitate.*
  feine Knistergeräusche, *fine rales.*
  hauchende (54) Geräusche, }
2053 Blasegeräusche, }
  *blowing sounds.*
2054 schnurren, *to hum, buzz, hiss.*
  sonore und schnurrende Geräusche, *sonorous and sibilant rales.*
2055 Ton, *m., sound, tone.*
  der laute, pfeifende Ton, *the loud wheezing.*
2056 summen, *to hum, buzz.*
  ein summendes Geräusch, *a humming sound.*
2057 knittern or knastern (2052). *to crackle.*
2058 Knattergeräusch, *n., crackling sound.*
2059 krähen, *to crow.*
  die Inspiration hat einen krähenden Ton, *the inspiration is crowing.*
2060 seufzen, *to sigh, groan.*
  das Athmen ist unregelmässig und seufzend, *the breathing is irregular and heaving.*
2061 niesen, *to sneeze.*
2062 häufiges Niesen, *frequent sneezing.*
2063 Ausfluss, *m.* (210), *flow, flux.*
  Ausfluss aus der Nase, *flow from the nose.*
2064 heiser, *a., hoarse.*
2065 Heiserkeit, *f., hoarseness.*
2066 heiser werden, *to become or get hoarse.*
2067 **husten, to cough,** *have a cough.*
2068 hüsteln, *to cough a little.*
2069 Husten, *m., cough.*
2070 Hustenanfall, *m., a fit of coughing.*
2071 keichen, } *to pant, gasp,*
  keuchen, } *wheeze.*
2072 Keichhusten, *m.* } *whooping*
  or } *cough.*
2073 der blaue Husten, }
2074 sticken, *choke, smother, stifle.*
2075 Stickhusten, *m., choking cough.*
  ein trockner (1868) Husten, *a dry cough.*
  ein feuchter (1372) Husten, *a moist cough.*

2076 bellen, *to bark.*
 der Husten ist rauh und bellend, *the cough is harsh and barking.*
 ein krampfhafter Husten, *a spasmodic cough.*
2077 metallisch, *a.*, *metallic.*
 der Husten ist laut, metallisch, und heiser, *the cough is loud, ringing (brazen) and hoarse.*

2078 **auswerfen,** *to expectorate,* *to expel, cough up.*
2079 Auswurf, *m.*, *expectoration, spitting, sputum.*
2080 **speien, spie,**
      **gespieen,** } *to spit.*
2081 spucken,
2082 Spucke, *f.*, *spittle.*
2083 ausspeien, } *to spit out.*
2084 ausspucken,
2085 blutspeien, *to spit blood.*
2086 Ausspeien, *n.*, *spitting.*
2087 klar, *a.*, *clear.*
 ein klarer Auswurf, *a clear expectoration.*
 zäh, *a.*, *tough, viscous, viscid.*
 ein zäher Auswurf, *a viscid expectoration.*
 zäher Schleim (543), *glairy mucus.*
2088 rostfarben, *a.*, *rust-colored.*
 rostfarbiger Auswurf, *a rusty colored sputum.*
2089 reichlicher Auswurf, *copious expectoration.*

2089 das Ausspeien von hellrothem Blut, *spitting up of florid blood.*
2090 **ergiessen, ergoss, ergossen,** *to effuse.*
2091 Bluterguss, *m.*, *effusion of blood.*
2092 gelbgestreift, *a.*, *streaked with yellow.*
 ein Auswurf von grüner, gelbgestreifter Farbe, *a sputum greenish in color, streaked with yellow.*
2093 der Gestank, *fetor, offensive smell.*
 der ausserordentliche Gestank des Auswurfs, *the extreme fetor of the expectoration.*
2094 schmutziggrau (1886), *dirty-grayish.*
2095 entschieden, *a.*, *and adv. decidedly.*
2096 eiterig (1922), *purulent, mattery.*
 der Auswurf hat ein schmutziggraues, entschieden eiteriges Aussehen (1589), *the sputum has a dirty-grayish, decidedly purulent aspect.*
2097 Münze, *f.*, *coin.*
2098 münzenartig, *a.*, *money-like, nummular.*
 ein Auswurf von runder, münzenartiger Form, *a sputum of nummular (or money-like) form.*

2099 Streifen, *m.*, *stripe, streak,* from:
2100 streifen, *to form streaks or stripes.*
2101 Blutstreifen, *bloody streak.*
 der Auswurf enthält Blutstreifen, *the expectoration is streaked with blood.*

## 6. Das Wärmemass, *Temperature.*

2102 Temperatur (tü'r), *f., temperature, amount of heat and cold.*
2103 Grad, *m., degree.*
2104 Wärmegrad, *degree of heat.*
2105 Körperwärme, *f., the body heat.*
 hohe Temperatur, *high temperature.*
 sehr hohe Temperatur, *very high temperature.*
 hohe Temperatur in der Achselhöhle, *increased axillary temperature* (408).
 die Temperatur erhöht (1756), *the temperature increased.*
2106 die Temperatur fällt, *the temperature is lowered.*
2107 herabgehen, *to go down, decrease.*
2108 nach und nach, *by degrees, gradually.*
2109 **die** Norm, *the normal standard.*
 die Temperatur geht nach und nach unter die Norm herab, *the t. decreases gradually below the normal standard.*
2110 Höhe, *f., height.*
2111 in die Höhe gehen, *to rise.*
 sie geht plötzlich (1724) in die Höhe, *it rises suddenly.*
2112 Frost, *m., cold, chill, chilliness.*
2113 Schauer, *m., chill, shivering, shuddering.*
2114 Frostanfall, *m., chill, rigor.*
2115 abwechseln, *to alternate, intermit.*
 abwechselnde Röthe (1764) und Frost, *alternate flushing and chilliness.*
 die Spitze (274) der Nase, der Finger und Zehen kalt, *the ends of the nose, fingers, and toes cold.*
2116 frösteln, *to be chilly, shiver.*
2117 Frösteln, *n., chill, shiver.*
2118 überlaufen (721) *v. a., to run over, spread over, overspread.*
 Hitze (or Frostschauer) überlief ihn, *he was seized by heat (with a shivering).*
 fröstelnde Empfindung, abwechselnd mit überlaufender Hitze, *chilly sensations which alternate with flushes of heat.*

2118 die Temperatur steigt (675) schnell, *the t. rapidly rises.*
2119 erreichen, *to attain, arrive at, reach.*
2120 stätig, *a., steadily.*
2121 das Fallen, *falling, decline.*
2122 **greifen, griff, gegriffen,** *to gripe, grasp, lay hold of.*
2123 begreifen, (1) *to feel, touch;* (2) *to contain, comprehend.*
2124 begriffen sein in, *to be at or about, to be engaged in.*
    die Temperatur erreicht ihre äusserste Höhe, *the temperature attains its maximum height.*
    sie ist in stätigem Fallen begriffen, *it steadily falls.*
    sie sinkt bis auf die Norm oder unter dieselbe herab, *it sinks to the normal standard or below it.*

## 7. The Tongue.

**die Zunge ist,** *the tongue is:*
    heiss, *hot.*
    kalt, *cold.*
    trocken, *dry.*
    feucht, *moist.*
2125   rein, *a., clean.*
2126   belegt, *coated.*
    sehr belegt, ⎫
    stark belegt, ⎬ *thickly coated, clammy.*
    dick belegt, ⎭
2127 weisslich, *a., whitish.*
2128 bedecken, *to cover.*
2129 der Käse (*gen.* -s), *cheese.*
2130 käseähnlich, *a., curd-like.*
2131 Beleg, *m., exudation, fur.*
    mit einem weisslichen käseähnlichen Belege bedeckt, *covered with a whitish, curd-like exudation.*
2132 **sie ist,** *it is:*
    pelzig, *a., furred.*
2133   rissig, *a., chapped, sprung.*
2134   gespalten (849), *fissured.*
    fortwährend trocken, *continually dry.*
    trocken und braun, *dry and brown.*
2135   dunkelbraun, *a., dark-brown.*
    weiss, *a., white.*
    blass, *a., pale.*
    gelb, *a., yellow.*
2136   gallig, *a., bilious.*
    gross und schlaff (1862), *large and flabby.*
    roth und geschwollen (1870), *red and swollen.*
2137 entzündet, *inflamed.*
2138 vergrössert, *enlarged.*
2139 einseitig (1729) gefurcht, *furred on one side.*

sie ist, *it is:*
2140 russartig belegt, *sooty, dusky.*
2141 geschworen, or mit Geschwüren behaftet, *ulcerated.*
2142 verhärtet, *indurated.*
2143 der Überzug der Zunge, *fur, coat.*
2144 abstossen, *to thrust, or push off*
2145 hervorragen, *to project, be prominent.*
2146 Papille, *f., papilla.*
der Beleg wird abgestossen, *the fur is cast off.*
ihre Papillen ragen hervor, *its (the tongue's) papillæ become prominent.*
die Zungenränder (1438) sind roth, *the borders of the tongue are red.*
2147 Himbeerzunge, *strawberry tongue, raspberry t.*

## 8. *Appetite and Thirst.*

2148 der Appetit, *appetite.*
2149 Appetit bekommen (322), *to get an appetite.*
2150 den Appetit reizen (825), *to whet, provoke the appetite.*
2151 hungrig sein, *to be hungry.*
2152 **verlieren, verlor, verloren,** *to lose.*
den Appetit verlieren, *to lose the appetite.*
2153 Appetitlosigkeit, *f., want of appetite.*
2154 Appetitmangel, *m., loss of appetite.*
2155 stören, *to disturb, trouble, interrupt.*
2156 **schwinden, schwand, geschwunden,** *to disappear, fail.*
der Appetit ist, *the appetite is:*
gestört, *impaired.*
geschwunden, *completely lost.*
der Appetit ist, *the appetite is:*
2157 mangelhaft, *a., deficient.*
2158 launisch, *a., depraved, perverted.*
2159 capriciös, *a., capricious.*
2160 gierig, *a., ravenous, craving.*
2161 Heisshunger, *m., canine appetite.*
einen Heisshunger haben, *to have a ravenous desire for food.*
2162 Ekel, *m., disgust, loathing.*
2163 Abneigung, *f., repugnance.*
2164 vertragen (360), *to bear, tolerate; agree with.*
absoluter Ekel gegen jede Nahrung (1057), *absolute disgust for eating.*
Abneigung gegen gewisse Nahrungsmittel (1059), *repugnance to taking certain articles of food.*

2164 Nahrungsmittel, die der Patient nicht vertragen kann, *articles of food that disagree with the patient.*
2165 fortwährend, *a.* and *adv., constant.*
2166 Leere (688), *f., emptiness.*
2167 verlangen, *to desire, call for, crave.*
2168 das Verlangen, *desire, longing, craving.*
2169 Säure, *f., acid.*
2170 Gemüse, *n., vegetables (cooked for eating).*
ein Verlangen nach Säuren und Gemüsen, *a craving for acids and for vegetables.*
Empfinden von fortwährender Leere und von Hesshunger, *feeling of a constant emptiness and of hunger.*
2171 der Durst, *thirst.*
2172 durstig, *a., thirsty.*
2173 Durstgefühl, *n., feeling of thirst.*
**der Durst ist,** *the thirst is :*
2174     ausserordentlich, *a., extreme.*
2175     unaufhörlich, *a., unceasing.*
2176     unstillbar nach kalten Getränken, *unquenchable for cold drinks.*
2177 Getränk, *n., drink, beverage.*
2178 Durstsucht, *f., morbid thirst.*
2179 Durstmangel, *m., adipsia, deficient sensation of thirst.*
2180 bitter, *a., bitter.*
ein bitterer Geschmack im Munde, *a bitter taste in the mouth.*

## 9. *Vomiting and Nausea.*

2181 **brechen, brach, gebrochen,** *to vomit, spit up; to break.*
2182 sich erbrechen, } *to vomit.*
2183 sich übergeben, }
2184 sich erbrechen (or übergeben) wollen, *to feel sick, to retch.*
2185 übel, *a.* (2042), *ill, qualmish*
2186 Übelkeit, *f.* } *rising of the stomach.*
2187 Brechlust, *f.* }
2187 Brechlust haben, } *to feel qualmish*
Übelkeit haben, }
es wird mir übel, *I feel my stomach rise.*
2188 Neigung, *f., inclination, desire.*
Neigung zum Erbrechen, *nausea, squeamishness.*
2189 Brechreiz, *m., provocation or tendency to vomit.*

2190 sich würgen, *to make an effort to vomit.*
2191 Würgen, *n., retching, qualm.*
2192 **liegen, lag, gelegen,** *to lie, be recumbent.*
    das liegt mir schwer im Magen, *that presses on my stomach.*
2193 blähen, *to inflate, cause flatulency.*
2194 blähend, *flatulent, causing flatulency.*
2195 Wind, *m., wind, flatulence.*
    Winde verursachend (1671), *causing flatulency.*
2196 Blähung, *f., flatulence.*
2197 Flatulenz, *f., wind.*
2198 das Vollsein, *fullness, distention.*
    Gefühl von Vollsein, *a feeling of distention.*
2199 faul, *a., rotten, foul, putrid.*
2200 **das** Ei (*pl.* Eier), *egg.*
    Geschmack und Geruch (1420) von faulen Eiern, *taste and odor of rotten eggs.*
2201 aufstossen, *to rise up, belch, eructate.*
2202 das Aufstossen, *eructation, belching.*
2203 rülpsen, *to belch, eructate.*
2204 das Rülpsen, *belching, eructation.*
2205 faulicht, *a., putrescent, putrid.*
2205 das faulichte Aufstossen, *sulphurous eructation.*
    saures Aufstossen, *sour eructation.*
    intensive Säure (2169) des Magens, *intensive acidity of the stomach.*
2206 Hitze, *f., heat, warmth.*
    brennende (1691) Hitze im Magen, *burning heat in the stomach.*
2207 das Magenbrennen,  ⎫ *heart-*
    or              ⎬ *burn,*
2208 das Sodbrennen,    ⎭ *pyrosis*
2209 Magensäure, *f., acidity of stomach.*
2210 Überschuss, *m., excess.*
    Überschuss an Säure im Magen, *excessive acidity of stomach.*
    Gefühl von Brennen im Magen, *a sensation of burning in stomach.*
    Gefühl von Unbehagen (1602) nach dem Essen, *uneasiness after meals.*
2211 Magenweh, *n., gastrodynia, stomach-ache.*
    marternde (1716) Brechneigung, *distressing nausea.*
2212 Anfall, *m., seizure, attack, fit.*
    qualvolle (1705) Anfälle von Würgen (2191), *distressing fits of retching.*

das **Erbrochene** ist,
*the ejected matter is:*
Nahrung oder Flüssigkeit mit Speichel (981) und Schleim vermischt, *food or liquid mixed with saliva and mucus.*
2213 vermischen, *to mix.*
2214 gähren, *to ferment.*
2215 Gährung, *f., fermentation.*
halb verdaute (1047) Nahrung in einem Zustande (1513) saurer Gährung, *half digested food in a state of acetous fermentation.*
Schleim, zuweilen in grosser Menge, *mucus, occasionally in large quantities.*
eine dünne, wässerige (708) Flüssigkeit, die dem Speichel ähnlich sieht, *a thin watery fluid looking much like saliva.*

das **Erbrochene** ist,
*the ejected matter is:*
2215 Galle (1099), *gall, bile.*
biliöses Erbrechen, } *bilious*
or } *vomit-*
2216 Gallenbrechen, } *ing.*
Eiter (1922), *matter, pus.*
Blut (683), *blood.*
2217 (zersetzen, *to decompose, putrefy, alter.*)
Mengen zersetzten Blutes, *quantities of altered blood.*
2218 Kaffeesatz, *m., coffee-grounds*
2219 (Kaffeesatzartig, *a., coffee-ground-like.*)
eine Kaffeesatzartige Flüssigkeit, *a fluid looking like coffee-ground.*
2220 Koth, *m., excrement, fæces.*
2221 Fäcalmasse, *f., stercoraceous matter.*
2222 das Kothbrechen, }
2223 das Miserere* (rā′rĕ), }
*vomiting of stercoraceous matter.*

## 10. Defæcation.

2224 Stuhlgang, *m., stool, alvine evacuation.*
2225 Stuhl, *m., chair, stool.*
2226 zu Stuhle gehen, *to go to stool.*
2227 nicht zu Stuhle kommen können, *to be constipated.*
2228 Darmthätigkeit, *f., action of the bowels.*
Darmthätigkeit gestört (2155), *the action of the bowels is deranged.*
Unbehaglichkeit im Leibe, *uneasy feeling in the abdomen.*

* Literally: "Have mercy!" beginning of the 51st Psalm in Rom. Cath. Church.

2229 kollern, *to roll, rumble.*
2230 das Kollern, *borborygmus.*
Kollern in den Eingeweiden (919), *rumbling in the intestines.*
2231 erregen, *to excite, stimulate.*
heftige, rollende (1316) Bewegung des erregten Darmes, *a violent rolling motion of the excited intestine.*
2232 Stuhlmangel, *constipation.*
2233 Stuhlverstopfung, f., *sluggish action of the bowels.*
2234 Leibesverstopfung, *costiveness.*
2235 hartnäckige (1634) Verstopfung, *bowels very constipated.*
2236 ein harter (1858) Stuhlgang, *costiveness.*
2237 einen harten Leib haben, *to be costive.*
2238 hartleibig, a., *costive.*
2239 Diarrhoe, f. ⎫
2240 Durchfall, m. ⎬ *Diarrhœa.*
2241 die schnelle (1629) Katharine, ⎭
2242 den Durchfall haben, *to have the diarrhœa.*
Neigung (2188) zur Diarrhöe, *tendency to diarrhœa*
2243 das Abweichen, *relaxation, looseness, diarrhœa.*
hartnäckige Diarrhöe, *obstinate diarrhœa.*

2244 Entleerung, f., *evacuation, stool.*
2245 Drang, m., *impulse, pressure.*
2246 drängen, *to throng, press.*
fortwährender (2165) Drang zum Stuhlgang, *a continual desire to go to stool.*
heftiges Drängen, *violent straining.*
2247 Stuhldrang, *tenesmus.*
2248 kritisch, a., *critical.*
2249 Darmentleerung, f., *discharge from the bowels.*
2250 Darmblutung, f., *bleeding from the intestinal canal.*
2251 farblos, a., *colorless.*
2252 leichtgelb gefärbt, *tinged with yellow.*
2253 trüb, a., *turbid.*
2254 reiswasserartig, a., *resembling rice-water.*
2255 Reiswasserstuhl, *rice-water stool.*
2256 Menge, f., *quantity, plenty.*
2257 **gleichen, glich, geglichen,** *to resemble.*
2258 das Eiweiss, *the white of an egg.*
2259 rahmähnlich, a., *creamy.*
**wie die Entleerungen sind,** *how the alvine evacuations are:*
kritische Entleerungen, *critical discharges.*
unregelmässige E., *irregular d.*

wie die Entleerungen sind, *how the alvine evacuations are:*

2259 unwillkührliche(553)E., *involuntary d., incontinence of stools.*
wässrige (708) Stuhlgänge, *watery stools.*
farblose E., *colorless d.*
leichtgelb gefärbte E., *d. tinged with yellow.*
sehr dünne und wässrige E., *very thin and watery d.*
entschieden gelbe E., *decidedly yellow d.*
trübe E., *turbid d.*
reiswasserartige E., *d. resembling rice water.*
dem Weissen vom Ei gleichende E., *d. resembling the white of an egg.*
aus Excrementen und Eiter bestehende E., *d. composed of fæces and pus.*
aus Galle bestehende E., *d. consisting of bile.*
Stühle mit Eiter in grosser Menge, *stools with pus in large amount.*

wie die Entleerungen sind, *how the alvine evacuations are:*

2259 Stühle mit vielem Schleim, *discharges with much mucus.*
Stühle mit weissen, rahmähnlichen Streifen, *stools with whitish creamy streaks.*
mit einem Überschuss an Galle, *with an excess of bile.*
grüngefärbte E., *d. of green color.*
2260 gelbbraune E., *yellowish-brown d.*
gelbe E., *yellow d.*
gallige E., *bilious d.*
2261 thonartige E., *clayey d.*
2262 fast weisse E., *almost white d.*
2263 schwarze E., *black d.*
2264 theerartige E., *d. of tarry appearance.*
Entleerungen von reinem Blute, *discharges of pure blood.*
2265 mit blutgestreiftem Schleim, *with bloodstreaked mucus.*
2266 fleischwasser ähnliche E., *d. resembling the washings of meat.*
2267 geléeartige E., *d. like jelly.*

**wie die Entleerungen sind,** *how the alvine evacuations are:*

2268 fettige Stühle, *fatty stools.*
2269 Entleerungen von speichelartigen Massen (82), *discharges of matter like saliva.*
2270 Excremente hart und knollig, *fæces hard and knotty.*
 weisse, gewöhnlich harte Kothmassen, *whitish stools, generally hard.*
2271 (widerlich, *a., offensive.*)

**wie die Entleerungen sind,** *how the alvine evacuations are:*

 Stühle mit ausserordentlich widerlichem Geruche, *stools with an extremely offensive odor.*
 Stühle mit saurem Geruch, *stools of a sour smell.*
2272 (die Form, *form.*)
2273 (platt, *a.* and *adv., flat, level.*)
2274 (drücken, *to press, squeeze.*)
 Excremente platt gedrückt, *flattened stools.*

## 11. *Excretion of the Urine.*

2275 absondern, *to secrete, set apart.*
2276 den Harn absondern, *to urinate, discharge urine.*
2277 Wasser lassen, *to make water.*
2278 pissen, *to urinate.*
2279 Pisse, *f., urine.*
2280 die kalte Pisse, *urine of vesical catarrh.*
2281 harnartig, *a., urinous.*
2282 Bestandtheil (1917), *m., component.*
2283 Harnbestandtheile, *constituents of the urine.*
2284 das Salz ( *pl.* -e), *salt.*
2285 Harnsalze, *urinary salts.*
2286 harnsaure Salze, *urates.*
2287 Phosphor, *m., phosphorus.*
2288 Harnphosphor, *uretic phosphorus.*
2289 Schwefel, *m., sulphur.*
2290 Schwefelsäure, *f., sulphate.*
2291 Harnsatz, *m., hypostasis. sediment, deposit.*
2292 abnorme Harnbestandtheile, *abnormal substances in the urine.*
2293 oxalsaure Salze, *oxalates.*
2294 Zucker, *m., sugar.*
2295 Überfluss, *m., superfluity, excess.*
 Harnsäure in grossem Überfluss, *uric acid in large excess.*

2296 Satz, *m.*, *deposit, sediment.*
ein starker Satz im Urin, *a heavy deposit in the urine.*
2297 Sand, *m.*, *sand.*
2298 sandig, *a.*, *sandly, gritty.*
eine sandige Masse, *a gritty substance.*
2299 Gries, *m.*, *gravel, sand.*
2300 Harngries, *m.* ⎫ *arena, gravel,*
or ⎬
2301 Harnsand, ⎭ *urine sand.*
2302 besehen, *to look on, inspect, examine.*
2303 besichtigen, *to inspect, view, examine.*
2304 beschauen, *to view, inspect, examine.*
den Harn besehen, *to inspect, examine the urine.*
2305 Harnbesichtigung, ⎫
2306 Harnbeschauung, ⎬
*inspection of urine, uroscopy.*

**die Farbe des Urins ist,** *the color of the urine is :*

2307 sehr hell, *very light.*
2308 hellgelb, *bright yellow.*
2309 grünlich-gelb, *greenish-yellow.*
braun, *brown.*
2310 (Schaum, *m.*, *foam, scum.*)
2311 braunschäumend, *having a brownish foam.*
2312 schmutzig blau, *dirty blue.*
schwarz, *black.*

**die Farbe des Urins ist,** *the color of the urine is :*

2312 blass, *pale.*
roth, *red.*
2313 von dunkler (710) Färbung, *of a dark hue.*
2314 hochgestellt, *high-colored.*
2315 seifenartig, *soapy-looking.*
dick und trüb, *thick and dark.*
2316 rauchig, *a.*, *smoky.*
2317 wolkig, *a.*, *cloudy.*
2318 Rauch, *m.*, *smoke, reek.*
2319 Wolke, *f.*, *cloud.*
2320 specifisch, *a.*, *specific.*
2321 Gewicht, *n.*, *weight, heaviness, gravity.*
2322 aussehen (1307), *to look, appear* (1589).
der Urin hat ein hohes specifisches Gewicht, *the urine is of high specific gravity.*
er hat eine rauchige Farbe, *it is of a smoky (dingy) color.*
er sieht wolkig und trübe aus, *it looks cloudy and dark-colored.*
2323 anwesend, *a.*, *present.*
2324 Blutkügelchen, *n.*, *blood globule.*
2325 sparsam, *a.*, *scanty.*
Blut und Eiter sind im Harn anwesend, *blood and pus are present in the urine.*

2325 Blutkügelchen sind im Harn, *blood-globules are in the urine.*
die Harnabsonderung ist sparsam, *the secretion of urine is scanty.*
2326 Harnverhaltung, ⎫
2327 Harnverstopfung, ⎬
2328 Harnzwang, ⎭
*ischuria, retention of urine.*
2329 häufig, *a., frequent.*
2330 Harndrang, *m.,* strangury.
häufiges Uriniren, oft mit Schmerz an der Eichel (1188) verbunden (1760), *frequent micturition, often attended with pain at the end of the penis.*
2331 Polinurie, *polyuria.*

2332 unwillkührlicher (553) Harnabgang, *involuntary (or spontaneous) discharge of urine, eneuresis.*
2333 excessive Ausscheidung (974) von Urin, *excessive flow of urine.*
2334 Strom, *m., stream.*
der Urin fliesst in einem ununterbrochenen (2002) Strom, *the urine flows in a continuous stream.*
2335 Tropfen, *m., drop.*
2336 tropfenweise, *drop by drop.*
der Urin wird tropfenweise gelassen und dabei besteht eine brennende Empfindung am Halse der Blase, *the urine is voided drop by drop, and is accompanied by a scalding sensation at the neck of the bladder.*

## 12. *Attitude and General Condition of the Body.*

2337 ausstrecken, *to stretch out, extend.*
ausgestreckt liegen, *to lie stretched out, extended.*
mit ausgestreckten Gliedern liegen, *to lie with extended limbs.*
im Bette krank liegen, *to lie ill in bed.*
2338 einziehen (174), *to draw in, inflect; bend, retract.*
mit eingezogenen Gliedern liegen, *to lie with flexed limbs.*

2339 vorbeugen, *to bend forward.*
liegen mit dem Körper nach vorn vorgebeugt, *to lie with the body bent forward.*
2340 aufrecht, *adv., upright, erect.*
im Bette aufrecht sitzen (456), *to sit upright (or erect) in bed.*
unbeweglich (141) in einer Lage (1755) liegen, *to lie immovably in same posture.*

2341 beständig, *a. and adv.*, *constant*.
2342 Seite, *f.*, *side, flank*.
    beständig auf dem Rücken oder auf einer Seite liegen, *to lie constantly on the back or on one side*.
    auf jeder Seite liegen, *to lie on either side*.
    gewöhnlich (1650) auf der leidenden Seite (1551) liegen, *to lie ordinarily on the affected side*.
    auf dem Gesichte (241) liegen, *to lie on the face*.
2343 frei, *a. and adv.*, *free, unconfined, at liberty*.
    sich im Bette frei bewegen (139) können, *able to move freely about in bed*.
    in einer und derselben Lage liegen müssen, *to lie fixed in one and the same position*.
    auf dem Rücken mit angezogenen (562) Schenkeln liegen, *to lie on the back with his thighs flexed*.
2344 **werfen, warf, geworfen,** *to throw*.
    liegen mit dem Kopfe nach hinten geworfen, *to lie with his head thrown back*.
2345 ruhig, *a. and adv.*, *quiet*.
2346 unruhig, *a. and adv.*, } *restless*.
    rastlos,
    ruhig liegen, *to lie quietly*.
2346 unruhig liegen, or rastlos liegen, } *to lie restlessly*.
2347 bekunden, *to make known, show*.
2348 grosse Rastlosigkeit bekunden, *to show intense restlessness*.
    ausser (922) dem Bette sein, *to be out of bed*.
2349 **stehen, stand, gestanden,** *to stand*.
2350 **gehen, ging, gegangen,** *to go, walk*.
2351 einhergehen, *to walk along*.
    schnell (1629) gehen, *to walk quickly*.
2352 rüstig, *a. and adv.*, *brisk, vigorous*.
    rüstig einhergehen, *to walk briskly*.
    aufrecht stehen oder einhergehen, *to stand or walk along erect*.
2353 Schritt, *m.*, *step*.
2354 fest, *a. and adv.*, *firm, solid*.
    mit festen Schritten einhergehen, *to walk along with firm steps*.
2355 unfähig (44), *unable*.
2356 sich bücken, *to bow, bend over*.
2357 gebückt, *bowed, bent*.
2358 vorwärts, *adv.*, *forward, toward the front*.

2358 nach einer Seite gebeugt stehen oder gehen, *to stand or walk with the body bent to one side.*

gebückt *or* nach vorwärts gebeugt stehen oder gehen, *to stand or walk with the body bent forward, or stoopingly.*

2359 fort, *adv. and prefix, forth, onward, forward.*

er bewegt sich nur mühsam (2039) fort, *he moves with difficulty.*

2360 seine Bewegungskraft ist beeinträchtigt (2031), *his powers of locomotion are impaired.*

2361 Gang, *m., gait, walk, step.*

**sein Gang ist,** *his gait is:*
schnell, *quick.*
langsam, *slow.*
mühsam, *painful.*
ungleich, *unequal.*
2362 steif, *stiff.*
2363 (hinken, *to limp, halt.*) hinkend, *limping, halting.*
2364 sicher, *steady.*
2365 unsicher, *uncertain, unsteady.*
2366 (taumeln, *to reel, totter.*) taumelnd, *tottering.*
2367 schwankend (2000), *staggering.*
2368 wackelig, *shaky.*

**sein Gang ist,** *his gait is:*
2369 (wackeln, *to waver, totter.*)
2370 beschwerlich, *laborious.*
2371 zitternd, *trembling.*
2372 (stolpern, *to stumble.*) stolpernd, *stumbling.*

2373 erheben, *to raise, lift.*
2374 sich erheben, *to raise one's self, rise, arise.*
2375 Mühe, *f., difficulty, trouble.*
2376 unmöglich, *a., impossible.*
2377 sich ausruhen, *to rest one's self.*
2378 berauben, *to rob, deprive.*
2379 beraubt sein einer (*genit.*) Sache, *to be deprived of a thing.*
2380 erschlaffen, *to relax, enervate.*
2381 Zuckung (1777), *convulsive movement, convulsion, fit.*
2382 lahm, *a. and adv., lame.*
2383 lahm gehen, *to halt.*
2384 gleich, *a. and adv., alike, equal.*
2385 unvollkommen, *a., imperfect.*
2386 beide, *both.*

er kann sich nur mit Mühe vom Sitze (1751) erheben, *he can rise from his seat only with difficulty.*

es ist ihm unmöglich sich zu bücken, *it is impossible for him to stoop forward.*

2386 er muss sich oft ausruhen, *he must often rest himself.*

er ist aller Kraft (17) zur Bewegung (523) beraubt, *he is deprived of all power of motion.*

seine Muskeln sind erschlafft, *his muscles are relaxed.*

sie sind von krampfhaften (1710) Zuckungen bewegt (139), *they are agitated with convulsive movements.*

seine Glieder sind steif, *his limbs are stiff.*

er geht lahm auf einer Seite, *he halts on one side.*

sein Gang ist gleich unvollkommen auf beiden Seiten, *his gait is equally imperfect on both sides.*

## 13. *General Aspect of the Patient.*

das allgemeine Aussehen (1589), *the general aspect.*
2387 Grösse, *f.*, *size, bulk.*
2388 Dickleibigkeit, *f.*, *bulkiness,*
2389 Umfang, *m.*, *size, extent, bulk.*
von grossem Umfang, *bulky.*
2390 **nehmen, nahm, genommen**, *to take.*
2391 zunehmen, *to take in addition, to increase.*
2392 das Zunehmen, } *increase.*
2393 Zunahme, *f.* }
die Zunahme an Umfang, ⎫
das Zunehmen des Umfangs, ⎬
*increase of bulk.* ⎭
2394 fettleibig, *a.*, *obese, fat.*
2395 Fettleibigkeit, *obesity.*
2396 Schmeer, *sebum, fat, grease.*
2397 Schmeerbauch, *paunch belly.*
2398 **schwellen, schwoll, geschwollen,** *to swell.*
2399 anschwellen (1761), *to swell.*
2400 **die** Geschwulst, *swelling, inflation, tumor.*
angeschwollene Fussgelenke, *swollen ankles.*
eine harte, weiche (385) Geschwulst, *a hard, soft tumor.*
2401 **treiben, trieb, getrieben**, *to drive.*
2402 auftreiben, *to drive up, distend, swell.*
2403 aufgetrieben (*p. part.*), *turgid.*
2404 **blasen, blies, geblasen**, *to blow.*
2405 aufblasen, *to blow up, inflate, swell.*
aufgeblasen (1871), *puffed up, swelled, emphysematous.*
2406 aufdunsen, *be swelled or puffed up.*
aufgedunsen (1872).

2407 Aufgedunsenheit, *f.*  
2408 Aufgeblasenheit, *f.*  
    *puffiness, emphysema, turgescence.*  
2409 eitern, *to suppurate.*  
2410 Eiterung, *suppuration.*  
2411 geschwürig, *a., ulcerous.*  
2412 geschwürig werden, *or*  
2413 **schwären, schwor, geschworen,**  
    *to ulcerate, fester.*  
2414 Geschwür, *n., ulcer.*  
    er hat ein Geschwür am Beine, *he has an ulcer on the leg.*  
2415 Beule, *boil.*  
2416 Eiterbeule, *abscess.*  
2417 Furunkel, *m., furuncle, boil.*  
2418 Blutgeschwür, *n.* } *furuncle*  
2419 Blutschwär, *n.*  
2420 Wassergeschwulst, *f., œdema, anasarca.*  
    der allgemeine Zustand (1513) der Haut ist ödematös, *the general condition of the skin is œdematous.*  
    die Haut ist an Händen und Füssen geschwollen, *the skin is tumefied on the hands and feet.*  
    der Bauch ist aufgeschwollen, *the abdomen is swollen.*  
2421 **kennen, kannte, gekannt,** *to know.*  
2422 erkennen, *to recognize, distinguish.*  

2423 Vergrösserung, *f.* (2138), *enlargement.*  
    die Vergrösserung des Kopfes, *enlargement of the head.*  
2424 abnehmen (2390), *to decrease, diminish.*  
2425 das Abnehmen } *diminution*  
2426 die Abnahme *decrease.*  
2427 Körperumfang, *m.* (2389), *size, extent of body, bulk.*  
    die Gesichtszüge sind so geschwollen, dass sie kaum zu erkennen sind, *the features are so tumefied as to be hardly recognizable.*  
    sein Gesicht schwillt immer mehr an, *the swelling of his face increases very much.*  
2428 wölben, *to arch, vault.*  
2429 Wölbung, *f., curved elevation, vaulting.*  
2430 die Hervorwölbung der Brust, *the prominence (lit. the curved elevation) of the chest.*  
    an Gewicht (2321) zunehmen, *to gain flesh.*  
    an Gewicht abnehmen, *to lose flesh.*  
    das Abnehmen an Körperumfang, *diminution in bulk.*  
2431 mager, *a., lean, thin, meagre.*  
2432 Magerkeit, *leanness, meagreness.*

2433 abmagern, *to become lean or thin, to emaciate.*
2434 abzehren, *to fall away.*
2435 Abmagerung, *emaciation.*
2436 Abzehrung, *waste, falling away, emaciation.*
2437 Auszehrung, *consumption.*
    die Abzehrung *or* Auszehrung bekommen(322),*to fall into a consumption.*
2438 abgemagert, } *a., emaciate.*
2439 abgezehrt,
2440 Marasmus, *m., marasmus.*
    ein Gefühl von fortwährend zunehmender Schwäche, *a feeling of steadily augmenting debility.*
2441 ermatten, *to grow or to make tired.*
2442 Ermattung, *lassitude, exhaustion.*
2443 müde, *a., tired, weary, fatigued.*
2444 ermüden, *to weary, grow tired.*
2445 Ermüdung, *weariness, lassitude.*
2446 danieder, *adv., down there, down.*
2447 danieder liegen, *to lie down, be broken, to succumb.*
2448 das Daniederliegen der Kräfte, *prostration.*
2449 zerschlagen (665), *to beat in pieces, use up.*
2450 zerschlagen, *a., bruised all over, worn out.*

2451 das Gefühl von Zerschlagenheit in allen Gliedern, *the feeling of extreme lassitude.*
2452 fühllos, *a., senseless.*
    taub (1851) *or* fühllos machen, *to benumb.*
2453 Betäubung, *f.* } *numbness.*
2454 Taubgefühl, *n.*
2455 Taumel, *m.* (2366), *reeling, tottering.*
    ein entschiedenes Gefühl von Taumel, *a decided impression of vertigo.*
    an allen Gliedern lahm sein, *to be quite out of joint.*
2456 starr, *benumbed, rigid, fixed.*
2457 Erstarrung, }
2458 Unempfindlichkeit,
    *torpor, torpidity.*
2459 die dumpfe (1688) Starrheit, }
2460 Lethargie, *f.*
    *stupor, lethargy.*
2461 Cachexie, }
2462 die ungesunde (1533) Natur des Körpers,
    *cachexia, unsound nature (condition) of the body.*
    das Glied ist abgezehrt (2439), *the limb has shrunken.*
2463 Erschlaffung (2380), *relaxation, debility, palsy.*
    Erschlaffung der Gliedmassen, *relaxation of the limbs.*

2464 **sinken, sank, gesunken,** *to sink.*
2465 einsinken, *to sink in.*
2466 Brustwand, *f.*, *chest-wall.*
das Einsinken der Brustwand, *the sinking in of the chest-wall.*
2467 einfallen (1583), *to fall in or down, to sink, flatten.*
die Oberfläche (1763) der Brust ist eingefallen, *the surface of the chest is flattened.*
die Ausdehnung (744) der Brust ist mangelhaft (2157), *the expansion of the thorax is defective.*
2468 schrumpfen, *to shrink, shrivel* (Eng. obs. *shrimp*).
2469 Schrumpfung, *f.*, *shrinking.*
2470 Brustseite, *chest-wall.*
Schrumpfung der Brustseiten, *the shrinking of the thoracic walls.*
2471 flachgedrückt (178, 2274), *depressed.*
2472 abgeflacht, *flattened.*
2473 hervorstehend, *prominent, bulging.*
2474 Muskelzucken, *muscular tremor or jerking.*
2475 Hautröthe, *erythema.*
Hautröthe mit Zucken (1951) und Brennen (1821), *erythema with itching and burning.*
2476 **schlafen, schlief, geschlafen,** *to sleep.*
2477 Schlaf, *m.*, *sleep.*
2478 einschlafen, *to fall or get asleep*, fig. *to be benumbed, asleep.*
der Fuss ist mir eingeschlafen, *my foot is benumbed, asleep.*
Gefühl des Kribbelns (1706) und Einschlafens in den Extremitäten, *a sense of tingling and of numbness in the extremities.*
Kribbeln in den Gliedern, *tingling in the limbs.*
2479 das Zittern (2371), *tremor, trembling.*
2480 knirschen, *to gnash, grate.*
2481 das Zahnknirschen, *grinding of the teeth.*
die Daumen (442) sind eingezogen, *the thumbs are drawn inward.*
2482 **schreien, schrie, geschrieen,** *to cry, scream, shriek.*
2483 Schrei, *m.*, *cry, scream, shriek.*
2484 ausstossen (736), *to thrust or push out; burst out with, utter abruptly.*
der Patient stösst Schreie aus, *the patient utters outcries, screams.*
fortwährend rollende Bewegung des Kopfes, *continuous rolling of the head.*
2485 schwammig, *a.*, *spongy, soft, fungous.*

2486 Schwamm, *m.*, *sponge*, *fungus*.
das Zahnfleisch (320) ist angeschwollen, schwammig, und bei der leisesten Berührung (1744) blutend, *the gums are swollen and spongy, and bleed on the slightest touch*.
2487 Klaue, *f.*, *claw*.
2488 klauenförmig, *a.*, *like a claw*.
2489 kerben, *to notch, jag, indent* (*related with Engl.* **kerf, carve**).
Nägel (951) klauenförmig gebogen und leicht gekerbt, *nails curved like claws and slightly cracked*
2490 Karbunkel, *carbuncle*.
Neigung zu Geschwüren und Karbunkeln, *a tendency to boils and carbuncles*.
fortwährend zunehmende Abmagerung (2435) des Körpers, *a steadily progressing waste of the body*.

**das Gesicht ist,** *the countenance is:*
ödematös, *œdematous*.
2491 dunkelgefärbt, *dark colored*.
2492 blauroth, *dusky red*.
gelb, *yellow*.
livid, *livid*.
2493 leichenhaft, *cadaverous*.
fahl, *sallow, of a sallow hue*.

**das Gesicht ist,** *the countenance is:*
2494 bleich (1881) und wachsähnlich, *pallid and waxlike*.
2495 todtenbleich, *deathlike*.
2496 entsetzlich abgemagert (2438), *horribly (or distressingly) emaciated*.
2497 spitz (274), *pointed*.
2498 eingesunken (*or* eingefallen) und spitz, *pinched and sharpened*.
2499 mit eingefallenen (2467) Backen, *hollow cheeked*.
2500 unverändert, *a.*, *unaltered*.
2501 vollblütig, *a.*, *flushed, full-blooded*.

2502 **der Gesichtsausdruck ist,** *the expression of countenance is:*
ruhig, *quiet*.
unruhig, } *unquiet, wild,*
rastlos, } *uneasy, restless*
2503 mürrisch, *peevish*.
leidend, *suffering*.
2504 niedergeschlagen, }
2505 deprimirt, }
*depressed*.
2506 ängstlich, *anxious, distressed*.
verzweifelt, *desperate*.
2507 erschreckt, *terrified*.
2508 erregt, *irritable*.

der Gesichtsausdruck ist, *the expression of countenance is:*
2509 aufgeregt, *excited.*
gereizt (825), *irritated.*
2510 leicht reizbar, *irritable.*
2511 verzerrt, } *distorted.*
2512 verdreht,
2513 verwirrt, *confused.*
2514 achtlos, } *listless.*
2515 sorglos,
2516 theilnahmslos, *impassive.*
2517 wild und ängstlich, *fierce and anxious.*
2518 dem einer Leiche ähnlich, *like that of a corpse.*
starr, *fixed, rigid, motionless.*
2519 stumpf, *obtuse, dull.*
2520 unstätt, *wandering.*
2521 ausdruckslos, *unmeaning.*
2522 einfältig, *silly.*
2523 rasend, *maniacal.*
2524 irrsinnig, *insane.*
leer (688), *vacant.*
2525 nichtssagend, *dull.*
2526 stupid und rauh (1866), *stupid and harsh.*

die Stirn gerunzelt (1319), *there is a frown on the brow.*
die Lippen blau, *the lips blue, livid.*
trocken (1868), *parched.*
mit Schaum (2310) bedeckt, *covered with foam.*

2527 die Sehkraft gestört, *the sight impaired.*

die Augen sind, *the eyes are:*
tief eingefallen, } *injected.*
2528 injicirt,
2529 glänzend aber wässerig (708), *brilliant but watery.*
geschlossen (405), *closed.*
2530 halb (246) offen, *half open.*
2531 schielend, *squinting.*
2532 von einem dunkeln Ringe umgeben, *surrounded by a dark ring.*
trüb, *dull, dim.*
2533 thränend, *watery.*
2534 stier, *staring.*
2535 perlenartig, a., *pearly.*
2536 mit gläsernem Aussehen, *with a glassy look.*

die Augäpfel sind, *the eyeballs are:*
starr, *fixed.*
rollend, *rolling.*
2537 oscillirend, *oscillating.*
in fortwährender Bewegung, *in constant motion.*
2538 unbeweglich, *immovable.*
2539 dilatirt, *dilated.*
2540 contrahirt, *contracted.*
2541 eng (998) zusammengezogen, *closely contracted.*

|  | die Augäpfel sind, *the eyeballs are :* | 2542 | die Augenlider sind halb offen, *the eyelids are half open.* |
|---|---|---|---|
| 2542 | unbeeinflusst vom Licht, *uninfluenced by light.* | | geschwollen, *puffy.* |

## 14. Derangement of general Sensation.

2543 Sinnestäuschungen,
2544 Sinnesstörungen,
2545 Hallucinationen,
*deceptions of the senses, illusions, hallucinations.*

2546 Sehstörungen, *perversion of the sense of vision.*
2547 Mücke, *f.*, *gnat, midge.*
2548 Fliege, *f.*, *fly.*
2549 Spinne, *spider.*
2550 Flocke, *flake, flocculus.*
2551 Mückensehen, *n.*
2552 Fliegensehen, *n.*
2553 Spinnensehen, *n.*
2554 Flockensehen, *n.*
2555 Funkensehen (59), *n.*
*muscæ volitantes, myodesopsia, photopsia.*
2556 Flimmerbewegungen, *ciliary motions.*
2557 Fleck, *m.*, *spot.*
2558 Fleckensehen, *n.*, *the fancied perception of black spots moving around.*
2559 Halbsehen, *n.*, *seeing but half of an object.*
2560 Doppelsehen, *n.*, *double vision.*
2561 Gehörtäuschungen, *illusions of the sense of hearing.*

2562 brausen, *to roar, bluster.*
2563 sausen, *to tinkle, tingle.*
2564 Brausegeräusche (655), *noise in the ears.*
2565 Ohrenbrausen,
2566 Ohrensausen, *buzzing in the ears,*
2567 Ohrensingen, *tinnitus*
2568 Ohrenklingen, *aurium.*
2569 Ohrentönen,
2570 Geruchsphantasmen, *phantasms of the sense of smelling.*
2571 Geschmacksphantasmen, *phantasms of the sense of taste.*
2572 tasten, *to touch.*
2573 subjective Tastempfindungen, *subjective sensations of the sense of touch.*
2574 Verstummung, *f.*, *loss of speech.*
2575 Beschwerde beim Articulieren, *defective articulation.*
2576 Stimmbeschwerde, *phonopathia.*

**die Stimme ist,** *the voice is :*
heiser, *hoarse.*

die Stimme ist, *the voice is:*
2576 rauh, *husky.*
näselnd (1489), *nasal.*
2577 (erlöschen (*incorrect for* **erleschen**), **erlosch, erloschen,** *to be extinguished.*)
2578 gänzlich erloschen, *completely lost.*
2579 die Stimme, *the voice:* (**klingen, klang, geklungen,** *to sound.*)
klingt gebrochen (2181), *sounds broken.*
klingt fremdartig heiser und schwach, *sounds strangely hoarse and faint.*
2580 fremdartig, *a. and adv., strange.*

### 15. *Alteration in the Mental Condition.*

2581 Schläfrigkeit, *drowsiness.*
2582 tief, *a., deep.*
tiefer Schlaf,
2583 Stupor, *m.*
2584 Schlafsucht, *f.*  } *somnolence stupor, lethargy.*
2585 Somnolenz, *f.*
2586 Coma, *n., coma.*
2587 Narkose, *narcosis.*
2588 der soporöse Zustand, *soporous state.*
lange andauernder (1727) Schlaf, *continued and profound sleep, the sleep of prostration.*
Neigung zu fortwährendem (2165) Schlafe, *tendency to somnolency.*
unruhiger (2346) Schlaf,
2589 Schlaflosigkeit, *f.*  }
2590 Insomnie, *f.*
*insomnia, wakefulness, sleeplessness.*
2591 Agrypnie, *agrypnia.*
2592 Schlafscheu, *f., hypnophobia.*
2592 an häufiger (2329) Schlaflosigkeit leiden, *to be tormented with sleepless nights.*
2593 **wissen, wusste, gewusst,** *to know.*
2594 bewusst, *a., conscious.*
2595 bewusstlos, *a., unconscious.*
2596 das Bewusstsein, *consciousness.*
2597 Bewusstlosigkeit, *f., unconsciousness, insensibility.*
2598 Ohnmacht, *f., swoon, fainting, syncope.*
2599 in Ohnmacht fallen, *to swoon, to be seized or taken by a fainting fit.*
2600 ohnmächtig sein, *to be in a swoon.*
2601 ohnmächtig werden, *to faint, go into a fit.*
2602 vollständig, *a., complete, total, perfect.*

2603 vollständiger Verlust des Bewusstseins, *complete loss of consciousness.*
2604 träumen, *to dream.*
2605 das Träumen, *dreaming.*
2606 **das** Alpdrücken, }
2607 Alp, *m.*
*nightmare, incubus.*
2608 das Aufschrecken, oder }
2609 Auffahren im Schlafe, *starting up out of one's sleep.*
2610 das Schlafwandeln, }
2611 das Nachtwandeln,
2612 der Somnambulismus, *somnambulism, act or practice of walking in sleep.*
2613 Gehirnsymptome (*plur.*), *cerebral symptoms.*
2614 phantasieren, } *to be deliri-*
2615 delirieren, *ous, to rave.*
2616 Delirium, *n.*, *delirium.*
stilles (1654) Delirium, *quiet delirium.*
furibundes (1655) D., *furious delirium.*
stumpf, *a.*(2519), *blunt, dull.*
2617 stumpfsinnig, *a.*, *dull-witted, stupid.*
2618 Stumpfsinn, *m.*, *imbecility.*
2619 Stumpfsinnigkeit, *f.*, *dullness, stupidity.*
Verlust der Sensibilität, }
2620 Empfindungslosigkeit,
2621 Gefühllosigkeit,
2622 Anästhesie, *anæsthesia.*

2623 stimmen, *to be in tune, to tune, to voice* (*organs*).
2624 verstimmen, *to put out of tune,* fig. *to put out of humor.*
2625 Stimmung, *f.*, *disposition, frame of mind.*
2626 Verstimmung, *f.*, *ill temper, uneasiness of the mind.*
2627 Gemüth, *n.*, *mind, soul.*
2628 Gemüthsstimmung, *f.*, *frame of mind.*
2629 geistig, *a.*, *mental.*
geistige Verstimmung, *mental depression.*
2630 schlecht, *bad, ill, wicked.*
er fühlt sich (1984) in schlechter Stimmung, *he feels himself out of spirits.*
gereizte (825) Stimmung, *peevishness of temper.*
gedrückte or niedergedrückte Stimmung, *depressed in spirit.*
erregte (2231) Stimmung, *irritability of temper.*
2631 Laune, *f.*, *humor, whim, temper, mood, caprice.*
2632 launenhaft, *a.*, *capricious, whimsical, peevish.*
2633 Launenhaftigkeit, *f.*, *capriciousness.*
grosse Launenhaftigkeit, *great irritability of temper.*
launische Stimmung, *id.*
veränderte Stimmung, *altered spirits.*

2634 unbekümmert (um), *listless.*
2635 sich kümmern um, *to care for, to concern one's self.*
sich um nichts kümmern, *to be listless.*
2636 gleichgültig (gegen), *indifferent (to), careless, apathetic.*
gleichgültig sein gegen Alles, *to be without animation.*
2637 Abspannung, *f., languor.*
2638 Ausdruck, *m., expression.*
das Gesicht zeigt den Ausdruck von Abspannung, *the countenance expresses languor.*
2639 Beklemmung, *f., oppression, anguish.*
Gefühl der Beklemmung, *a sensation of discomfort.*
2640 aufregen, *to agitate.*
sehr aufgeregt sein, *to be much agitated.*
2641 Aufregung, *f., excitement, exaltation.*
grosse geistige Aufregung, *great mental excitement.*
Aufregung des Gemüths, *exaltation of mind.*
2642 Gemüthsbewegung, *mental emotion.*
2643 das Vermögen, *capability, power.*
2644 Empfindungsvermögen, *n., sensation.*
2645 Fähigkeit, *f.* (44), *faculty, ability.*
2646 **denken, dachte, gedacht,** *to think.*
2647 Denkfähigkeit, *f., faculty of thought.*
2648 Intelligenz, *f., intellection.*
2649 Gedächtniss, *n., memory.*
2650 urtheilen, *to judge.*
2651 Urtheilskraft, *f.* (17), *judgment.*
2652 Function, *f., function.*
2653 Störung, *f.* (2155), *disturbance, disorder, derangement.*
2654 intellectuell, *a., intellectual.*
Störung der intellectuellen Functionen, *disorder of the intellectual functions.*
2655 **der** Wille *or* der Willen (*gen.* Willens), *will, voluntas.*
2656 Willensstörung, *f., impairment of the faculty of volition.*
2657 Sensibilitätsstörung, *impairment of sensation.*
2658 Gefühlsvermögen, *n., sensation.*
gestörtes Gefühlsvermögen, *deranged sensation.*
2659 Trägheit, *f., indolence, laziness.*
2660 Geistesträgheit, *f., intellectual indolence, torpor.*
2661 Geistesstörung, *f., alienation of mind.*
2662 Verstand, *m., understanding, intellect.*

2663 Begriff, *m.*, *conception, idea.*
2664 **der** Gedanke (*gen.* **-ns,** *pl.* -n), *thought.*
2665 verwirren, *to confuse, confound.*
2666 Verwirrung, *f.*, *confusion, disorder.*
2667 Verworrenheit, *f.*, *confusion.*
2668 verwirrt, *a.*, *deranged, distracted.*
2669 Verwirrung der Begriffe, *confusion of the conceptions.*
Verwirrung des Verstandes, *alienation of mind.*
2670 Geistesverwirrung, *confusion, derangement, aberration of mind, craziness.*
2671 zerrütten, *to disorder, destroy, to shatter.*
2672 Geisteszerrüttung, *f.*, *disorder of mind.*
2673 Confusion, *f.*, *confusion.*
Schwere (1832) und Confusion der Gedanken, *heaviness and confusion of thought.*
2674 Geistesstumpfheit, *f.*, *dullness of intellect.*
2675 besinnungslos, *a.*, *senseless.*
2676 Besinnungslosigkeit, *suspension of consciousness.*
2677 Beeinträchtigung (2031), *impairment.*

2677 allmählige (1653) Beeinträchtigung der Intelligenz, *gradual impairment of intelligence.*
gestörte Intelligenz, *deranged intellection.*
Sensibilität erhöht (1756), *sensibility increased.*
2678 verringern, *to diminish.*
Sensibilität verringert, *sensibility diminished.*
2679 Verschlechterung (2630), *deterioration.*
Verschlechterung des Gedächtnisses, *deterioration of memory.*
Schwäche (1542) des Gedächtnisses, *weakening of the memory.*
2680 zeitweise, *adv.*, *from time to time, at times.*
zeitweiser Verlust des Gedächtnisses, *attacks of loss of memory.*
2681 Wortgedächtniss, *n.*, *verbal memory.*
Verlust des Wortgedächtnisses, *loss of verbal memory.*
abnorme Sensationen im Kopfe, *abnormal sensations in the head.*
2682 Irrreden, *mental wandering.*
lautes und heftiges Delirium, *noisy and violent delirium.*
2683 flüchtig, *a.*, *fugitive, fleeting, transient.*

2683 flüchtiges Delirium, *fleeting delirium.*
2684 Zerstreutheit, *absence of mind.*
2685 Trübsinn, *m.*, *dejection, melancholy.*
2686 Wahnsinn, *m.*, *insanity, madness.*
2687 wahnsinnig, *a.*, *mad, insane.* wahnsinnig werden, *to become insane.*
2688 verfallen, *to fall into, run into, decline, sink.* in Wahnsinn verfallen, *to go or run mad.*
2689 Verzuckung, *f.*, *contortion, convulsion.* Verzuckungen bekommen, *to fall into convulsions.*
2690 mondsüchtig, *lunatic, moonstruck.*
2691 rasen, *to rave, rage.*
2692 Raserei, *f.*, *frenzy.*
2693 Augenblick, *m.*, *moment, interval.*
2694 lichte Augenblicke, *lucid intervals.*
2695 verrückt, *a.*, *deranged, crazy.* verrückt, *or* wahnsinnig werden, *to run or to go mad.*
2696 toben, *to rave, rage.*
2697 toll, *a.*, *mad, frantic.*
2698 Tollheit, *f.* } *folly, madness, lunacy.*
2699 Tollwuth, *f.*
2700 Tobsucht, *f.*, *frenzy, insanity.*
2701 der, die Wahnsinnige, *madman, mad woman, maniac.*
2702 der, die Verrückte, }
2703 der, die Tolle, *deranged man, deranged woman, lunatic.*
2704 Trunkenheit, *f.*, *drunkenness.*
2705 Trunkenbold, *m.*, *drunkard.*
2706 berauscht, } *intoxicated,*
2707 betrunken, *inebriated.*
2708 Säufer, *m.*, *drunkard, tippler.*
2709 Säuferwahnsinn, *m.* }
2710 Zitterwahnsinn, *m.* *delirium tremens, madness of drunkards.*

## B. NAMES OF DISEASES.

*Some of the terms given under Symptoms as well as many of the following, serve to designate both primary diseases and symptoms of other diseases.*

2711 Hirnkrankheit, } Gehirnkrankheit, *disease of the brain, brain disease.*
2712 Hirnverletzung, *injury of the brain.*
2713 verletzen, *to injure, hurt.*
2714 Entzündung, *f.*, *inflammation, phlegmon.*
2715 Hirnentzündung, *inflammation of the brain.*

2716 Hirndruck, *m.* (1745), *compressio cerebri.*
2717 Erweichung(385),*softening.*
2718 Hirnerweichung, *softening of the brain.*
2719 Verhärtung (1858), *hardening, induration.*
2720 Hirnverhärtung, *induration of the brain.*
2721 Hirneiterung (2410), *abscess of the brain,*
2722 Wassersucht, *f., dropsy.*
2723 Hirnwassersucht, }
    Gehirnwassersucht, }
    *dropsy of the brain, hydrocephalus.*
2724 erschüttern, *to shake, toss.*
2725 Hirnerschütterung, *concussion of the brain.*
2726 Hirnanämie, *f., cerebral anæmia.*
2727 Eingenommenheit des Kopfes, *heaviness of the head.*
2728 klopfen, *to knock, pulsate violently.*
    Klopfen im Innern des Schädels, *throbbing in the interior of the cranium (or skull).*
2729 auseinander, *asunder.*
    auseinander treiben, *to drive asunder.*
2730 vergrössern, *to enlarge.*
2731 verkleinern, *to diminish.*
2732 zusammenpressen (568), *to compress.*

2732 das Gefühl von Vergrösserung (2423) und Auseinandertreiben des Kopfes, *the feeling of enlargement and driving asunder of the head.*
    das Gefühl von Verkleinerung und Zusammenpressen des Kopfes, *the feeling of diminution (atrophy) and compression of the head.*
    gestörte Bewegung, *deranged motion.*
2733 lähmen (2382), *to lame, paralyze, palsy.*
2734 Lähmung, *lameness, paralysis.*
2735 Paralyse, *f., general paralysis.*
2736 Parese (rā'), *f., partial paralysis.*
2737 halbseitig, *a.* (246), *unilateral.*
    halbseitige Lähmung, }
2738 Halbhähmung, }
    *unilateral palsy.*
2739 Hemiplegie, *hemiplegia.*
    Lähmung beider Seiten, }
2740 Querlähmung (625), }
2741 Paraplegie, *f.,* }
    *palsy occurring on both sides, paraplegia.*
    locale (1661) Lähmungen, *local palsies.*
2742 Lähmungskrankheiten, *paralytic diseases.*

2743 Krampfkrankheiten (1765), convulsive diseases.
2744 tonisch, a., tonic.
2745 klonisch, a., clonic.
2746 tonischer, or Starrkrampf (2456), tonic spasm, tetanus, spasm with rigidity.
2747 clonischer, or Stosskrampf (736), clonic spasm.
2748 Krampfanfälle (2212), spasmodic or convulsive attacks.
den Krampf bekommen (322), to be taken with a cramp.
2749 lachen, to laugh.
2750 das Lachen, laughter, risus.
2751 Krampflachen, n., convulsive laugh, sardonic laugh.
2752 Gehirnhaut, f., membranes covering the brain, meninges.
2753 Entzündung der Gehirnhäute, acute Meningitis, inflammation of the membranes of the brain.
2754 Enzündung der Häute des Gehirns und des Rückenmarks, cerebro-spinale Meningitis, inflammation of the membranes of the brain and of the spine.

2755 geistesschwach, a., feeble-minded, imbecile.
2756 geistesstumpf, a., torpid, imbecile.
2757 der, die Geisteskranke, the diseased in mind, imbecile.
2758 der, die Irre (2524), the lunatic, insane.
2759 Irrsinn, m., mental alienation, insanity.
2760 Monomanie, f., monomania.
2761 die Schwermuth, melancholia, dejection of spirits.
2762 Schlagfluss (665), apo-
2763 Apoplexie, plexy.
2764 Schlaganfall, apoplectic seizure, fit of apoplexy.
2765 rühren, to touch, strike up.
vom Schlage gerührt, seized with apoplexy.
2766 Sonnenstich, m., sun-stroke, insolation.
2767 Catalepsie, f. catalepsy,
2768 Starrsucht, f. trance.
2769 starrsüchtig, a., cataleptic.
2770 Muskelstarre, f., muscular rigidity.
2771 Epilepsie, f. epilepsy, fall-
2772 Fallsucht, f. ing sickness.
2773 epileptisch, a., epileptic.
epileptischer Anfall, epileptic fit.
2774 St. Veits Tanz, St. Vitus dance (chorea St. Viti).
2775 Gesichtskrampf, m., facial spasm.
krampfhaftes Lachen (2751).

2776 Hysterie, *f.*, *hysteria, hysterics.*
2777 Zufall, *m.*, *accident;* (med.) *fit, attack.*
   hysterische (1665) Zufälle, *hysteric fits, attacks.*
   hysterische Krämpfe, *hysteric cramps.*
2778 Hirntumor, *m.*, *tumor of the brain, cerebral tumor.*
2779 Blödsinn, *m.*, *imbecility of mind, idiocy.*
   paralytischer Blödsinn, } progressive Paralyse, } *general paralysis.*
2780 Neuralgie (1776) des Gesichts, *tic douloureux, facial neuralgia.*
2781 Halsleiden, *throat trouble.*
   Steifheit (1759) der Halsmuskeln, *stiffness of the muscles of the throat.*
   Schmerz im Halse, *pain in the throat.*
2782 Schlingbeschwerde, *dysphagia.*
2783 Kehlentzündung, }
2784 Kehlkopfentzündung, } *laryngitis.*
2785 Kehlkopfbräune, *quinsy.*
2786 Mandelbräune, *cynanche tonsillaris.*
2787 Ödem (dä'm) der Glottis, }
2788 Kehlkopfödem, } *œdema of the glottis.*
2789 Kehlgeschwulst, *tumor in the throat.*
2790 falscher Croup, *false croup.*

2791 wahrer Croup, *true croup.*
2792 häutig, *a.*, *skinned, cuticular.*
   die häutige Bräune, *croup, laryngitis crouposa.*
2793 Kehldeckelentzündung, *epiglottitis.*
2794 Kehlschwindsucht, }
2795 Kehlkopfschwindsucht, } *laryngeal phthisis.*
   Krankheiten der Knorpel und des Perichondrium, *diseases of the cartilages and of the perichondrium.*
2796 Carcinom, *n.*, *carcinoma.*
   Carcinome des Kehlkopfs, *cancerous growths of the larnyx.*
2797 Polyp, *m.*, *polypus.*
   Polypen des Kehlkopfs, *polypi of the larynx.*
2798 Katarrh, *m.*, *catarrh.*
   acuter Katarrh der Nasenschleimhaut, *acute catarrh of the nasal mucous membrane.*
2799 Schnupfen, *m.*, *cold (in the head), coryza.*
2800 Nasenbluten, } Epistaxis, } *bleeding from the nose, rhinorrhagia, epistaxis.*
2801 Nasenpolyp, *nasal polypus.*
2802 Stinknase, *f.*, *ozæna.*
   Entzündung der Luftwege, }
2803 Bronchialkatarrh, *m.* } *acute bronchitis.*

2804 capilläre Bronchitis, f., capillary bronchitis.
2805 Bronchienerweiterung, f. }
Bronchienectasie, f. } bronchial dilatation.
2806 Windgeschwulst, f. }
2807 Luftgeschwulst, f. }
2808 Emphysem (sä'm), n. }
2809 Lungenemphysem, } emphysema, pulmonary emphysema.
2810 Brustbeschwerde, f., chest disease.
2811 Brustbeklemmung, f., oppression of the chest.
2812 Brustgeschwür, n., empyema.
2813 Brustkrebs, m., cancer of the chest.
2814 Brustwassersucht, f., hydrothorax, dropsy in the chest.
2815 Brustkrampf, m. }
2816 Lungenkrampf, m. }
2817 Asthma, } spasms in the chest; spasms of the lungs; asthma.
2818 Lungensucht, f. }
2819 Lungenschwindsucht, }
2820 Phthisis, f. } consumption of the lungs, phthisis pulmonalis.
2821 vertrocknen, to dry up.
2822 Vertrocknung, f., drying up. Vertrocknung der Lunge, phthisis pulmonalis.
2823 gallopierende Lungensucht, rapid decline, galloping consumption.
2824 lungensüchtig, a., phthisical, consumptive.
2825 Lungenbrand, pulmonary gangrene.
2826 Lungenentzündung, }
2827 Brustentzündung, } Pneumonia, pneumonia.
2828 katarrhalisch, a., catarrhal. katarrhalische Lungenentzündung, catarrhal or lobular pneumonia.
2829 croupöse Lungenentzündung, acute pneumonia.
2830 Lungenödem, pulmonary œdema.
2831 Lungenlähmung, paralysis of the lungs.
2832 Lungenabscess, pulmonary abscess.
2833 Lungenapoplexie, pulmonary apoplexy.
2834 Lungenblutsturz, } hæmo-
2835 Bluthusten, } ptysis.
2836 Lungenblutung, hemorrhage from the lungs.
2837 Lungenerweichung, breaking down of the lung tissue.
2838 Lungenschrumpfung, }
2839 Lungenatrophie, } pulmonary atrophy.
2840 Lungenwassersucht, hydropneumonia.
2841 Lungengeschwür, pulmonic ulcer.
2842 Brustfellentzündung, acute pleurisy.

2842 Luft (1005) in der Brust- ⎫
 höhle,        ⎬
2843 Luftbrust,      ⎭
 Pneumothorax,
 *pneumothorax.*
 chronische (1630) Brustfellentzündung, *chronic pleu-*
 *risy.*
2844 Circulation, *f.*, *circulation.*
2845 Circulationsstörung, *de-*
 *rangement of circulation.*
2846 Arterienerweiterung, ⎫
2847 Pulsadergeschwulst,  ⎬
 Aneurysma,    ⎭
 *aneurism, varix.*
2848 Herzschwäche, ⎫
2849 Herzfehler,   ⎭
 *defect or disease of the*
 *heart.*
2850 Herzleiden, *cardiac affec-*
 *tion.*
2851 Herzklopfen, *palpitation of*
 *the heart.*
2852 Herzzittern, *trepidatio cor-*
 *dis.*
2853 Herzlähmung, *paralysis of*
 *the heart.*
2854 Herzbeklemmung, ⎫
2855 Herzbeschwerung, ⎭
 *oppression of heart, angor,*
 *anxietas præcordium.*
2856 Herzbrennen, *heart-burn.*
2857 Herzkrampf, ⎫ *spasm of the*
 or    ⎬ *heart, angina*
2858 Herzbräune, ⎭ *pectoris.*
 Entzündung der innern
 Herzhaut, *acute endocar-*
 *ditis.*

2859 Herzentzündung, *carditis.*
2860 Herzbeutelentzündung, *peri-*
 *carditis.*
 reizbares (2510) Herz, *irri-*
 *table heart.*
2861 Herzmuskelentzündung, ⎫
2862 Herzfleischentzündung, ⎭
 *myocarditis.*
2863 Vermehrung, *increase, aug-*
 *mentation.*
2864 Vermehrung der Herz- ⎫
 muskulatur (535) or  ⎬
2865 Herzhypertrophie,  ⎭
 *hypertrophy of the heart.*
2866 Herzerweiterung, *dilatation*
 *of the heart, cardiectasis.*
2867 Herzhöhlenerweiterung, *dil-*
 *atation of ventricle.*
2868 Herzverengerung, *narrowing*
 *or coarctation of the heart.*
2869 Herzerweichung, *cardiac*
 *softening.*
2870 Herzverhärtung, *induration*
 *of the heart.*
2871 Herzverknöcherung, *ossifi-*
 *cation of the heart.*
2872 Herzschrumpfung, ⎫
2873 Herzatrophie,   ⎭
 *atrophy of heart.*
2874 Herzabscess, *abscess of the*
 *heart.*
2875 Verfettung, *f.*, *adipose de-*
 *generation.*
 Verfettung des Herz- ⎫
 fleisches or    ⎬ *fatty*
2876 Fettherz,     ⎭ *heart*
2877 Herzgeschwulst, *f.*, *heart*
 *tumor.*

2878 Herzgeschwür, *n.*, *ulcer of the heart.*
2879 Herzpolyp, *polypus of the heart.*
2880 Blausucht, *cyanosis.*
2881 Herzblausucht, *cardiac cyanosis.*
2882 Herzbeutelwassersucht, *hydro-pericardium.*
2883 Klappenkrankheiten, *valvular affections.*
2884 Lageveränderung, *displacement, change of position.*
Lageveränderung des Herzens, *displacement of the heart.*
Wassersucht bei Herzleiden, *cardiac dropsy.*
2885 Gaumenblutung, ⎫
2886 Mundfäule, ⎬
Mundfäulniss, ⎭
*ulcerated gums, stomacace.*
2887 Wasserkrebs, *gangrenous stomatitis, noma.*
2888 Mundentzündung, ⎫
2889 Mundschleimhautentzündung, ⎬ *stomatitis.*
2890 Mundkrebs, *cancer of the mouth.*
2891 Mundschwamm, ⎫
Mundschwämmchen, or ⎬
2892 Soor, ⎭
*thrush, aphthæ, muguet.*
2893 Zahnfleischentzündung, *inflammation of the gums.*
2894 Zahnfäule, *caries, or rottenness of the teeth.*
2895 das Wackeln (2369) der Zähne, *looseness of the teeth.*
2896 **fressen, frass, gefressen,** *to eat (of animals),* fig. *to eat away, corrode, waste.*
2897 anfressen, *to canker, gangrene.*
2898 der angefressene Zahn, *decayed tooth.*
2899 Zahnweinstein, *m.*, *tartar of the teeth.*
2900 Zahnentzündung, *odontitis.*
2901 Zahnhusten, *cough during teething.*
2902 Zahnruhr, *diarrhœa from teething.*
2903 Zahnkrampf, *teething convulsions.*
2904 Rachenkatarrh, *pharyngeal catarrh.*
2905 Gaumengeschwulst, *f.*, *palatine swelling, p. tumor.*
2906 Zungengeschwulst, *tumor of the tongue.*
2907 Zungengeschwür, *ulcer of the tongue.*
2908 Zungenentzündung, *glossitis.*
2909 Zungenkrebs, *cancer of the tongue.*
2910 Zungenkrampf, *spasm of the tongue.*
2911 Zungenlähmung, *paralysis of the tongue.*
2912 Zungenvergrösserung, *macroglossia.*

2913 Zungenvorfall, *prolapsus linguæ.*
2914 bösartige (1647) Rachenbräune, *angina maligna.*
2915 Schlundkopfentzündung, }
2916 Schlundkopfbräune, } *pharyngitis.*
2917 Schlundkopfkrampf, *pharyngeal spasm.*
2918 Schlundkopflähmung, *pharyngoplegia, pharyngeal paralysis.*
2919 Schlundkopfpolyp, *polypus of the pharynx.*
2920 Schlundkopfschwindsucht, *pharyngeal phthisis.*
2921 Halsbräune, } *quinsy,*
    or } *tonsil-*
2922 Halsentzündung, } *litis.*
2923 Speiseröhrenbräune, }
2924 Speiseröhrenentzündung, } *œsophagitis.*
2925 Speiseröhrenentartung, *degeneration of the œsophagus.*
2926 Speiseröhrenkrampf, *spasm of the œsophagus.*
2927 Speiseröhrenlähmung, *paralysis of the œsophagus.*
2928 Strictur, *f., stricture.*
    Strictur *or* Verengerung (1002) der Speiseröhre, *stricture of the œsophagus.*
    Erweiterung (1003) der Speiseröhre, *dilatation of the œsophagus.*

2929 ein Straussenmagen, *m., stomach of an ostrich (apt to digest anything).*
2930 Magendrücken, *n., cardialgia, stomach-ache.*
2931 Verdauungsbeschwerde, }
    Dyspepsie, } *dyspepsia.*
2932 Magenkatarrh, *catarrh of the stomach, attack of indigestion.*
2933 schwerer (1610) Katzenjammer, *the indisposition after a drunken debauch* (lit. *cat-misery*).
2934 Magengeschwulst, *gastric tumor.*
2935 Magenkrebs, *cancer of the stomach.*
2936 Magengeschwür, *n., gastric ulcer.*
2937 peptisch, *a., peptic.*
    peptisches Geschwür, *peptic ulcer.*
2938 Magenentzündung, *gastritis.*
2939 Magenerweiterung, *f., dilatation of the stomach.*
2940 Magenerweichung, *softening of the stomach.*
2941 Magenverhärtung, *induration of the stomach.*
2942 Magenlähmung, *gastric paralysis.*
2943 Magenschwindsucht, *f., atrophy of the stomach.*
2944 Magenbruch, *m., rupture of the stomach.*

2945 Seekrankheit, *f.*, *sea-sickness.*
2946 Brechreiz (2189) verursachen (1671), *to turn the stomach.*
2947 das Übelsein, *nausea, qualm.*
2948 sich erbrechen wollen, *to heave the gorge.*
2949 flau, *a.*, *weak, feeble.*
mir wird ganz flau, *I feel unwell.*
2950 grimmen (1124), *to give severe pain.*
2951 Kolik (1816), *f.* ⎱
2952 Darmkolik, ⎰
2953 Darmgrimmen, *n.*
*the gripes* (1124), *colic, gripes.*
2954 einfache (1652) Kolik, ⎱
2955 Krampfkolik, ⎰
*simple colic.*
2956 Windkolik, *flatulent colic.*
2957 Gallenkolik, ⎱ *bilious*
biliöse Kolik, ⎰ *colic.*
2958 biliös, *a.*, *bilious.*
nervöse (804) Kolik, *nervous colic.*
2959 Metallkolik, *metallic colic.*
2960 Bleikolik, *lead colic.*
2961 Kupferkolik, *copper colic.*
2962 sympathetisch, *sympathetic.*
sympathetische Kolik, *sympathetic colic.*
2963 rheumatische or entzündliche (1636) Kolik, *rheumatic or inflammatory colic.*
2964 Perforation, *f.*, *perforation.*
2964 Perforation des Darms, *perforation of the intestines.*
2965 Darmentzündung, ⎱
Enteritis, ⎰
*acute enteritis.*
2966 Darmkatarrh, *intestinal catarrh, muco-enteritis.*
2967 Darmfellentzündung, ⎱
2968 Bauchfellentzündung, ⎬
Peritonitis, ⎰
*peritonitis.*
2969 Kindbettfieber, *n.*, *acute puerperal fever.*
Peritonitis des Kindbettfiebers (1280), ⎱
puerperale Peritonitis, ⎰
*puerperal peritonitis.*
particlle *or* locale Peritonitis, *partial or local peritonitis.*
chronische Peritonitis *chronic peritonitis.*
Typhlitis, ⎱
Entzündung des Blinddarms (1117), ⎰
*typhlitis, inflammation of the cæcum.*
Perityphlitis, *perityphlitis.*
Ausdehnung (744) des Cæcum (1117), *distension of the cæcum.*
Carcinom (2796) des Cæcum, *cancer of the cæcum.*
2970 Blinddarmbruch, *hernia cæcalis.*
2971 Darmgeschwür, *intestinal tumor.*

2972 Darmkrebs, *cancer of the intestines.*
2973 Darmerweichung, *softening of the intestines.*
2974 Darmschwindsucht, *intestinal phthisis.*
2975 einklemmen, *to squeeze in.*
2976 einschieben, } *to push in, to*
2977 einstülpen, } *intercalate.*
2978 verschlingen, *to intertwine, twist.*
2979 Darmeinklemmung,
2980 Darmeinschiebung,
2981 Darmeinstülpung,
2982 Darmverschlingung, *invagination or intussusception of the intestines.*
2983 Darmverengung, *stricture of the intestines.*
2984 Darmverhärtung, *induration of the intestines.*
2985 Darmverschliessung, *intestinal obstruction.*
2986 Darmgicht, *ileus volvulus.*
2987 habituell, *a., habitual.*
habituelle Verstopfung, *habitual constipation.*
acute Diarrhöe, *acute diarrhœa.*
biliöse Diarrhöe, *bilious diarrhœa.*
fettige (2268) Diarrhöe, *fatty diarrhœa.*
2988 die Ruhr,
2989 Darmruhr, *f.,* } *dysentery.*
. Dysenterie,
2990 Brechdurchfall. } *cholera*
Brechruhr. } *morbus.*

2990 asiatische Cholera, *cholera asiatica.*
2991 Hämorrhoiden, *hemorrhoids, piles.*
2992 Gelbsucht, } *jaundice.*
Icterus, }
2993 Blutüberfüllung der Leber,
2994 acute Leberhyperämie, *acute congestion of the liver.*
2995 Leberverstopfung, *hepatic obstruction.*
2996 eitrige Leberentzündung, *acute hepatitis.*
Entzündung des Leberüberzugs (176), *perihepatitis.*
2997 Pigmentleber, *pigmented liver.*
Entzündung der Gallenblase (1104) und Gallengänge (1102), *inflammation of the gall-bladder and gall-ducts.*
acute gelbe Atrophie, *acute yellow atrophy.*
2998 chronische Hyperämie, *chronic congestion of the liver.*
2999 Lebervergrösserung, }
Hypertrophie der Leber, } *hypertrophy of the liver.*
3000 Leberabscess, *abscess of the liver.*
3001 Fettleber, *fatty liver.*
3002 Speckleber, } *waxy liver.*
3003 Wachsleber, }

3004 Leberkrebs, *cancer of the liver.*
3005 syphilitisch, *syphilitic.*
syphilitische Leberentzündung, *syphilitic liver.*
Echinococcus der Leber, }
3006 Säuferleber,
Cirrhose,
*hob-nail liver, cirrhosis.*
Entzündung und Trombose der Pfortader (776), *inflammation of the portal vein.*
chronische einfache Atrophie der Leber, *chronic simple atrophy of the liver.*
3007 Leberschwindsucht, *hepatic consumption, consumption in consequence of liver disease.*
3008 Lebererweichung, *softening of the liver.*
3009 Abdominalvergrösserung, *abdominal enlargement.*
allgemeine oder partielle Abdominalvergrösserung, *general or partial abdominal enlargement.*
3010 Bauchwassersucht, } *ascites.*
Ascites,
3011 Tympanie, *tympanites.*
chronische Tympanie,
3012 Trommelsucht (Windsucht),
Tympanitis,
*chronic tympanitis.*
3012 Geschwülste in der Bauchhöhle ;
Tumoren im Abdomen, im rechten, linken Hypochondrium,
*abdominal tumors in the right, left hypochondrium.*
Entzündung der Milz (1105),
Splenitis,
*inflammation of the spleen, splenitis.*
chronische Vergrösserung (2423) der Milz, *chronic enlargement of the spleen.*
3013 Wandermilz, *movable spleen.*
epigastriche Geschwulst, *epigastric tumor.*
3014 Bauchspeicheldrüsengeschwulst, *pancreatic tumor.*
3015 Bauchspeicheldrüsenkrebs, *pancreatic cancer.*
3016 Nabelgegend, *f., umbilical region.*
Geschwulst in der Nabelgegend, *tumor in the umbilical region.*
3017 Tuberkulose, *tuberculous disease.*
3018 mesenterial, *mesenteric.*
3019 Lendengegend, *f., lumbar region.*
3020 Hüftgegend, *f., iliac region.*
3021 hypogastrisch, *a., hypogastric.*

3022 Pulsation, *f.*, *pulsation.*
Tuberkulose der Mesenterialdrüsen, *tuberculous disease of the mesenteric glands.*
3023 Wanderniere, *movable kidney.*
Geschwülste in der Lendengegend, *tumors in the lumbar region.*
Krebs der Lymphdrüsen, *cancer of the lymphatic glands.*
Geschwülste in der Hüftgegend, *tumors in the iliac region.*
3024 Eierstocksgeschwülste, ⎫
3025 Ovarialtumoren, ⎭ *ovarian tumors.*
Geschwülste in der hypogastrischen Gegend, *tumors in the hypogastric region.*
Pulsation der Aorta, *aortic pulsation.*
3026 Nierenentzündung, *nephritis.*
Neuralgie der Niere, *neuralgia of the kidney.*
3027 Nierenkrebs, *cancer in the kidney.*
3028 Nierenstein, *renal calculus.*
acute Bright'sche Krankheit, *acute Bright's disease.*
3029 Blutharnen, *haematuria.*
3030 Eiterharnen, *pyuria.*
3031 Harnvergiftung, *uræmia.*
chronische Bright'sche Krankheit, *chronic Bright's disease.*

3032 Fettniere, *fatty kidney.*
vergrösserte Niere, *enlarged kidney.*
chronisch-entzündete Niere, *chronically inflamed kidney.*
3033 die Wachsniere, *the waxy kidney.*
die geschrumpfte (2468) Niere, *contracted kidney.*
3034 Blasenentzündung, *f.*, *acute cystitis.*
3035 Nierenabscess, *abscess of the kidney.*
3036 Nierenbeckenentzündung, *pyelitis.*
Erweiterung des Nierenbeckens, *hydronephrosis.*
die Lähmung der Blase, *paralysis of the bladder.*
3037 Blasenkrampf, *enuresis.*
3038 Steinkrankheit der Blase, *stone in the bladder.*
3039 Harnsteine, *urinary calculi.*
3040 Steinkolik, *or* ⎫ *nephritic*
3041 Nierenkolik, ⎭ *colic.*
3042 Harnblasengeschwür, *ulcer of the bladder.*
Geschwülste der Blase, *tumors of the bladder.*
3043 Harnblasenerweichung, *softening of the bladder.*
3044 Blasenkrebs, *m.*, *cancer of the bladder.*
3045 Harnblasenschwindsucht, *tuberculous disease of the bladder.*

3046 die Harnruhr, \} *diabetes.*
　　 Diabetes,
3047 Zuckerharnruhr, *f.*
3048 Honigharnruhr, *f.*
3049 Meliturie, *f.*
3050 Glycosurie, *f.*
　　 *diabetes mellitus.*
3051 die zuckerfreie Harnruhr, *diabetes insipidus.*
3052 Harnröhrenentzündung,
3053 Harngangentzündung,
3054 Harnleiterentzündung,
　　 *urethritis.*
3055 Harnröhrenverengung, *stricture of the urethra.*
3056 Harnlosigkeit, *suppression of urine.*
3057 Harnsperre, \} *ischuria,*
3058 Harnstrenge, / *dysuria.*
3059 Harnbrennen, *n.*, *ardor urinæ, strangury.*
3060 Harnröhrengeschwüre, *urethral ulcers.*
3061 unwillkührlicher (558) Harnabfluss, *incontinence of urine, enuresis.*
3062 Herzwassersucht, *f.*, *cardiac dropsy.*
3063 Nierenwassersucht, *f.*, *renal dropsy.*
3064 Leberwassersucht, *f.*, *hepatic dropsy.*
3065 Blutarmuth, *f.* \} *anæmia.*
3066 Anämie, *f.* /
3067 Bleichsucht, *f.*, *chlorosis.*
3068 Addinson'sche Krankheit, *Addinson's disease.*
3069 perniciös, *pernicious.*
　　 perniciöse Anämie, *pernicious anæmia.*
3070 Leukämie, \} *leucocy-*
3071 Weissblütigkeit, / *thæmia.*
3072 Eitervergiftung, \} *pyæmia.*
3073 Pyämie, /
3074 Skorbut, *m.*, *scurvy.*
3075 Blutfleckenkrankheit, \}
　　 Purpura, /
　　 *purpura.*
3076 acuter und chronischer Rheumatismus, *acute and chronic rheumatism.*
3077 Gehirnrheumatismus, *cerebral rheumatism.*
3078 Gelenkentzündung, *arthritis.*
3079 Muskelrheumatismus, *muscular rheumatism.*
3080 Hexenschuss, *m.*, *lumbago.*
3081 schief, *a.*, *oblique; awry, distorted.*
　　 der schiefe Hals, \} *wry-neck,*
　　 Torticollis, / *torticollis*
3082 **die** Gicht, *gout.*
3083 Fussgicht, *f.*, *podagra, gout in the foot.*
3084 Handgicht, *f.*, *chiragra, gout in the hand.*
3085 Kniegicht, *f.*, *gonagra.*
　　 die Gicht in der Hüfte, *hipgout.*
3086 Gichtfieber, *n.*, *arthritic fever.*
3087 **das** Fieber, *fever.*

3088 Fieberanfall, *m.*, *febrile attack.*
3089 das täglich wiederkehrende Fieber, *quotidian fever.*
3090 das einmal täglich sich einstellende Fieber, *simple quotidian fever.*
3091 das zweimal täglich sich einstellende Fieber, *double quotidian fever.*
3092 das dreitägige Fieber, *tertian fever.*
3093 das viertägige Fieber, *quartan fever.*
3094 Fieber haben, *or* fiebern, *to have fever, be in a fever.*
3095 ein Fiebergefühl verspüren, *to feel feverish.*
3096 einen Fiebergeruch verbreiten, *to emit a feverish odor*
3097 vor Fieber zittern, *to shake with the ague.*
Fieberschauder, *or*
3098 Fieberschauer haben, *to have ague-fits.*
3099 vor Frost zittern, *to shiver with cold.*
3100 Wechselfieber, *n.*, *intermittent fever.*
das katarrhalische Fieber, *catarrhal fever.*
3101 Heufieber, *n.*, *hay fever, hay asthma.*
3102 Hustenfieber, *n.*, *cough- (or catarrhal) fever.*
3103 Schweissfieber, *n.*, *sweating fever.*
3104 Schnupfenfieber, *n.*
3105 Grippe, *f.*, *influenza, catarrhal f.*
3106 Brechfieber, *n.*, *fever attended with vomiting.*
3107 Frühlingsfieber, *n.*, *spring fever.*
3108 Herbstfieber, *n.*, *autumnal fever.*
3109 Gallenfieber, *n.*, *bilious fever.*
3110 Schleimfieber, *n.*, *mucous fever.*
3111 Entzündungsfieber, *n.*, *inflammatory fever.*
3112 Wundfieber, *n.*, *traumatic fever.*
3113 Schleichfieber, *slow, lingering fever.*
das hectische (1885) Fieber, *hectic fever.*
das syphilitische (3005) Fieber, *syphilitic fever.*
das gelbe Fieber, *yellow fever.*
congestives Fieber, *pernicious malarial fever.*
verlarvtes Fieber, *masked fever.*
3114 **lassen, liess, gelassen,** *to let, leave.*
3115 nachlassen, *to abate, subside, cease.*
3116 das Nachlassen des Fiebers, *the abatement of the fever.*
3117 schütteln, *to shake, move back and forth, agitate.*
3118 Schüttelfröste (2112), *rigors, ague.*

3118 die Schüttelfröste der Eitervergiftung, *the rigors (ague) of pyæmia.*
3119 Abdominaltyphus, *typhoid fever.*
3120 Nervenfieber, *nervous fever.*
3121 Flecktyphus (2557),
3122 Exanthematischer Typhus,
  *typhus fever.*
3123 Hungertyphus, } *hunger ty-*
3124 Hungerpest, } *phus.*
3125 Rückfallstyphus (1524) febris recurrens, *relapsing fever.*
3126 **die** Pest, *pest, pestilence, plague.*
3127 das Scharlach,
3128 Scharlachfieber,
3129 Scharlachfriesel,
  *scarlet fever, scarlatina.*
3130 **das** Friesel, *miliary eruption.*
  Friesel mit Fieber, *miliary fever.*
3131 rother Hund, *roseola.*
3132 Masern (*plur.*), *measles, morbili.*
3133 Röthcln (*plur.*), *rose rash.*
3134 Feuermasern, *roseola.*
3135 Pocken (*pl.*) } *variola,*
3136 Blattern (*pl.*) } *small-pox.*
3137 pockenartig, *a., varioloid.*
3138 Pockenfieber, *n., variolous fever.*
3139 Narbe, *f.* } *cicatrix,*
3140 das Wundenmal, } *scar.*
3141 Pockennarbe, *f.* } *pock-mark*
3142 Blatternarbe, *f.* } *or pit.*
3143 pockennarbig, *a.*
3144 blatternarbig, *a.*
  *pock-pitted or marked.*
3145 Windpocken,
  fliegende (1708) } *chicken-*
  Pocken, } *pox.*
3146 Kuhpocke, *f.* } *vaccina.*
3147 Schutzpocken, }
3148 (schützen, *to protect, to guard.*)
3149 impfen, *to inoculate, vaccinate.*
3150 Pockenimpfung, *f.*
3151 Impfung, *f.* } *vaccination.*
3152 das Impfen,
3153 Impfstoff, *m., the vaccine matter.*
3154 Wasserpocken (*pl.*) }
3155 Spitzpocken (*pl.*) }
  *varicella, chicken-pox.*
3156 das Varioloid, *varioloid.*
3157 variolös, *a., variolous.*
3158 Rose, *f., rose.*
3159 Rothlauf, *m., erysipelas.*
3160 Kopfrose, *f., erysipelas of the head.*
3161 Gesichtsrose, *f., facial erysipelas.*
3162 Bläschenrothlauf, *m., erysipelas vesiculosum seu bullosum.*
3163 Blasenrose, *f., erysipelas bullosum.*
3164 **das** Wunderysipel, *wound erysipelas.*

3165 die Nessel, *nettle.*
3166 Nesselausschlag, *m.* } *nettle-*
3167 Nesselsucht, *f.* } *rash.*
3168 Nesselfieber, *n.* }
3169 Nesselfriesel, *n.* }
  *febris urticata, urticaria.*
3170 das Eczem, *eczema.*
3171 nässen, *to wet, moisten.*
3172 Bläschenflechte (1945),
  *herpes.*
3173 die nässende Bläschenflechte,
  *eczema rubrum.*
3174 Gesichtsgrind (1938), *m.,*
  *porrigo larvalis.*
3175 Borke, *f., scurf, scab, crust.*
3176 Milchschorf (1926), *m.* }
3177 Milchborke, *f.* }
  *impetigo, crusta lactea.*
3178 Gürtel, *m., girdle, zone, shingles.*
3179 Gürtelausschlag, *m.* }
3180 Gürtelflechte, *f.* }
3181 Gürtelrose, *f.* }
  *herpes zoster, zona, cingulum.*
3182 Blasenausschlag, *m., pemphigus.*
3183 Wasserblasen, } *water*
  kleine (1988) Blasen, } *blebs.*
3184 Finne, *f.* } *acne, blotch,*
3185 Hautfinne, *f.* } *pimple.*
3186 finnig, *a., pimpled.*
3187 Finnenausschlag, *acneform dermatosis.*
3188 Mitesser, *m., comedone.*
3189 Kupferrose, *f., acne rosacea.*

3190 Burgundernase, }
3191 Säufernase, }
  *Burgundy nose, gutta rosacea.*
3192 Bartfinne, *f., acne mentagra, sycosis.*
3193 Pustelflechte, *f., impetigo, ecthyma.*
3194 Schmutzflechte, *f.* } *rupia.*
3195 Schmutzgrind, *m.* }
3196 Kleie, *f., bran.*
3197 Kleienflechte, *f.* }
3198 Kleiengrind, *m.* }
3199 Kleienschwinde, *f.* }
3200 Kleiensucht, *f.* }
  *dandruff, pityriasis.*
3201 Schwinde, *f., tetter.*
  die rothe Kleienflechte, *pityriasis rubra.*
3202 Schuppenausschlag (1931), *m., squamous dermatosis.*
3203 Schuppenflechte, *psoriasis.*
3204 Schuppengrind, *m., psoriasis nummularis, crusta lamellosa.*
3205 Schuppicht, *a., scaly, squamose.*
3206 Fischschuppenkrankheit, *fish-skin disease, ichthyosis.*
  die schuppichten Hautausschläge (1901), *the scaly eruptions.*
3207 Sommersprossen (*pl.*), *ephelid, freckle, summer rash.*
3208 sommersprossig, *a., freckled, freckle-faced.*

3209 Bronzekrankheit, *bronzed skin.*
3210 Leberflecken (*pl.*), *liver-spots, chloasma hepaticum.*
3211 **das** Mal, *mole, mark.*
3212 Muttermal, *mother-spot, macula, nævus.*
　farbiges (689) Muttermal, *nævus vasculosus, teleangiectiasis.*
3213 Feuermal, *nævus flammeus, seu vasculosus.*
3214 Albinismus, *albinism.*
3215 Neubildung, *neoplasm, tumor.*
3216 Wolf, *m., wolf, lupus.*
　der fressende (2896) Wolf, *lupus vorax.*
　die fressende Flechte, *eating tetter.*
　die scherende (286) Flechte, *herpes tonsurans, ringworm.*
3217 Aussatz, *m., lepra.*
3218 Hautkrebs, *m., carcinoma cutis, epithelioma.*
3219 Elephantenaussatz, *elephantiasis Græcorum, lepra.*
3220 Elephantiasis, *elephantiasis Arabum, pachydermia.*
3221 vertheilen, *to distribute, divide.*
3222 Vertheilung, *dispersion, resolution.*
　Vertheilung einer Geschwulst, *resolution of a tumor.*

3223 aufbrechen, *to break open, burst.*
　das Geschwür bricht auf, *the abscess is bursting.*
3224 Reife, *f., ripeness, maturity.*
　die Reife eines Geschwürs, *the maturity of an abscess.*
3225 Balggeschwulst, *f.* ⎫
3226 Sackgeschwulst, *f.* ⎭ *encysted tumor.*
3227 Nagelgeschwür, *n.* ⎫ *whit-*
3228 Fingerwurm, *m.* ⎭ *low.*
3229 ein böser Finger, *a sore finger.*
　das entzündliche (1636) Geschwür, *acute abscess.*
3230 **die** Fistel, ⎫ *fistula.*
3231 Hohlgeschwür, *n.* ⎭
3232 Fistelgeschwür, *n., a fistulous ulcer.*
3233 Schwären (2413), *m., (used indifferently for) abscess, ulcer, boil.*
　das fressende Geschwür, *ringworm.*
3234 schwammicht, *a., spongy, fungous.*
　das schwammichte Geschwür, *fungous ulcer.*
3235 wildes Fleisch, *fungus, proud flesh, dead flesh.*
　Syphilis, *f.* ⎫ *syphilis, vene-*
3236 Lustseuche, ⎭ *real disease.*
3237 venerisch, *a., venereal.*
　die venerische Krankheit, ⎫
3238 Venerie, *f.* ⎭ *venereal disease.*

3238 der syphilitische Hautausschlag, *syphilitic eruption.*
3239 der harte Schanker, *hard chancre.*
3240 das weiche Schankergeschwür, *soft chancre, chancroid.*
3241 brandig, *a.*, *gangrenous, chancroid.*
der brandige (gangränöse) Schanker, *phagedenic chancroid.*
3242 Feigwarze, *f.* }
die breite Kondylome, } *condyloma.*
3243 breit, *a.*, *broad, wide.*
3244 die Schleimpapel, *mucous patch (plaques muqueux).*
3245 syphilitische (3005) Augenentzündung, *syphilitic ophthalmia.*
3246 syphilitische Hodenanschwellung, *syphilitic orchitis.*
syphilitische Knoten (1907), *syphilitic nodes.*
das venerische Geschwür, *venereal tumor.*
3247 Venusbeule, *f.*, *bubo.*
3248 Venuskrone, *f.*, *venereal crown.*
3249 Tripper, *m.*, } *gonorrhœa.*
Gonorrhöe, *f.* }
der entzündliche Tripper, *inflammatory gonorrhœa, blennorrhœa.*
die Verengerung der Vorhaut (1189), *stenosis of the prepuce.*
3249 abnorme Verengerung der Vorhaut, Phimose, *f.* } *phimosis.*
3250 Kragen, *m.*, *collar.*
3251 spanisch, *Spanish.*
3252 der spanische Kragen, } Paraphimose, } *paraphimosis.*
3253 menstruieren, *to menstruate.*
3254 Menstruation, *f.*, *menstruation.*
3255 Reinigung, *f.*, *purging.*
3256 monatlich, *a.*, *monthly.*
monatliche Reinigung, *menstrual flux, monthly courses.*
3257 Monatsfluss, *m.*, *monthly flow.*
3258 das Monatliche, } *menses.*
3259 die Menses (*pl.*) }
3260 Katamenien (*pl.*), *menstruation.*
3261 Blutgang, *m.*, *flow of blood.*
3262 die Regel, } *menses,*
3263 Periode, } *period.*
3264 Menstruationsstörung, *f.*, *derangement of menstruation.*
3265 Amenorrhöe, *amenorrhœa.*
3266 vikariierende Menstruation, *vicarious menstruation.*
3267 Blutkrämpfe (*pl.*), *uterine colic.*
3268 Menstrualkolik, *dysmenorrhœa.*
3269 Fluss, *m.*, *flow, flux; river.*

3270 der weisse Fluss,  
3271 die Leukorrhöa,  
    *flux albus, leucorrhœa.*  
3272 **das** Weisse, *the whites.*  
3273 Ordnung, *f.*, *order.*  
    in Ordnung, *in order, correct, regular.*  
    eine Frau, die ihre Regel hat, *a woman having the menses*  
    eine Frau, bei der die monatliche Reinigung in Ordnung (in Unordnung) ist, *a regularly (an irregularly) menstruating woman.*  
    die gestörte (2155) Reinigung, *dysmenorrhœa.*  
3274 Mutterkrankheit,  
3275 Mutterbeschwerde, *hysteria.*  
3276 Mutterkrampf, *m.*, *uterine spasm.*  
3277 Mutterkrebs, *m.*, *cancer of the womb.*  
3278 Mutterscheidenentzündung, *f.*, *vaginitis.*  
3279 Mutterblutfluss, *m.*  
3280 die Lochien (*pl.*) *lochia.*  
3281 Gebärmutterschmerz, *m.*, *hysteralgia.*  
3282 senken, *cause to sink, let down, lower.*  
    Senkung, *f.*, *sinking, lowering.*  
3283 Muttervorfall, *m.*  
    Senkung,  
3284 das Heraustreten der Gebärmutter, *prolapse of the womb, hysterocele.*

3285 Mutterwuth, *f.*  
3286 Mannestollheit, *f.* *nymphomania.*  
3287 die Schüttelwehen (*pl.*), *tremulous labor pains.*  
3288 die Nachwehen (*pl.*), *after-pains.*  
3289 **die** Missgeburt, *miscarriage; monster.*  
3290 Zangenentbindung, *f.*  
3291 Zangengeburt, *f.* *forceps delivery.*  
3291 Kaisergeburt, *delivery by Cæsarean section.*  
3292 Kaiserschnitt (298), *f.*, *Cæsarean section; hysterotomy.*  
3293 Zwilling, *m.*, *twin.*  
    Zwillinge gebären, *to bear twins.*  
3294 Frühgeburt,  
    zu früh gebären,  
    vorzeitig (1272) gebären, *premature birth, to be delivered prematurely.*  
3295 unrichtig, *a. and adv., not right, wrong.*  
    es ist ihr unrichtig gegangen, *she has miscarried.*  
3296 abtreiben (2401), *to expel.*  
    ein Kind ——, *to cause an abortion or miscarriage.*  
3297 **das** Mutterfieber,  
    Wochenfieber, *puerperal fever.*  
3298 Milchfieber, *milk fever.*  
3299 **die** Milchruhr, *galactorrhœa.*

3300 Milchkanalentzündung, *f.*, *inflammation of the milk-ducts.*
3301 Milchverhaltung, *f.*, *galactischesis.*
das Wundsein der Brustwarzen (vom Säugen), *sore nipples.*
3302 schlimm, *a.*, *sore, bad.*
die schlimme Brust, *sore breast.*
3303 verwunden, *to wound.*
3304 Verwundung, *wounding, hurting.*
3305 beschädigen, } *to injure,*
3306 lädieren, } *hurt.*
3307 Beschädigung, *f.* ⎫ *injury,*
3308 Verletzung, *f.* ⎬ *lesion,*
3309 Läsion, *f.* ⎭ *hurt.*
3310 die Tödtlichkeit einer Wunde, *a fatal wound.*
eine fressende Wunde, *an ulcerating wound.*
eine eiternde Wunde, *a suppurating wound.*
die Wunde eitert, fliesst stark, *the wound discharges freely.*
die eitrige Beschaffenheit, *purulence, purulency.*
3311 Eiterherd, *m.*, *the focus of suppuration.*
3312 untersuchen, *to examine.*
3313 eine Wunde untersuchen, *to probe a wound.*
3314 ätzen, *to cauterize.*
eine Wunde ätzen, *to cauterize a wound.*

3315 verbinden, *to apply a bandage, to dress.*
die Wunde verbinden, *to dress a wound.*
3316 reinigen, *to clean.*
die Wunde reinigen, *to clean or wash a wound.*
3317 bähen, *to foment.*
die Wunde bähen, *to foment or bathe a wound.*
das wilde (*or* schwammige) Fleisch einer Wunde, *proud flesh.*
3318 **hauen, hieb, gehauen,** *to hew, cut, chop.*
3319 Hiebwunde, *f.*, *slash, cut by a blow.*
3320 Schnittwunde, *f.*, *incised wound.*
3321 **stechen, stach, gestochen,** *to sting, prick, pierce.*
3322 Stichwunde, *f.*, *a punctured wound.*
3323 Meissel, *f.*, *chisel.*
3324 Meisselwunde, *f.*, *wound from a chisel.*
3325 **schiessen, schoss, geschossen,** *to shoot.*
3326 Schusswunde, *f.*, *gun-shot wound.*
3327 quetschen, *to bruise, contuse.*
3328 Quetschung, *f.*, *bruise, contusion.*
3329 Quetschwunde, *f.*, *a contused wound.*
3330 Prellschuss, *m.*, *a gun-shot contusion.*

3331 Streifschuss, *m.*, *a furrowed gun-shot wound.*
3332 das Wundliegen, } *excoriation.*
3333 das Aufliegen,
3334 wundliegen, } *to excoriate.*
3335 durchliegen,
sich wundliegen, }
3336 sich aufliegen, *to lie one's self sore.*
3337 unterlaufen, *to run under (the skin).*
3338 mit Blut unterlaufen, *suffused with blood.*
der blutunterlaufene Fleck, *a suggillation.*
3339 aufspringen, }
3340 schrundig (1929) werden, } *to chap.*
aufgesprungene Lippen, *chapped lips.*
3341 **reissen, riss, gerissen,** *to tear, pull.*
3342 Riss, *m.*, *laceration, tear.*
3343 Risswunde, }
gerissene Wunde, } *laceration, lacerated wound.*
einfache Wunde, *simple wound.*
3344 complicirt, *complicated.*
complicirte Wunde, *complicated wound.*
3345 vergiften, *to poison, infect.*
vergiftete Wunde, *poisoned wound.*
3346 Längenwunde, *f.*, *longitudinal wound.*
3347 Querwunde, *f.*, *transverse wound, cross-wound.*
die schiefe Wunde, *oblique wound.*
die oberflächliche Wunde, *superficial wound.*
die tiefe Wunde, *deep wound.*
3348 die penetrierende Wunde, *penetrating wound.*
3349 Wundrand, *m.*, *edge or margin of a wound.*
3350 klaffen, *to gape.*
das Klaffen der Wundränder, *the gaping of the edges (lips) of a wound.*
3351 zertheilen, *to divide; dissipate, discuss.*
3352 Zertheilung, *resolution.*
3353 verwachsen, *to unite, coalesce.*
3354 Verwachsung, *coalescence, intergrowth.*
3355 vernarben *and* sich vernarben, *to cicatrize, heal over.*
3356 Vernarbung, *cicatrization.*
3357 Blutung, *f.*, *bleeding.*
3358 Pfropf, *m.*, *stopper, plug; thrombus.*
3359 Blutpfropf, *m.*, *thrombus, embolus.*
Compression, *f.*, *compression.*
3360 unterbinden, *to tie under; to ligate, bind up.*
3361 Unterbindung, *f.*, *ligation; bandaging, ligature.*
3362 zusammenziehen, *to contract, constrict.*

3363 kleben, *to cleave, stick.*
klebende Mittel, *agglutinants.*
3364 Glüheisen, *n., cautery iron.*
3365 Brandschorf, *m., gangrenous eschar.*
3366 entfernen, *to remove, put far off.*
3367 fremd, *a., foreign, strange.*
3368 Entfernung der fremden Körper aus der Wunde, *removal of foreign bodies from a wound.*
Amputation, *f., amputation.*
Granulation, *f., granulation.*
3369 Bausch, *m.* (dim. Bäuschchen), *(little) pad, compress.*
Charpie, *charpie, lint.*
3370 Charpiebäuschchen, *compress of lint.*
3371 wuchern, *to grow exuberantly, proliferate.*
3372 Wucherung, *proliferation; extuberance.*
3373 Fleischwärzchen, *verruca carnea, granulations.*
3374 zersplittern, *to split, splinter.*
3375 Zersplitterung der Knochen, *splintering of bones.*
3376 zerreissen, *to rend, tear, lacerate.*
3377 Zerreissung der Gefässe, *laceration of vessels.*
3378 zerschmettern, *to crush.*
3379 Zerschmetterung der Gelenke, *crushing or comminution of joints.*

3380 schinden, *to flay, skin.*
3381 aufreissen (3341), *tear up, crack; chap.*
sich die Haut aufreissen, *to rub off one's skin.*
3382 Brandwunde, *f., burn, scald.*
3383 sich verbrühen, *to scald one's self.*
3384 stillen (1297), *to still, staunch, arrest, stop.*
das Blut stillen, *to staunch the blood.*
das Stillen des Blutes, *staunching of the blood.*
3385 besorgen, *to care for, manage, carry out.*
eine Wunde besorgen, *to take care of a wound.*
3386 sondieren, *to sound, probe, examine.*
eine Wunde sondieren, *to probe a wound.*
3387 nässen (3171), }
3388 aussickern (1921), } *to run, discharge.*
3389 Auswuchs, *m., excrescence, protuberance, tubercle.*
der schwammichte (3234) Auswuchs, *fungus, proud flesh, dead flesh.*
3390 wegnehmen (2300), *to take away.*
3391 Grund, *m., ground, bottom, basis, foundation.*
bis auf den Grund, *to the (very) bottom.*

3391 bis auf den Grund wegnehmen, *to extirpate, eradicate.*
3392 einander, *adv., one another.*
3393 nähern, *to bring near, close.* die Ränder(131)einerWunde einander nähern, *to close the lips of a wound.*
3394 Steigerung (675), *increase, exacerbation.*
3395 Brand, *m., burn; gangrene, mortification.*
3396 Knochenbrand, *m., necrosis.*
3397 heisser Brand, *acute gangrene.*
3398 kalter Brand, *sphacelus.*
3399 feuchter Brand, *moist gangrene.*
3400 trockner Brand, *dry gangrene.*
3401 knochenbrandig, *a., necrotic.*
3402 Brandjauche, *gangrenous sanies.*
3403 Brandblase, *blister.*
3404 Hospital, *n., hospital.*
3405 Hospitalbrand, *hospital gangrene.*
3406 brandig (3241) werden, *to gangrene, mortify.*
3407 Bruch (324), *m., fracture; hernia.*
3408 Knochenbruch, *m., fracture.*
3409 der lange Knochenbruch, *longitudinal fracture.*
3410 der quere Knochenbruch, *transverse fracture.*
3411 der schiefe Knochenbruch, *oblique fracture.*
3412 Längenbruch, *m., longitudinal fracture.*
3413 Querbruch, *transverse fracture.*
3414 Splitter (3374), *m., splinter; shiver.*
3415 Splitterbruch, *comminuted fracture.*
3416 einrichten, *to reduce a fracture.*
3417 Einrichtung, *f.* (eines gebrochenen Knochens), *reduction, setting of a fracture.*
3418 Einrenkung, *f., reduction.*
3419 wiedereinrichten, *to reset, reduce.*
3420 wiedereinsetzen, *to reset, reduce.*
3421 Wiedereinrichtung, }
3422 Wiedereinrenkung, } *reduction of a dislocation.*
3423 verrenken, *to dislocate.*
3424 Verrenkung, *f., dislocation.*
3425 verdrehen (2512), *to twist out of shape, distort.*
3426 Verdrehung, *f., distortion.*
3427 verstauchen, *to sprain, strain.*
3428 Verstauchung, *spraining, wrenching.*
3429 verkrümmen(1107), *to crook, curve, spoil in bending.*
3430 Verkrümmung, *curvature, excurvation.*

3431 verkrüppeln, *to be crippled; to make a cripple, mutilate.*
3432 Verkrüppelung, *crippling, mutilating.*
schlimme (3302) Augen, *sore eyes.*
3433 die Bindehaut, *conjunctiva.*
Entzündung der Bindehaut, *conjunctivitis.*
Entzündung der Augenlider, *inflammation of the eyelids.*
Entzündung der Hornhaut, *inflammation of the cornea*
Entzündung der Gefässhaut und Nervenhaut, *inflammation of the choroid and of the retina.*
3434 Augentriefen (1981), *n.* }
3435 Augenfluss, *m.*
3436 Triefauge, *n.*
*lippitudo.*
3437 krankhaft, *a., diseased, morbid; abnormal.*
3438 krankhafter Thränenfluss, *a watering of the eyes, epiphora.*
3439 Thränenfistel, *f.* }
3440 Augenthränenfistel, *f.*
*fistula lachrymalis.*
3441 Augenkrebs, *m., cancer of the eye, scirrhophthalmia.*
3442 Gerstenkorn, *a sty (on the eyelid).*
3443 der weisse Fleck auf der Hornhaut, *film, speck, leucoma.*

3444 Augenliderkrampf, *blepharospasm.*
3445 Stumpfsichtigkeit, *dullness of vision.*
3446 Kurzsichtigkeit, *myopia.*
3447 Weitsichtigkeit, *far-sightedness.*
3448 Trübung (2253) der Linse, *opacity of the crystalline lens.*
3449 Staar, *m., cataract.*
3450 der graue Staar, *cataract.*
3451 der schwarze Staar, *amaurosis.*
Amblyopie, *or*
3452 unvollkommener schwarzer Staar, }
*ambliopia amaurotica.*
3453 Staarfell, *n.* } *capsular*
3454 Kapselstaar, *m.* } *cataract.*
3455 Linsenstaar, *m., lenticular cataract.*
3456 der harte Staar, *hard cataract.*
3457 der weiche Staar, *soft cataract.*
3458 der punktierte Staar, *punctured cataract, cataracta punctata.*
3459 der grüne Staar, *glaucoma.*
3460 der weisse Staar, *albugo, leucoma.*
3461 der häutige Staar, *membranous cataract.*
3462 der falsche Staar, *false cataract, cataracta spuria.*
3463 staarblind, *a., cataractous, blind from cataract.*

3464 den Staar stechen (3321), *to couch.*
3465 das Staarstechen, ⎫ *couch-*
3466 die Staaroperation, ⎭ *ing.*
3467 Blindheit, *f.*, *blindness.*
3468 stockblind, *a.*, *stone-blind, absolutely blind.*
3469 Stockblindheit, *f.*, *absolute blindness.*
3470 umstülpen, *to till or turn over.*
3471 Umstülpung der Lider, *turning over or eversion of the eyelids.*
3472 heraustreten, *to step out, come out, protrude.*
das Heraustreten des Augapfels, *protrusion of the eyeball.*
3473 übersichtig, *a.*, *hypermetropic.*
3474 Übersichtigkeit, *f.*, *hypermetropia.*
3475 schielen (2531), *to squint.*
3476 das Schielen, *squinting, strabismus.*
schlimme Ohren, *sore ears.*
3477 Ohrgeschwür, *n.*, *otitis.*
3478 Ohrenfluss, *m.*, *discharge from the ear.*
3479 stocktaub (1851), *absolutely deaf.*
3480 der (die) Harthörige, *a person hard of hearing.*
3481 Harthörigkeit, *f.*, *deafness, cophasis.*
3482 der (die) Taubstumme, *a deaf and dumb person.*
3483 Taubstummheit, *f.*, *deaf and dumbness.*
Verstopfung (2235) des Gehörgangs, *obstruction of the auditory canal.*
verhärtetes (2142) Ohrenschmalz (1391), *indurated cerumen.*
3484 das Nasentriefen, *discharge from the nostrils.*
3485 Zahnfleischgewächs, *n.*, *epulis.*
3486 Zahngeschwür, *n.* ⎫
3487 Zahnfleischgeschwür, *n.* ⎭ *gum-boil, parulis.*
3488 Zahnfistel, *f.*, *fistula of the gum.*
3489 Zahngicht, *f.*, *odontagra.*
3490 Zahnrose, *f.*, *erysipelas odontalgicum.*
3491 Zahnwurm, ⎫ *caries of the*
3492 Zahnfäulniss, ⎭ *teeth.*
3493 Gaumenabscess, *alveolar abscess.*
3494 das Gift (*pl.* -e), *venom, poison.*
3495 die Wuth, *rage, fury; madness.*
das Wuthgift, *a rabid virus.*
3496 Hund, *m.*, *dog.*
ein toller Hund, *a mad dog.*
3497 Tollsucht, *f.*, *insanity, madness.*
3498 die Wasserscheu, *hydrophobia.*
3499 das Blatterngift, *the variolous virus.*

3500 Schlangengift, *venom of serpents.*
3501 Wurstgift, *sausage-poison.*
3502 anstecken, *to infect.*
3503 Ansteckungstoff, m., *contagium.*
3504 Ansteckung, f., *infection.*
3505 Vergiftung (3345), *poisoning.*
  ein schleichendes (1637) Gift, *a slow poison.*
3506 **erfrieren, erfror, erfroren,** *to freeze.*
  erfrorne Glieder, *frozen limbs.*
3507 Frostbeule, } *chilblain,*
3508 Frostballen, } *pernio.*
3509 sich verbrennen (1821), *to burn one's self.*
  sich die Finger verbrennen, *to burn one's fingers.*
3510 Verbrennung, *combustio, burn.*
3511 Schiesspulver, n., *gun-powder.*
  Verbrennung mit Schiesspulver, *burning by gunpowder.*
3512 **das** Korn, *grain.*
3513 Pulverkörner, *grains of gunpowder.*
3514 Schmarotzer, } *parasites.*
3515 Parasiten, }
3516 thierisch, a., *animal.*
3517 thierische Schmarotzer, *animal parasites.*
3518 Schmarotzerthier, n., *parasitic animal.*
3519 pflanzlich, a., *vegetable.*
3519 pflanzliche Parasiten, *vegetable parasites.*
3520 Schmarotzerpflanze, f., *parasitic plant.*
3521 Schmarotzerschwamm, m., *parasitic fungus*
3522 Pilz, m., *fungus, favus.*
3523 Erdgrindpilz, m., *favus.*
3524 Flechtengrind, m., *psoriasis.*
3525 Krätzmilbe, f., *itch-tick.*
3526 **die** Laus (*pl.* **Läuse**), *louse.*
3527 **das** Kleid, *garment, dress.*
3528 Kleiderlaus, *pediculus vestimenti.*
3529 Filzlaus, *a crab, felt-louse, pediculus pubis.*
3530 **das** Ungeziefer, *vermin, noxious insects.*
  voll Ungeziefer sitzen, *to be full of lice.*
3531 **die** Niss (*pl.* Nisse), (Lausei, n.), *a nit.*
3532 Floh, m., *flea.*
3533 Flohstich, m., *a flea-bite.*
3534 Wanze, f., *bug.*
3535 Wurm, m. (*pl.* Würmer), *worm, vermis.*
3536 Helminthen, *helminths, worms.*
3537 Enthelminthen, }
3538 Eingeweidewürmer (919), }
3539 Binnenwürmer, }
  *entozoa (living within the body).*
  (Binnen- (*in comps.*) = *within, internal.*
3540 Plattwürmer (2273), *flatworms.*

3541 Bandwürmer, *tænia, tape-worms.*
3542 Egelwurm, or
3543 Leberegel, } *liver-worm, distomum hepaticum.*
3544 Bundwürmer, *round-worms, ascarides.*
3545 Faden, *m.*, *thread, filament.*
3546 Fadenwürmer, *thread-worms, filaria.*
3547 Spulwurm, *ascaris lumbricoides.*
3548 Springwurm, *oxyuris vermicularis.*
3549 Made, *f.*, *maggot, mite.*
3550 Madenwurm, *pin-worm, ascaris.*
3551 Mastdarmwurm (1129),
3552 Askaride,
    *ascaris.*
3553 Peitschenwurm, *or*
3554 Haarkopf,
    *tricocephalus dispar.*
3555 Trichine, *trichina.*
3556 Blasenwurm, *cysticercus.*
3557 Finnenwurm,
3558 Blasenschwanzwurm, } *cysticercus cellulosæ.*
3559 Hüllsenwurm, *echinococcus.*
3560 das Hühnerauge, *corn, clavus pedum.*
3561 **wachsen, wuchs, gewachsen,** *to grow, sprout.*
3562 einwachsen, *to grow into.*
3563 das Einwachsen, *the growing into.*
    das Einwachsen des Nagels ins Fleisch, *nails growing into the flesh.*
3564 Neid, *envy.*
3565 Neidnagel, } *hangnail,*
3566 Niednagel, } *paronychia.*
(*thus in French :* envie (aux doigts), *hangnail.* Nietnagel is incorrect.)

## III. *MEDICAL, SURGICAL, AND HYGIENIC APPLIANCES.*

### *1. Means of Transportation.*

3567 Wagen, *m.*, *carriage, wagon.*
3568 Schlitten, *m.*, *sleigh.*
3569 Trage, *f.*, *litter, hand-barrow.*
3570 Bahre, *f.*, *hand-barrow.*
3571 Tragbahre, *litter.*
3572 das Bett (*pl.* -en), *bed.*
3573 Tragbett, *n.*, *litter, portable bed.*
3574 Stuhl, *m.*, *chair.*
3575 Sessel, *m.*, *chair, settle.*
3576 Tragstuhl, *m.*, *portable chair.*
3577 Krankenträger, *m.*, *litter, stretcher; stretcher-bearer.*
3578 das Brett, *board, plank.*
3579 Matraze, *f.*, *mattress.*

3580 Strohsack, *m.*, *straw-bed, pad of straw.*
3581 Streckbett (560), *n.*, *stretcher.*
3582 Streckstuhl, *m.*, *stretch or extension chair.*
3583 Fussbett, *n.*, *cradle (for a broken foot).*
3584 Fussmaschine, *f.*, *apparatus for deformed feet.*
3585 Luftbett, *n.*, *air-bed.*
3586 Krankenstuhlwagen, *m.*, *invalid wheel-chair.*
3587 Ambulance, *f.*, *ambulance.*

## 2. Bandages, Bands.

3588 Binde, *f.*, *bandage.*
3589 grosse Binde, *handkerchief bandage.*
3590 Hobelspänbinde, *shaving bandage.*
3591 Kopfbinde, *head-band.*
3592 Sternbinde, }
das Stirnband, } *frontal, head bandage.*
3593 Aderbinde, *f.*, *bandage to tie up a vein with.*
3594 Nasenbinde, *nose bandage, accipiter.*
3595 Kinnbinde, }
3596 das Kinntuch, } *chin-piece, bandage for the chin.*
3597 Bauchbinde, *abdominal bandage.*
3598 Leibbinde, *body bandage.*
3599 Nabelbinde, *naval band or bandage.*
3600 Nabelbruchband, *n.*, *truss for umbilical hernia.*
3601 Oberarmbinde, *bandage for the upper arm.*
3602 Brustbinde, *bandage for the chest.*
3603 Aufhebebinde, *suspensory, truss.*
3604 Scheibenbinde (492), *circular bandage.*
3605 Kniebinde, *knee-cap (bandage).*
3606 Fussbinde, *bandage for the foot.*
3607 Bruchband, *n.*, *truss.*
3608 Bruchbinde, *bandage for a fracture or hernia.*
3609 Bauchbruchband, *n.*, *abdominal bandage or truss for hernia.*
3610 überbinden, *to bind over, tie, wrap.*
3611 umbinden, *to bind or tie round.*
3612 Überbinde, *f.*, *upper bandage.*
3613 Unterband, *n.*, *under bandage, subligation.*
3614 Verband, *m.*, *dressing, bandage.*
3615 Schnürverband, *laced bandage, compressive bandage.*
3616 Wattenverband, *padded bandage or dressing.*

3617 Hangband, n. } sling.
3618 Hangbandage, f.
3619 Tragband, n. } sling, truss, suspensory bandage.
3620 Tragbinde, f.
3621 Tragbeutel, m.
3622 Hauptbinde, bandage for the head.
3623 vierköpfig, a., four-headed.
3624 die vierköpfige Hauptbinde, four-headed or four-tailed bandage of the head.
3625 Schleife, noose, loop.
3626 Schlinge, sling, hanging bandage.
3627 Schärpe, sling.
3628 Rollbinde, roller bandage.
3629 Querbinde, transverse band.
3630 Ligatur, f., ligature.
3631 kreuzweis, crucial.
3632 Knotenbinde (1907), star bandage, packer's bandage.
3633 Compressionsverband, m., compress, bolster.
3634 Lassband, n. } bandage after bleeding.
3635 Lassbinde, f.
3636 Wachstaffet, oiled silk.

## 3. Surgical Instruments.

3637 das Messer, knife.
3638 Lanzette, f. } lancet.
    Lancette, f.
3639 Ritzmesser, lancet, scarificator.
3640 Schnittmesser, bistoury.
3641 Incisionsmesser, scalpel.
3642 Incisionslanzette, abscess lancet.
3643 Schere, f., scissors, shears.
3644 Incisionsschere, surgeon's scissors.
3645 Schlitzmesser, lancet, bistoury.
3646 Zergliederungsmesser, dissecting knife.
3647 Schneidemesser, cutting knife, chopping blade.
3648 Stecher, m., picker, pierce.
3649 Stichlanzette, f., thumb-lancet.
3650 Säge, f., saw.
3651 Knochensäge, bone-saw.
3652 Knochenschere, bone-forceps.
3653 Kopfsäge, skull-saw.
3654 Ringmesser, ring or annular knife.
3655 Kreissäge, circular saw.
3656 Kettensäge, chain saw.
3657 Schabemesser, n. } raspatory.
3658 Schabeeisen, n.
3659 Knochenschneidewerkzeug, n., osteotome.
3660 Skalpell, n., scalpel.
3661 die Nadel, needle.
3662 Heftnadel, f., suture needle.
3663 Sichelnadel, sickle needle.
3664 Punktiernadel, acupuncture needle.
3665 Sonde (3386), sound, probe.
3666 Sondiernadel, stylet.
3667 hohle Sondiernadel, needle probe, grooved probe.

3668 Nadelsonde, *exploring needle.*
3669 Leitsonde, *or* }
3670 Leitungssonde, } *grooved sound or staff.*
3671 Schlundsonde, *pharyngeal catheter.*
3672 Wachssonde, }
3673 Wachsröhrchen, } *catheter.*
3674 Katheter, *m.* }
3675 Wundeisen, *n., probe.*
3676 Impfnadel, *f., vaccinating needle.*
3677 Impfinstrument, *n., instrument for vaccinating.*
3678 Schielnadel, *f., strabismus needle.*
3679 Gummiröhrchen, *n., bougie.*
3680 Höllensteinhalter, } *caustic-*
3681 Aetzstifthalter, } *case.*
3682 Schwammhalter zum Aetzen des Schlundes, *sponge-holder for cauterizing the pharynx.*
3683 ausziehen (174), *pull out, extract.*
3684 Zahninstrumente (*pl.*), *dental instruments.*
3685 Zahnauszieher, *m.* } *tooth*
3686 Zahnbrecher, *m.* } *drawer.*
3687 Zahneisen, *n., forceps.*
3688 Zahnbrecheisen, *n., instrument for drawing teeth.*
3689 Zahnschlüssel, *m., key, turnkey.*
3690 Zahnfeile, *f., raspatory.*
3691 Zahnputzer, *scraper.*
3692 **das** Pflaster, *plaster.*

3693 Wundfäden (*pl.*), *m., lint, charpie.*
3694 Charpie, *f., charpie.*
3695 Patent-Charpie, } *patent*
3696 Charpiewatte, } *lint.*
3697 zupfen, *to pluck, tug, unravel.*
3698 gezupfte Charpie, *picked lint.*
3699 schaben, *to scrape.*
3700 geschabte Charpie, *scraped lint.*
3701 Wundpflaster, *n., plaster.*
3702 Wundpulver, *n., wound powder.*
3703 Heftpflaster, *n., sticking plaster.*
3704 Heilpflaster, *n., healing plaster.*
3705 das englische Pflaster, *court plaster.*
3706 Zange, *f., tongs (pair of).*
3707 Zängelchen, *n., (pair of) pincers, tweezers, nippers.*
3708 Pincette, *f., pincette, surgeon's forceps.*
3709 Splitterzange, *f., splinter forceps; parrot-beak.*
3710 Sperrpincette, *forceps with a catch.*
3711 Pincettenschere, *forceps-scissors.*
3712 Rupfzange, *tweezers.*
3713 Krückenzange, *crutch-shaped forceps, scraping forceps*
3714 Kugelzange, *bullet forceps, ball extractor.*
3715 Knochenzange, *bone forceps.*
3716 Zahnzange, *tooth forceps.*

3717 Schnäpper, *fleam, spring lancet.*
3718 Feile, *f., file.*
3719 Knochenfeile, *bone-file.*
3720 Zungenschaber, *m., tongue scraper.*
3721 Knochenschaber, *bone scraper.*
3722 Schlundkopföffner, *pharyngeotome.*
3723 Schlundstosser,
3724 Schlundkopfstosser, *probang.*
3725 Kugelbohrer, *m., ball gimlet.*
3726 Schädelbohrer, *trephine, perforator.*
3727 Trepanschlüssel, *m., trepan-key.*
3728 Zugbohrer, *m., trephine.*
3729 spritzen, *to inject.*
3730 einspritzen, *to inject.*
3731 **Spritze** (die kleinere), *syringe.*
3732 Einspritzer, *m., syringe.*
3733 Injectionsspritze, *syringe.*
3734 die Spritze von Hartgummi, *syringe of hard india-rubber.*
3735 Spritzröhrchen, *n., injection-pipe.*
3736 subcutan, *a., hypodermic, subcutaneous.*
3737 Injection, *f., injection.* Spritze zu subcutanen Injectionen, *hypodermic syringe.*
3738 klystieren, *to apply a clyster.*
3739 das Klystier, *clyster, enema, lavement.*
3740 Klystierröhre, *f.*
3741 Klystierschlauch, *m.* *clyster-pipe, tube, hose.*
3742 Klystierspritze, *squirt, syringe.*
3743 Klystierpumpe, *clyster-pump, enema apparatus.*
3744 Magenspritze, *stomach*
3745 Magenpumpe, *pump.*
3746 Milchpumpe, *breast or nipple glass.*
3747 Nasenspritze, *nasal syringe.*
3748 Ohrenspritze, *ear*
3749 Ohrspritze, *syringe.*
3750 Mutterspritze, *uterine syringe*
3751 **der Spiegel,** *mirror,*
3752 das Speculum, *speculum.*
3753 Mundspiegel, *speculum or dilator oris, stomatoscope.*
3754 Rachenspiegel, *pharyngeal mirror or speculum.*
3755 Kehlkopfspiegel, *laryngoscope.*
3756 Schlundkopfspiegel, *pharyngoscope.*
3757 Nasenspiegel, *rhinoscope.*
3758 Ohrenspiegel, *speculum*
3759 Ohrspiegel, *auris.*
3760 Mutterspiegel, *vaginal speculum.*
3761 Aderlasszeug, *n., bleeding instruments.*
3762 Schröpfkopf, *m., glass, cucurbita.*
3763 Schröpfschnäpper *m., spring-lancet, scarificator.*

## 4. Baths.

3764 das Bad, bath.
3765 Heilbad, n., medical or mineral bath.
3766 Warmbad, n, warm spring, thermal waters.'
3767 Laubad, tepid bath.
3768 Sitzbad, hip bath, slipper-bath.
3769 Gliedbad, bath for a limb.
3770 Handbad, hand-bath.
3771 Fussbad, foot-bath.
3772 Schweissbad,
3773 Schwitzbad, } sweating bath, steam bath, hot-air bath.
3774 Dampfbad,
3775 Dunstbad, } vapor bath.
3776 Qualmbad,
3777 Douche, f.
3778 Douchebad, } douche, douche-bath.
3779 Giessbad,
3780 Traufbad,
3781 Sprudelbad,
3782 Tropfbad, } shower-bath.
3783 Staubbad,

3784 Schauerbad,
3785 Sturzbad, } shower-bath.
3786 Sturzbad von oben, shower-bath from above.
3787 Sturzbad von der Seite, shower-bath from the side.
3788 Überraschungsbad, plunge-bath.
3789 (überraschen, to surprise, overtake.)
3790 Übergiessung, douche.
3791 Schwefelbad, sulphur-bath.
3792 Luftbad, air-bath, exposure to the atmosphere.
3793 Sonnenbad, sun-bath.
3794 Sandbad, sand-bath or heat.
3795 Rauchbad, vapor-bath.
3796 Kräuterbad, bath of medicinal herbs.
3797 Medicinalbad, medicated bath.
3798 Salzbad, salt water bath.
3799 Seebad, sea-bathing.
3800 Waschbad, wash, bath taken for cleanliness.

## 5. Medical Officials and Establishments.

3801 die Kunst (pl. Künste), art.
3802 Kunde, f., knowledge, science.
3803 Heilkunst,
Heilkunde, }
Medicin,
medicine (science of medicine), medical science, therapeutics.

3804 medicinal, a., medical.
3805 Arzt, m.
Doctor (pl. -en) } physician, doctor.
3806 ärztlich, a., medical.
3807 behandeln, to treat.
3808 Behandlung, treatment.
ärztliche Behandlung, medical treatment.
3809 practiciren, to practise.

3810 practisch, *a.*, *practical; practising.*
der practische Arzt, *practising physician.*
der practicierende Arzt, *practitioner.*
einen Artz zu Rathe ziehen, *to ask a doctor's advice, to consult a doctor.*
3811 das Amt, *office, board.*
3812 Gesundheitsamt, *n.* }
3813 Medicinal-Collegium, *n.* } *the board of health.*
3814 Polizei, *f.*, *police.*
3815 Gesundheitspolizei, *sanitary police.*
3816 Mediciner, *physician; medical student.*
3818 Militärarzt, *military surgeon.*
3819 Militäroberarzt, *surgeon general.*
3820 Wundarzt, } *surgeon.*
3821 Chirurg, }
3822 special, *a.*, *special.*
3823 Specialarzt, *specialist.*
3824 Augenarzt, } *oculist.*
Okulist, }
3825 Ohrenarzt, *aurist.*
3826 Zahnarzt, } *dentist.*
Dentist, }
3827 Geburtsarzt, }
3828 Geburtshelfer, }
3829 Accoucheur, }
*accoucher, man-midwife, obstetrician.*
3830 Hebamme, } *midwife.*
3831 Bademutter, }

3832 Thierarzt, }
3833 Veterinärarzt, }
*veterinary surgeon, horse-doctor.*
3834 Thierarzneikunde, *veterinary science.*
3835 geschickt, *a.*, *clever, skillful, qualified.*
ein sehr geschickter Doctor, *a very clever doctor.*
3836 Allopath, *m.*, *allopathic doctor.*
3837 Homöopath, *m.*, *homœopathist, homœopathic doctor.*
3838 der Rath, (*title*) *councillor,* (*counsellor*).
3839 der Geheimrath, *privy councillor.*
3840 Bandagist, *m.* }
3841 Bruchbandmacher, *m.* } *the truss-maker.*
3842 Hüneraugenoperateur, *m.*, *corn-doctor, corn-cutter, chiropodist.*
3843 Bader, *m.* }
3844 Bartdoctor, *m.* } *barber.*
3845 Barbier, *m.* }
3846 Krankenwärter, }
3847 Krankenwärterin, *f.* }
3848 Krankenpfleger (in). }
*tender (of sick persons), nurse, sick-nurse.*
3849 Apotheker, *m.*, *apothecary, druggist, chemist.*
3850 Apothekergehilfe, *apothecary's assistant.*
3851 Apothekerlehrling, *apothecary's apprentice.*

3852 Provisor, *head-clerk, provisor, manager.*
3853 Droguist. *m., druggist.*
3854 Apothekerwaaren, } *drugs.*
3855 Droguen,
3856 Droguenhandlung, *druggery or druggist's shop.*
3857 Arzenei, *f.* } *medicine.*
    Medicin, *f.*
3858 Arzeneipflanzen (*pl.*) }
    Arzeneikräuter (*pl.*) *officinal or medicinal herbs.*
3859 Apotheke, *apothecary's shop, drug-store.*
3860 Apothekerladen, }
3861 die Officin, *druggist's shop.*
3862 Laboratorium, *laboratory.*
3863 Feldapotheke, *field-dispensary.*
3864 Reiseapotheke, *travelling medicine-chest.*
3865 Hausapotheke, *family medicine chest.*
3866 Schiffsapotheke, *a ship's medicine chest.*
3867 die Anstalt, *establishment, institution.*
3868 die Heilanstalt, *sanitary or medical establishment.*
3869 die Klinik, *clinic.*
3870 Polyklinic, *ambulatory clinic.*
    Hospital, } *hospital.*
3871 Spital,
3872 Krankenhospital, }
3873 Krankenhaus, *infirmary, hospital.*
3874 Lazareth, *lazaretto hospital.*
3875 fliegendes Lazareth, *flying lazaretto, ambulance.*
3876 Feldlazareth, *field hospital.*

## 6. Articles of Food and Drink.

3877 **essen, ass, gegessen,** *to eat.*
3878 **fressen, frass, gefressen,** *to eat greedily, devour, gluttonize* (2896).
3879 **schlingen, schlang, geschlungen,** *to swallow eagerly, devour.*
3880 schlucken, *to swallow.*
3881 aufessen, *to eat up, consume.*
3882 sich überessen, *to over-eat one's self.*
3883 das Essen, *eating, meal, repast.*
3884 frühstücken, *to breakfast.*
    speisen (764).
    Speise (765).
3885 die Esslust, } *appetite.*
    Appetit (2148), *m.*
3886 verschlingen, *to swallow up, gulp down, devour.*
3887 verschlucken, *to swallow up, absor*
3888 sich verschlucken, *to swallow the wrong way.*

3889 befriedigen, *to appease, satisfy, set at rest.*
3890 die Esslust, *or* ⎫
3891 den Appetit, *or* ⎬ befriedigen *or* stillen,
3892 den Hunger ⎭
  *to stay the appetite, the hunger.*
3893 sich sättigen, *to satisfy one's self or one's hunger.*
3894 **geniessen, genoss, genossen,** *to enjoy; to take or taste food or drink.*
3895 zu sich nehmen (2390), *to take food, drink.*
3896 tüchtig, *a., proper, excellent; hearty.*
3897 tüchtigen Appetit haben, *to have a hearty appetite.*
3898 viel vertragen können, *able to digest much.*
3899 wenig vertragen können, *able to digest little.*
3900 Alles verdauen (1047) können, *able to digest everything.*
3901 **trinken, trank, getrunken,** *to drink, to drink like a beast.*
3902 **saufen, soff, gesoffen,** *to drink; to drink hard, tipple.*
3903 den Durst (2171) stillen, *or* ⎫
3904 löschen, ⎭
  *to quench the thirst.*
3905 Bier trinken, *to drink beer.*
3906 schnappsen, *to drink drams.*
3907 zu viel, *too much, to excess.*
3908 über den Durst *or* zu viel trinken, *to drink beyond the thirst, to drink to excess.*
3909 sich betrinken, ⎫
3910 sich besaufen, ⎬ *to get drunk.*
3911 sich berauschen, ⎭
3912 betrunken sein, *to be drunk, intoxicated.*
3913 Etwas im Kopf haben, *to be rather drunk.*
3914 gucken, *to look, peep.*
3915 **das** Glas, *glass.*
3916 zu tief ins Glas geguckt haben, *to have been too familiar with the bottle, to have got tipsy or drunk.*
3917 Speisen (765), *dishes, viands.*
3918 **das** Mittel, *medium, expedient, means.*
3919 Lebensmittel (*pl.*), *victuals, provisions, means of subsistence.*
  Nahrungsmittel (1059).
3920 Mahlzeit, *f., meal.*
3921 **das** Gastmahl, *banquet.*
3922 **das** Frühstück, *breakfast.*
3923 das Gabelfrühstück (1021), *lunch.*
3924 Mittagsmahl, ⎫
3925 Mittagsessen, ⎬ *dinner.*
3926 Mittagsbrod, ⎭
3927 Abendbrod, ⎫
3928 Abendessen, ⎬ *supper.*
3929 Nachtmahl, ⎭
3930 Kost, *f., food, fare, diet, regimen.*

3931 Hausmannskost, *plain fare.*
3932 Brod, *n.* (*pl.* -e), *bread.*
3933 Weissbrod, *white bread.*
3934 Weizenbrod, *wheat bread.*
3935 Schwarzbrod, *brown bread.*
3936 Brodkrumme, *f., crumb.*
3937 Brodrinde, *f., crust.*
3938 ein Bissen Brod, *a bit (morsel) of bread.*
3939 **backen, buk, gebacken,** *to bake.*
3940 rösten, *to toast (bread); roast, burn (coffee).*
3941 hausbackenes Brod, *home-baked bread.*
3942 neubackenes Brod, *fresh bread.*
3943 altbackenes Brod, *dry (stale) bread.*
3944 geröstetes Brod, *toast-bread.*
3945 Milchbrod, *French roll (milk bread).*
3946 **die** Semmel, *roll.*
**die** Butter (1334), *butter.*
3947 Butterbrod, *n., bread and butter.*
3948 belegen (30), *to lay over, cover with.*
3949 belegtes Butterbrod, *sandwich.*
3950 **das Backwerk,** }
3951 **das Gebäck,** } *pastry.*
3952 Kuchen, *m., cake.*
3953 Zwieback, *m.* }
3954 Biscuit, } *biscuit.*
3955 **das** Fleisch, *meat.*
3956 Rindfleisch, *n., beef.*

3957 Kalbfleisch, *n., veal.*
3958 Hammelfleisch, *n., mutton.*
3959 Schweinefleisch, *n., pork.*
3960 **das** Wildpret, *venison, game.*
3961 Suppe, *f., soup.*
3962 Suppenfleisch, *n.* }
3963 (kochen, *to cook, boil*), }
3964 gekochtes Fleisch, } *boiled meat.*
3965 Fleischbrühe, *f., broth.*
3966 **braten, briet, gebraten,** *to roast.*
3967 Braten, *m., roast meat, roast.*
3968 Rinderbraten, *m.* } *roast-*
or Rindsbraten, } *beef.*
3969 Kalbsbraten, *roast-veal.*
3970 Hammelbraten, *roast-mutton*
Sauce, *f., sauce.*
3971 Brühe, *f., gravy.*
3972 Hammelkeule, *leg of mutton.*
3973 Hammelcotelett, *n.* }
3974 Hammelrippchen (*pl.*) } *mutton-chop.*
3975 Kalbscotelett, *n., veal-cutlet.*
3976 Rauchfleisch (2318), *smoked beef.*
3977 pökeln, *to pickle, to salt.*
3978 Pökelfleisch, *corned beef.*
3979 **die** Wurst, *sausage.*
3980 Schinken, *ham.*
3981 Speck, *bacon.*
3982 Geflügel, *n., poultry.*
3983 **die** Gans, *goose.*
3984 Ente, *duck.*
3985 **das** Huhn, *fowl.*
3986 **das** Rebhuhn, *partridge.*

3987 Fasan (sah'n), m., *pheasant.*
3988 Taube, *pigeon, dove.*
3989 Truthahn, m., *turkey.*
3990 spicken, *to lard.*
3991 Spickgans, f., *smoked goose.*
3992 Gänsebraten, *roasted goose.*
3993 Gänsebrust, f., *(smoked) breast of a goose.*
3994 Entenbraten, *roasted duck.*
3995 Gänseklein, *giblets.*
3996 Füllsel, n., *stuffing.*
3997 Fisch, *fish.*
3998 Karpfen, *carp.*
3999 Barsch, *perch.*
4000 Hecht, m., *pike.*
4001 Schleie, *tench.*
4002 Weissling, *whiting.*
4003 Aal, *eel.*
4004 Lachs, *salmon.*
4005 Forelle, *trout.*
4006 Makrele, *mackerel.*
4007 Sardelle, *anchovy.*
4008 Häring, *herring.*
4009 Krebs (3655), *craw fish.*
4010 Hummer, *lobster.*
4011 **die** Auster, *oyster.*
4012 Schildkröte, *tortoise, turtle.*
4013 harte Eier (1211), *hard-boiled eggs.*
4014 weiche Eier, *soft-boiled eggs.*
4015 Eierkuchen, *omelet.*
4016 Pfannkuchen, *pancake, fritter.*
4017 gerührte Eier, *buttered eggs.*
4018 Rühreier (2765), *scrambled eggs.*
4019 Setzeier, *poached eggs.*
4020 Dotter, n., *the yolk.*
4021 Pastete (tā'te), *pastry, pie.*
4022 Gelée (Galerte), *jelly.*
4023 Gurken (*pl.*), *cucumbers.*
4024 Pfeffergurken, *pickled cucumbers (gherkins).*
4025 Eingemachtes (in Zucker), *preserve.*
4026 Eingemachtes in Salz (oder Essig), *pickles.*
4027 Küchenkräuter (*pl.*), *vegetables, greens.*
4028 Hülsenfrüchte (*pl.*), *legumes*
4029 Erbsen (*pl.*), *peas.*
4030 Schoten (*pl.*), *green peas.*
4031 Bohnen (*pl.*), *beans.*
4032 Linsen (1375) (*pl.*), *lentils.*
4033 Kohl, *cabbage.*
4034 Blumenkohl, *cauliflower.*
4035 Kraut, *cabbage.*
4036 Sauerkraut, *(sour) crout or krout.*
4037 Rüben (*pl.*), *turnips.*
4038 Möhren (*pl.*), } *carrots.*
4039 gelbe Rüben,
4040 Kartoffeln (*pl.*), *potatoes.*
4041 Reis, *rice.*
4042 Frucht, f. (*pl.* Früchte), *fruit.*
4043 Obst, n., *fruitage.*
4044 Apfel, m., *apple.*
4045 Birne, f., *pear.*
4046 Pflaume, f., *plum.*
4047 Kirsche, f., *cherry.*
4048 Apricose, *apricot.*
4049 Pfirsich, f., *peach.*
4050 Orange, f. } *orange.*
4051 Pomeranze, f.

4052 die Nuss (Nüsse), *nut.*
4053 Wallnuss, *f.*, *walnut.*
4054 Haselnuss, *f.*, *hazel nut.*
4055 Ananas, *f.*, *pine-apple.*
4056 Feige, *fig.*
4057 Torte, *tart.*
4058 Macaroni, *macaroni.*
4059 Makrone, *macaroon.*
4060 Bonbons, *bon-bon.*
4061 Gefrornes, *n.*, *or* ⎫ *ice,*
4062 Eis, *n.* ⎭ *ice-cream.*
4063 Zuckerwerk (2294), *sweetmeats.*
4064 Confect, *n.*, *confectionery.*
4065 Pfeffer, *m.*, *pepper.*
4066 Essig, *m.*, *vinegar.*
4067 Zimmt, *m.*, *cinnamon.*
4068 Ingwer, *m.*, *ginger.*
4069 Muscatnuss, *f.*, *nutmeg.*
4070 Wassersuppe (705), *water-porridge.*
4071 Haferschleim, *m.*, *gruel.*
4072 Sodawasser, *n.*, *soda-water.*
4073 Limonade, *f.*, *lemonade.*
4074 Himbeersaft, *m.* ⎫
4075 Himmbeerensyrup, *m.* ⎭ *raspberry syrup.*
4076 Sahne, *f.* ⎫ *cream.*
4077 Rahm, *m.* ⎭

4078 Kaffee, *m.*, *coffee.*
4079 Thee, *m.*, *tea.*
4080 Chocolade, *f.*, *chocolate.*
4081 grüner Thee, *green tea.*
4082 schwarzer Thee, *black tea.*
4083 Wein, *m.*, *wine.*
4084 Weisswein, *white wine.*
4085 Rothwein, *red wine.*
4086 Rheinwein, *Rhenish wine, Hock.*
4087 Moselwein, *Moselle.*
4088 Ungarwein, *Hungary wine.*
4089 Bordeaux, *Claret, Bordeaux.*
4090 Burgunder, *Burgundy.*
4091 Champagner, *Champagne.*
4092 Tokaeir, *Tokay.*
4093 Portwein, *Port.*
4094 Xercswein, ⎫ *Sherry.*
4095 Scherry, ⎭
4096 Äpfelwein, *cider.*
4097 geistige Getränke, *spirits.*
4098 Doppelbier, *n.*, *ale.*
4099 Branntwein, *m.*, *brandy.*
4100 Liqueur, *m.*, *liquor.*
4101 Rum, *m.*, *rum.*
4102 Punsch, *m.*, *punch.*
4103 Mineralwasser, *n.*, *mineral water.*

## 7. *Methods of Treatment.*

4104 heilen, *to cure, heal.*
4105 Heilung, *healing, cure, recovery.*
4106 Genesung (1517), *recovery.*
4107 wiederherstellen, *restore, recover, recruit.*
4108 Wiederherstellung, *restoration, recovery.*
4109 rückfallen, *to relapse.*
Rückfall, ⎫ *relapse* (1524).
4110 Recidiv, ⎭
4111 curieren, *to cure.*

4112 die Cur, *cure, course of medicine, medication.*
4113 gebrauchen, *to make use of, employ.*
4114 eine Cur gebrauchen, *to make use of (certain) remedies.*
4115 in der Cur sein, *to be under treatment.*
4116 allopathische Cur,
4117 homöopathische Cur,
4118 hydropathische Cur
   *allopathic, homœopathic, hydropathic cure.*
4119 Badecur,
4120 Brunnencur, } *use of mineral waters.*
4121 Kaltwassercur, *cold water cure or remedy*
4122 Milchcur, *milk-cure.*
4123 Molke, *f.*
   Molken, } *whey.*
4124 Molkencur, *whey-cure.*
   Traube, *f.* (1366), *bunch (of grapes).*
4125 Traubencur, *cure by grapes.*
4126 Eisen, *n., iron.*
4127 Eisencur, *cure by iron.*
4128 der Stahl, *steel.*
4129 Stahlcur, *cure by iron.*
4130 Quecksilber, *n., quicksilver.*
4131 Merkur, *m., mercury.*
4132 Quecksilbercur,
4133 Merkurialcur, } *cure by mercury.*
4134 Speichelcur (981), *salivation*
4135 Vorcur, *preparatory treatment.*
4136 Nachcur, *after-treatment.*
4137 magnetische Cur, *magnetic cure.*
4138 sympathetisch (904), *a., sympathetic.*
4139 sympathetische Cur, *sympathetic cure.*
4140 zaubern, *to practise magic, witchcraft.*
4141 Zauber, *magic, spell, witchcraft.*
4142 Zaubercur, *magic cure.*
4143 das Wunder, *wonder, miracle.*
4144 Wundercur, *miraculous cure*
4145 Hungercur, *hunger-cure, cure by fasting, starving system.*
4146 Hunger-Schwitzcur,
   Inanitionscur,
4147 Abmagerungscur (2435), } *cure by emaciation.*
4148 das Klima, *climate.*
4149 Klimacur, *f., climatic cure.*
4150 Kräutercur (4035), *cure by means of herbs.*
4151 Athmung (984), *f., breathing, respiration.*
4152 Athmungsheilung, *f., respiratory cure.*
4153 Lufttheilung (1005), *air-cure.*
4154 zuheilen, *to heal up, close.*
4155 Zuheilung, *f., healing up.*
4156 verharschen, *to grow hard, cicatrize, close.*
4157 Verharschung, *f., cicatrization.*

## 8. Remedies and Expedients.

4158 laben, *to refresh, recreate.*
4159 Labung, *f., refreshment, recreation.*
4160 Labemittel, *n.* }
4161 Labungsmittel, *n.* }
    *refreshing remedy.*
4162 Kräftigung (1555), }
4163 Stärkung (1558), }
    *strengthening, invigoration.*
4164 Kräftigungsmittel, *n.* }
4165 Stärkungsmittel, *n.* }
    *strengthening remedy.*
4166 **das** Heil (4104), *health, soundness.*
4167 Heilmittel, *n.* }
4168 Heilungsmittel (4105), *n.* }
    *remedy, cure.*
4169 turnen, *to perform gymnastic exercises.*
4170 **das** Turnen, *gymnastic art.*
4171 üben, *to exercise.*
4172 Übung, *f., exercise.*
4173 Turnübung, *f., gymnastic exercise.*
4174 gymnastisch, *a., gymnastic.*
    gymnastische Übungen, *gymnastic exercises.*
4175 Muskelübung, *muscular exercise, gymnastics.*
    Bewegung in freier (2343) Luft, *exercise (movement) in the open air.*
4176 regulieren, *to regulate.*
4177 die Diät (ä't), *diet.*

4178 knapp, *a., narrow, tight.*
4179 knappe Diät, *low regimen.*
4180 einwickeln, *to wrap (up or in).*
4181 Einwickelung, *f., wrapping up.*
4182 Watte, *f., wadding.*
    Einwickelung in Watte, *to wrap in wadding.*
4183 Kräuterküssen, *n., medicated cushion (stuffed with herbs).*
4184 Kräuterarzenei, *f., herb medicine.*
4185 Arzeneimittel, *n., remedy, medicine.*
4186 verordnen, *to order, prescribe.*
4187 Verordnung, *f., order, prescription.*
4188 Hausmittel, *n., family medicine; old woman's remedy.*
4189 **das** Recept, } *recipe.*
4190 Recipe, }
4191 abführen, }
4192 laxieren, } *to evacuate, purge.*
4193 purgieren, }
4194 Abführmittel, }
4195 Laxiermittel, } *purgative, evacuant.*
4196 Purgiermittel, }
4197 Brechmittel, } *emetic.*
4198 Vomiermittel, }
4199 einschläfern, *to lull (to sleep), to drowse.*

4200 Einschläferungsmittel, *narcotic, soporific.*
4201 Schlafmittel, *soporific.*
4202 Schlafarznei, *hypnotic.*
4203 Schlaftrank, *m.*, *narcotic draught.*
4204 Mittagsschläfchen, *siesta.*
4205 betäuben, *to stupefy, narcotize.*
4206 Betäubungsmittel, *stupefactive, narcotic.*
4207 verwahren, *to preserve from.*
4208 verhüten, *to prevent, ward off.*
4209 Preservativmittel, *n.* }
4210 Verwahrungsmittel, } *preservative.*
4211 Verhütungsmittel, }
4212 Linderungsmittel (1673), *n.*, *lenitive, palliative.*
4213 lindernd, *a.*, *lenitive, assuaging.*
4214 lindernde Mittel, *lenitives, anodynes.*
reizen (825), } *to irritate,*
4215 anreizen, } *stimulate.*
4216 Reizmittel, *n.*, *stimulant.*
4217 kühlen, *to cool, freshen, refrigerate.*
4218 Kühlmittel, *n.*, *a refrigerant.*
4219 erweichen, *to soften, mollify.*
4220 Erweichungsmittel, *emollient.*
4221 auflösen, *to loosen, dissolve, melt.*
4222 Auflösungsmittel, *an expectorant, diluent.*
4223 Atzmittel (3314), *caustic.*
4224 erfrischen, *to refresh, cool, refrigerate.*
4225 Erfrischungsmittel, *a refrigerant.*
4226 Brustmittel, *n.*, *pectoral.*
4227 Schweissmittel, *n.*, *diaphoretic, sudorific.*
4228 Niesemittel, *sternutatory.*
4229 Wundmittel, *traumatic.*
4230 Wurmmittel, *helminthic.*
4231 Schutzmittel, *prophylactical remedy.*
4232 Radikalmittel, *specific.*
4233 Bähung (3317), *fomentation*
4234 beräuchern, *to fumigate.*
4235 Beräucherung, } *fumigation.*
Beräuchern, *n.* }
4236 Einspritzung (3730), *injection.*
die Einspritzung unter der Haut, *a subcutaneous injection.*
4237 einträufeln, *to instil.*
4238 Einträufelung, } *instillation.*
Einträufeln, *n.* }
4239 einblasen, *to breathe into.*
4240 Einblasung, } *insufflation.*
Einblasen, *n.* }
4241 röthen, *to redden.*
4242 die Röthung der Haut, *rubification.*
4243 Reizung, *irritation.*
4244 bespritzen, *to syringe, inject.*
4245 Bespritzen, *n.* }
4246 Einspritzen, *n.* } *syringing, injecting.*
4247 Ausspritzen, *n.* }
4248 Haarseil, *n.*, *seton.*

4249 Haarseillegen, *n.*, *to put in a seton.*
4250 schröpfen, *to cup.*
4251 zur Ader lassen, *to bleed a person.*
4252 Senfpflaster, *n.*, *mustard plaster, sinapism.*
4253 Senfumschlag, *m.*, *mustard poultice.*
4254 Teig, *m.*, *dough, paste.*
4255 Senfteig,*m.*,*mustard plaster.*
4256 Umschlag, *m.*, *poultice, cataplasm.*
  ein Bad mit Senf, *a bath with mustard.*
4257 bestreuen, *strew over,spread.*
  mit Senf bestreuen (einen Umschlag), *to powder with mustard.*
4258 Pille, *f.*, *pill.*
4259 Pulver, *n.*, *powder.*
  Brei (1085), } *pulp, pap,*
4260 Mus, *n.* } *porridge.*
4261 Syrup, *m.*, *syrup.*
4262 Tinctur, *f.*, *tincture.*
4263 Kühltrank, *m.* }
4264 Krankenthee, *m.* } *cooling draught, diet-drink, ptisan.*
4265 Salbe, *f.*, *salve.*
4266 Waschung, *f.*, *lotion, wash.*
4267 Fettsalbe, *f.*, *ointment.*
4268 Gurgelwasser, *n.*, *gargle.*
4269 Stuhlzäpfchen, *suppository.*
4270 Augenwasser, *n.* } *collyrium*
4271 Augenmittel, *n.* }
4272 Augensalbe, *f.*, *eye-salve.*
4273 Augenpulver, *n.*, *eye-powder.*
4274 Zahnmittel, *n.*, *dentifrice.*
4275 Enthaarungsmittel, *n.*, *depilatory.*
4276 Schönheitsmittel, *n.*, *cosmetic.*
4277 Medicament,*n.*,*medicament, remedy.*
4278 Pastille, *pastile, a medicated lozenge.*
4279 Trank, *m.* }
4280 Tränkchen (*dim.*), *n.* } *beverage, potion.*
4281 Tropfen (*pl.*) (2335), *drops.*
4282 Balsam, *m.*, *balm.*
4283 Breiumschlag, *m.*, *poultice, cataplasm.*
4284 Panacee, *f.*, *panacea, all-healing remedy.*
4285 Gegengift (3494), *n.* }
4286 Theriak, } *antidote, counter-poison.*
4287 schmerzstillendes Mittel (1686), *anodyne.*
4288 Opiat, *n.*, *opiate.*
4289 Anästheticum, *anæsthetic.*
4290 Aether, *ether.*
4291 Schwefeläther, *sulphuric ether.*
4292 Irritans, *irritant.*
4293 Stimulans, *stimulant.*
4294 Holzessig (4066), *pyroligneous vinegar.*
4295 Himbeeressig, *raspberry vinegar.*
4296 Essigsäure, *acetic acid.*
4297 Carbolsäure, *carbolic acid.*

4298 Kohlensäure, *carbonic acid*.
4299 Citronensäure, *citric acid*.
4300 Milchsäure, *lactic acid*.
4301 Salpetersäure, *nitric acid*.
4302 Vitriolöl, *n.*, *oil of vitriol*.
4303 Weinsteinsäure, *tartaric acid*.
4304 Baldriansäure, *valerianic acid*.
4305 Alkohol, *m.*, *alcohol*.
4306 Aloë, *f.*, *aloes*.
4307 Alaun, *m.*, *alum*.
4308 destillieren, *to distill*.
destilliertes Wasser, *distilled water*.
4309 Kalk, *m.*, *lime*.
4310 Kalkwasser, *n.*, *lime water*.
4311 Chlorwasser, *n.*, *chlorine water*.
4312 Theer, *m.*, *tar*.
4313 Theerwasser, *tar water*.
4314 Schwefelwasser, *sulphurated water*.
4315 Silbernitrat, *n.*, *nitrate of silver*.
4316 Höllenstein, *nitrate of silver, lunar caustic*.
4317 Stinkasant, *m.* }
4318 Teufelsdreck, *m.* }
asafœtida, *devil's dung*.
4319 die spanische (3251) Fliege, *cantharides, blistering fly, Spanish fly*.
4320 Thierkohle, *f.*, *bone-black, (wiry black) animal charcoal*.
4321 Holzkohle, *f.*, *charcoal (wood-charcoal)*.
4322 Magentropfen, }
4323 Magenpulver, }
*stomachic drops or powder*.
4324 Brusttropfen, }
4325 Brustpulver, }
*pectoral drops or powder*.
4326 Brustsyrup, } *pectoral syrup*.
4327 Brustsaft, }
4328 Altheewurzel, *f.*, *marshmallow root*.
4329 Altheensyrup, } *althea syrup*.
4330 Eibischsyrup, }
4331 Fliederblume, } *elder flowers*.
4332 Hollunderblüthe, }
4333 Lindenblüthe, *linden flowers*
4334 Pomeranzenblätter, *orange leaves*.
4335 Bilsenkraut, *n.*, *hyoscyamus leaves, henbane leaves*.
4336 Melissenblätter, *balm-mint*.
4337 Krauseminze, *curled mint, balm-mint*.
4338 Pfefferminze, *peppermint*.
4339 Rosmarin, *rosemary*.
4340 Salbeiblätter, *garden-sage*.
4341 Sennesblätter, *senna leaves*.
4342 Reutenblatt, *n.*, *common rue*.
4343 Tabak, *m.*, *tobacco leaves*.
4344 Brustthee, *m.*, *pectoral tea*.
4345 Kräuterthee, *m.*, *tea of medicinal herbs*.
4346 Kamillenthee, *m.*, *camomile tea*.
4347 Fenchelthee, *fennel tea*.
4348 Fliederthee, *elder tea*.
4349 Pfefferminzthee, *peppermint tea*.

4350 Schlüsselblumenthee, *primrose or lungwort tea.*
4351 Flachssamenthee, *flax-seed tea.*
4352 Holzthee, *aperient roots.*
4353 erwärmen, *to make warm.*
Erwärmenderthee, *carminative species.*
4354 das Moos, *moss.*
4355 Isländisches Moos, *Iceland moss.*
4356 Fiebermittel, *febrifuge, anti-febrile.*
4357 Chinin, *n., quinine.*
4358 Chinarinde, } *Peruvian*
4359 Fieberrinde, } *bark, quinine.*
4360 Chinapulver, *n., Jesuit's powder.*
4361 Sassaparille, *f.* } *sarsaparilla.*
Sarsaparille, *f.* }
4362 Brausepulver (2652), *n., effervescent.*
4363 das englische Brausepulver, *soda.*
4364 das Seidlitzpulver, *Seidlitz's powder.*
4365 abführendes (4191) Brausepulver, *effervescent purgative.*
4366 Kinderpulver, }
4367 Beruhigungspulver, } *magnesia and rhubarb.*
4368 niederschlagen, *to strike down; precipitate, deposit.*
4369 das niederschlagende Pulver, *antispasmodic powder.*
4370 Mandelemulsion (1454), *f., almond emulsion.*

4371 Mandelmilch, *f., orgeat.*
4372 Thran, *m., train-oil, blubber*
4373 Leberthran, *m., cod-liver oil.*
4374 Klauenfett (927), *n., neat's-foot oil.*
4375 Wachholderbeere, *juniper.*
4376 Wachholderbeeröl (967), *n., oil of juniper.*
4377 Lavendel, *m., lavender.*
4378 Lavendelöl, *n., oil of lavender.*
4379 Leinsamen, *m., flax seed.*
4380 Leinöl, *n., linseed oil.*
4381 Majoranöl, *n.* } *oil of sweet*
4382 Mairanöl, } *marjoram.*
4383 Krauseminzöl, *n., oil of curled mint.*
4384 Muskatnussöl, *n., oil of nutmegs.*
4385 Olivenöl, *n.* } *olive oil,*
4386 Baumöl, *n.* } *sweet oil.*
4387 Mohn, *m., poppy (plant).*
4388 Mohnöl, *n., poppy oil.*
4389 Steinöl, *n.* } *petroleum,*
Petroleum, *n.* } *naphtha.*
4390 Fichtenöl, *n., resin oil.*
4391 Ricinusöl, *castor oil.*
4392 Rosmarinöl, *n., oil of rosemary.*
4393 Senföl, *n., oil of mustard.*
4394 Bernstein, *m., amber.*
4395 Bernsteinöl, *n., oil of amber.*
4396 Terpentin, *m., turpentine.*
4397 Terpentinöl, *n.* }
4398 Terpentinspiritus, *m.* } *oil of turpentine.*
4399 Mandelöl, *n., oil of almonds.*
4400 Anisöl, *n., anise-seed oil.*

4401 Bergamottenöl, n., *oil of bergamot.*
4402 Kalmus, m., *calamus (prop. a reed); sweet flag.*
4403 Kalmusöl, n., *oil of sweet flag.*
4404 Kampher, m., *camphor.*
4405 Kampheröl, n., *camphorated oil, liniment of camphor.*
4406 Nelke, f., *clove.*
4407 Nelkenöl, n., *oil of cloves.*
4408 Zimmtöl, n., *oil of cassia.*
4409 Crotonöl, n., *croton oil.*
4410 Thymianöl, n., *oil of thyme.*
4411 Baldrianöl, n. } *oil of*
4412 Valerianöl, n. } *valerian.*
4413 Hirschhorn, n., *hartshorn.*
4414 Hirschhornsalz, n., *salt of hartshorn.*
4415 Hirschhorngeist, m. }
4416 Salmiakgeist, } *spirits of hartshorn.*
4417 englisches Salz, *English salt, Epsom salts.*
4418 Glaubersalz, *Glauber's salt.*
4419 Hoffmannsche Tropfen, *Hoffman's drops.*
4420 peruvianischer Balsam, *balsam of Peru.*
4421 Wundbalsam, *vulnerary balsam.*
4422 Blei, n., *lead.*
4423 Bleisalbe, f., *lead salve.*
4424 Bleipflaster, n., *sticking plaster, lead plaster.*
4425 Pechpflaster, n., *pitch plaster.*
4426 Blasenpflaster, n., *blistering plaster, vesicatory.*
4427 Ziehpflaster, or } *drawing*
4428 Zugpflaster, } *plaster.*
4429 Spanischfliegenpflaster, n. }
4430 Spanischfliegensalbe, f. } *cantharides plaster; fly blister.*
4431 Quecksilberpflaster, n., *mercurial plaster.*
4432 Senfpflaster, n., *mustard plaster.*
4433 Gummipflaster, n., *compound litharge plaster.*
4434 Kampferspiritus, m. }
4435 Kamphergeist, m. } *tincture of camphor.*
4436 Seifenspiritus, m., *tincture of soap.*
4437 Senfspiritus, m., *spirit of mustard.*
4438 versüssen, *to sweeten.*
4439 Salpeterweingeist, *spirit of nitre.*
   der versüsste Salpeterweingeist, *sweet spirit of nitre.*
4440 Wurmsamen, } *worm-seed.*
4441 Zitwersamen, }
4442 Lavendelblüthe, *lavender flowers.*
4443 Hanfsamen, *hemp seed.*
4444 Koriandersamen, *coriander.*
4445 Petersiliensamen, *parsley seed.*
4446 Vanille, *vanilla.*
4447 Galläpfel (pl.), m., *galls.*
4448 Gelatine, *gelatin.*

4449 Glycerin, *n.*, *glycerine.*
4450 **der** Wermuth, *wormwood.*
4451 Bitterholz, *n.* } *quassia.*
4452 Quassiaholz, *n.* }
4453 Fenchelholz, *n.* } *sassafras.*
4454 Sassafras, *m.* }
4455 süss, *a.*, *sweet.*
4456 Süssholz, *n.*, *licorice root.*
4457 **das** Malz, *malt.*
4458 Honig, *m.*, *honey.*
4459 Moschus, *m.*, *musk.*
4460 Myrrhe, *myrrh.*
4461 Rhabarber, *m.*, *rhubarb.*
4462 Milchzucker, *m.*, *sugar of milk.*
4463 Mohnköpfe (*plur.*), *poppy heads.*
4464 Mohnsamen, *m.*, *poppy seed.*
4465 Seife, *f.*, *soap.*
4466 die medicinische Seife, *medicinal soap.*
4467 Oelseife, *hard soap.*
4468 Terpentinölseife, *turpentine liniment.*
4469 Wachsschwamm, *m.*, *prepared sponge.*
4470 Pressschwamm, *sponge tents.*
4471 Lakritzensaft, *m.*, *juice of licorice.*
4472 Traganth, *m.*, or } *tragacantha.*
Tragacanth, *m.* }
4473 Salep, *m.*, *salep*, *salop.*
4474 Mutterkorn, *n.*, *ergot.*
4475 Safran, *saffron.*
4476 Rainfarrn (*tanacetum*), *tansy.*
4477 Stärke, *f.*, *starch.*
4478 **das** Wachs, *wax.*
4479 **das** Mehl, *meal, flour.*
4480 Hafer, *m.*, *oats.*
4481 Waizen, *m.*, *wheat.*
4482 Hafermehl, *n.*, *oatmeal.*
4483 Weizenmehl, *n.*, *wheat flour.*
4484 Gerste, *f.*, *barley.*
4485 Rettich, *m.*, or } *radish.*
Rettig, *m.* }
4486 Meerrettig, *m.*, *horse radish.*
4487 Wolle, *f.*, *wool.*
4488 Baum, *m.*, *tree.*
4489 Baumwolle, *f.*, *cotton.*
4490 Kreide, *f.*, *chalk.*

### 9. Poisons.

das Gift (3494), *poison*, (*like Engl.* **gift**, *a fatal, deadly gift*).
4491 vergeben, *to give a fatal gift, to poison.*
4492 giftig, *a.*, *poisonous.*
4493 giftlos, *a.*, *void of poison.*
4494 Giftstoff (202), *m.*, *poisonous matter.*
4495 Giftmittel, *n.*, *antidote.*
4496 Giftkunde, }
4497 Giftlehre, } *toxicology.*
Toxikologie, }

4498 **Giftthiere,** or } venomous
 giftige Thiere, } animals.
4499 Giftschlangen (1110), venomous serpents.
4500 Klapperschlange, rattlesnake.
4501 Kupfer, copper.
4502 Kupferschlange, copper-snake.
4503 die Natter, adder.
4504 die Viper, viper.
4505 die Tarantel, tarantula.
4506 Kröte, toad.
4507 **Giftpflanzen,** venomous plants.
4508 Giftkräuter, venomous -herbs.
4509 Giftpilze, }
4510 Giftschwämme, } poisonous mushrooms.
4511 Gifthahnenfuss, m., crow-foot (plant), hemlock.
4512 Giftlattich, m., lettuce, strong-scented lettuce.
4513 Giftkresse, f., poisonous cress, cresses.
4514 Schierling, m., hemlock.
4515 Nachtschatten, m., nightshade.
4516 Stechapfel, m., thorn-apple, stramonium.
4517 Tollkirsche, f., belladonna, deadly nightshade.
4518 Fingerhut, m., digitalis, foxglove.
4519 **giftige Mineralien,** poisonous minerals.
4520 Arsenik, m., arsenic.
4521 Giftmehl, n., white arsenic, poisoned flour.
4522 Blausäure, f., prussic acid, hydrocyanic acid.
4523 essigsauer, a., acetous.
4524 (essigsaures) Kupferoxyd, (acetate of) black oxide of copper.
4525 Giftstoff, m., poisonous matter, virus.
4526 Grünspan, verdigris, acetate of copper.
4527 Strychnin, strychnine.
4528 Nicotin, nicotine.
4529 Opium, opium.
4530 Morphium, morphine.
4531 Dampf, m., vapor, steam.
 **giftige Dämpfe,** poisonous vapors.
 Miasma, miasma.
4532 Ansteckungsgift (3504), }
4533 Ansteckungsstoff, }
4534 Contagionsgift, } virus, contagium.
4535 Pestgift, pestilential poison.
4536 Leichengift, cadaveric or septic poison.
4537 Toxication, poisoning, infection.
4538 Blutvergiftung, poisoning of blood.
4539 Luftvergiftung, f., air-poisoning.
4540 verpesten, to infect.
4541 Verpestung, infection.

## 10. Articles of Furniture and of Clothing.

4542 das Zimmer, *chamber, room*.
4543 Stube, *f., room*.
4544 Schlafzimmer, *n., bedroom*.
4545 Krankenzimmer, *n., sick-room*.
4546 Schlafstube, *f., bedroom*.
4547 Krankenstube, *f., sick-room*.
4548 Lehnstuhl (3574), } *arm-*
4549 Lehnsessel (3575), } *chair*.
4550 Schlafstuhl, *m.* } *easy-*
4551 Schlafsessel, *m.* } *chair*.
4552 schaukeln, *to swing, rock*.
4553 Schaukelstuhl, *rocking-chair*
4554 Bettstelle, } *bedstead*.
4555 Bettstatt, *f.* }
4556 Wiege, *f., cradle*.
4557 Bettzeug, *n., bedding, bed-clothes*.
4558 Bettwäsche, *f., bed-linen, clothes*.
4559 Bettlinnen, *n., sheeting linen, sheeting*.
4560 Bettüberzug, *bed-cover, pillow-case*.
4561 das Laken, }
4562 Bettlaken, *n.* } *sheet*.
4563 Betttuch. *n.* }
4564 Bettdecke, *f., blanket, cover, coverlet*.
4565 das Federbett, *feather-bed*.
4566 das Unterbett, *under-bed*.
4567 Feder-Matratze (3579), *spring mattress*.
4568 das Küssen, *cushion*.
4569 Kopfküssen, *n., pillow*.
4570 Luftküssen, *n., air pillow*.
4571 Riechküssen, *n., perfume cushion, scent-bag*.
4572 Wollendecke, *f., blanket*.
4573 steppen, *to quilt, stitch*.
4574 Steppdecke, *f., quilt*.
4575 Oberdecke, *f., coverlet*.
4576 Fussdecke, *f., cover for the feet, floor-cloth, carpet*.
4577 Vorhang, *m.* } *curtain*.
4578 Gardine, *f.* }
4579 Bettvorhänge, }
4580 Bettgardinen, } *bed-hangings, bed-curtains*.
4581 Schirm, *m., screen*.
4582 Bettpfanne, }
4583 Bettwärmer, } *warming-pan*.
4584 Bettwärmflasche, }
4585 Moskitonetz, *n.* } *mosquito-*
4586 Fliegennetz, *n.* } *bar, fly-net*.
4587 Waschtoilette, *f.* } *wash-*
4588 Waschtisch, *m.* } *stand*.
4589 Waschbecken (449), *n., wash basin*.
4590 Waschschwamm (2486), *sponge*.
4591 Bürste, *f., brush*.
4592 Zahnbürste, *tooth-brush*.
4593 Haarbürste, *hair-brush*.
4594 Kamm, *m., comb*.
4595 kämmen, *to comb*.
4596 das Handtuch, *towel*.
4597 Abtritt, *m.* } *privy, water-*
4598 das Closet, } *closet*.

4599 Nachttopf, m. } chamber
4600 Nachtgeschirr, n. } pot.
4601 Nachtstuhl, m., close-stool.
4602 Stechbecken, n., bed-pan.
4603 Spucknapf (2031), m., spittoon.
4604 Lampe, f., lamp.
4605 Nachtlampe, f., night-lamp.
4606 Zündhölzchen, n. ⎫
4607 Streichhölzchen, n. ⎬ match.
4608 Schwefelholz, n. ⎭
4609 das Licht, } light,
4610 Kerze, f. } candle.
4611 Stearinkerze, stearine candle.
4612 Gaslampe, f., gas lamp.
4613 Docht, m., wick.
4614 eine Lampe anzünden, to light a lamp.
4615 eine Lampe (Licht) putzen, to trim a lamp (a candle).
4616 eine Lampe (Licht) auslöschen, to extinguish, to put out a lamp (candle).
4617 Leuchter, m., candlestick.
4618 Laterne, f., lantern.
4619 Lichtschirm, m., light-screen, shade.
4620 Wachsstock, m., wax-stand, wax taper.
4621 Ofen, m., stove.
4622 Ofenschirm, m., screen.
4623 Ofenthür, f., stove-door, vent-door.
4624 Ofenloch, n., mouth of a stove.
4625 Aschenkasten, m., ash-pan.
4626 Brennholz, n., fire-wood.
4627 Feuerherd, m., fireplace, hearth.
4628 Kohlen (plur.), coal.
4629 Besen, broom, **besom**.
4630 Schornstein, m., chimney.
4631 Fensterscheibe (492), f., (window) pane.
4632 Fensterladen, m., window-shutter.
4633 **das** Möbel, }
4634 die Mobilien (plur.) } furniture.
4635 **die** Bank (plur. **Bänke**), bench.
4636 **die** Fussbank, footstool.
4637 Schrank, m., cupboard, shrine, press.
4638 Kleiderschrank, m., clothes press, wardrobe.
4639 Wäschschrank, linen-press, clothes-press.
4640 Commode, commode, chest of drawers.
4641 Schublade, f., drawer.
4642 **das** Geheimfach, secret drawer.
4643 **das** Sopha, sofa.
4644 **das** Pult, }
4645 Schreibtisch, m. } desk, writing-desk.
4646 **das** Rouleau, blind.
4647 Stiefelzieher (174), bootjack.
4648 Küche, f., kitchen.
4649 Küchenschrank, m., larder, kitchen closet.
4650 Speiseschrank, pantry, safe.

4651 Herd, *m.*, *hearth.*
4652 Backofen, *m.*, *oven.*
4653 Rost, *m.*, *grate.*
4654 Keller, *m.*, *cellar.*
4655 Eimer, *m.*, *pail, bucket.*
4656 Krug, *m.*, *pitcher, jug.*
4657 Topf, *m.*, *pot.*
4658 Kessel, *m.*, *kettle.*
4659 Korb, *m.*, *basket.*
4660 **das** Fass, *cask.*
4661 Trichter, *m.*, *funnel.*
4662 kleine Flasche (1300), *flask.*
4663 Propfen, *cork.*
4664 Propfenzieher, *corkscrew.*
4665 Becher, *m.*, *cup, goblet.*
4666 Tischzeug, *n.*, *table linen.*
4667 Tischtuch, *n.*, *tablecloth.*
4668 Serviette, *f.*, *napkin.*
4669 Teller, *m.*, *plate.*
4670 Löffel, *m.*, *spoon.*
4671 Esslöffel, *tablespoon.*
4672 Theelöffel, } *teaspoon.*
4673 Kaffeelöffel, }
4674 Tasse, *f.*, *cup.*
4675 Untertasse, *saucer.*
4676 **die** Schüssel, *dish, bowl.*
4677 Kleidung, *f.*, *clothing.*
4678 Tracht, *f.*, *costume, mode of dress.*
4679 Gewand (*plur.* Gewänder), *n.*, *garment, dress.*
4680 Anzug, *m.*, *dress, attire, toilet.*
4681 Rock, *m.*, *coat.*
4682 Überrock, *m.*, *overcoat.*
4683 Weste, *f.*, *waistcoat.*

4684 Hosen (*pl.*) }
4685 Beinkleider (*pl.*) } *pantaloons, breeches.*
4686 Mantel, *m.*, *cloak.*
4687 Regenmantel, *m.*, *waterproof.*
4688 Pelz, *m.*, *fur.*
4689 Ärmel, *m.*, *sleeve.*
4690 Knopf, *m.*, *button.*
4691 Knopfloch, *n.*, *buttonhole.*
4692 **das** Futter, *lining.*
4693 Hut, *m.*, *hat.*
4694 Mütze, *f.* } *cap.*
4695 Kappe, *f.* }
4696 Halsbinde, *f.*, *necktie.*
4697 Halstuch, *n.*, *neckerchief.*
4698 Schuh, *m.*, *shoe.*
4699 Schuhband, *n.* }
4700 Schuhriemen, *m.* } *shoestring, shoelace.*
4701 Stiefel, *m.*, *boot.*
4702 Halbstiefel, *half-boot.*
4703 Überschuh, *m.*, *overshoe.*
4704 Strumpf, *m.*, *stocking.*
4705 Socken (*pl.*), *socks.*
4706 Unterhosen (*pl.*), *drawers.*
4707 Hosenträger (*pl.*), *m.*, *suspenders.*
4708 **das** Hemd(*plur.* -en),*shirt*
4709 **das** Nachtzeug, *night-dress.*
4710 das Nachthemd, *night-shirt.*
4711 Nachtjacke, *f.*, *night-jacket.*
4712 Nachtmütze, }
4713 Nachtkappe, } *night-cap.*
4714 Schlafmütze, }
4715 Schlafrock, *m.*, *dressing or night-gown, chamber-robe.*

4716 Pantoffel (*pl.* -n), *m.*, *slippers.*
4717 Frauenkleider, *female attire.*
4718 **das** Kleid, *dress.*
4719 Hauskleid, *n.*, *morning-dress.*
4720 Unterkleid, *n.*, *under-dress, undergarment.*
4721 Rock, *m.*, *skirt.*
4722 Unterrock, *m.*, *petticoat.*
4723 Schlafrock, *m.*, *wrapper.*
4724 **das** Mieder, } *bodice,*
4725 **das** Corset (sĕ't), } *corsets.*
4726 der Leib, or } *bodice.*
     Leibchen, *n.* }
4727 Schnürmieder, *n.* }
4728 Schnürleib, *m.* } *stays.*
4729 Schnürbrust, *f.* }
4730 Schnürriemen (*plur.*), *laces.*
4731 Jacke, *f.*, *jacket.*
4732 Schlafjacke, *night-jacket.*
4733 Schürze, *f.*, *apron.*
4734 Frauenhemd, *n.*, *chemise.*
4735 Haube, *f.*, *cap.*
4736 Nachthaube, *night-cap.*
4737 Morgenhaube, *morning-cap.*
4738 Gürtel, *m.*, *belt, sash.*
4739 Strumpfbänder (*plur.*), *garters.*
4740 Pelzkragen, *fur collar.*

4741 kleiden, *to clothe.*
4742 bekleiden, *to put clothes on.*
4743 umkleiden, *to put other clothes on, dress anew.*
4744 entkleiden, *to undress.*
4745 sich entkleiden, *to undress one's self.*
4746 anziehen, *to put, get on clothes*
4747 sich anziehen, *to dress one's self.*
4748 ausziehen, } *to undress.*
4749 sich ausziehen, }
4750 die Kleider ablegen, *to lay off clothes, undress.*
4751 entblössen *or* sich entblössen, *to divest one's self of clothes.*
4752 sich der Kleider entledigen, *idem.*
4753 auskleiden, *to undress.*
4754 aufbinden, *to untie.*
4755 zubinden, *to tie.*
4756 aufmachen, *to open, loosen.*
4757 zumachen, *to close, to button up.*
4758 aufknöpfen, *to unbutton.*
4759 zuknöpfen, *to button up.*
4760 schnüren, *to tie, lace.*
     sich schnüren, *to lace one's self.*
4761 zusammenschnüren (1700), *to lace, tie together.*
4762 einschnüren, } *to tie up, lace,*
4763 zuschnüren, } *to corset.*
4764 aufschnüren, *to unlace.*
4765 losbinden, *to unbind, untie.*
4766 losmachen, *to detach.*
4767 lose machen, }
4768 locker (1861) machen. }
4769 lockern, }
     *to make loose, to loosen.*
4770 lösen, *to loosen; dissolve.*
4771 emporheben, *to raise, lift up.*

4772 heruntermachen, }
4773 herunterlassen, } *to let down, untie.*
4774 sich überlegen, *to lay one's self over.*
4775 sich hinlegen, *to lie down.*
4776 sich aufsetzen, *to sit upright.*
4777 niederknien, *to kneel down.*
4778 (die Arme) kreuzen,
4779 (die Arme) übereinander legen, *to cross, to fold across, to superimpose, cross, double (one's arms).*
4780 auseinander (2729) spreizen, *to distend, spread out, open.*
4781 die Beine auspreizen, *or* ausspreiten, *to open one's legs.*
4782 Wasser in den Mund nehmen, *to take water into one's mouth.*
    es im Munde halten, *to keep it in one's mouth.*
4783 es nicht hinabschlucken, *or* hinunterschlucken, *not to swallow or gulp it down.*

# PART II.

## CONVERSATIONS.

# CONVERSATIONS.

## I. VOCABULARY CONTAINING THE NEW WORDS OCCURRING IN THE CONVERSATIONS.

### Conversations 1 to 4.

4784 dienen, *to serve*.
4785 sich fürchten, *to be afraid*.
4786 Beistand, *m., assistance*.
4787 nöthig, *a., necessary, needful*.
4788 nöthig haben, *to want, to need*.
4789 **rufen, rief, gerufen,** *to call*.
4790 rufen lassen, *to send for*.
4791 gar nicht, *not at all*.
4792 recht, *adv., quite*.
4793 ungewöhnlich, *a. and adv., uncommon*.
4794 **fangen, fing, gefangen,** *to catch*.
anfangen, *to commence, begin*.
4795 vorgestern, *adv., the day before yesterday*.
4796 zuweilen, *adv., sometimes*.
4797 schwindelig, *a., dizzy, giddy*.
4798 Schwindel, *m., dizziness, vertigo*.
4799 besonders, *especially, particularly*.
4800 versuchen, *to attempt, try*.
4801 **die** Lust (*pl.* **Lüste**), *inclination*.
4802 fieberisch, *a., feverish*.
4803 bitte (*for* ich bitte), *pray*.
4804 aufwachen, *to awake*.
4805 das Aufwachen, *awaking*.
4806 unrein, *a., unclean, foul*.
4807 einnehmen, *to take medicine*.
4808 verschreiben, *to prescribe*.
4809 Stunde, *f., hour*.
4810 sonst, *adv., besides, else*.
4811 beobachten, *to observe; to perform*.

4812 **bleiben, blieb, geblieben,** *to remain, stay.*
4813 Erkältung, *f.*, *catching cold.*
4814 sich hüten, *to take care or heed.*
4815 sich hüten vor, *to beware of, to guard against.*
4816 früh, *early.*
4817 wiederkommen, *to call again.*
4818 hoffentlich, *adv.*, *it is to be hoped, (as) I hope.*
4819 aufgelegt, *disposed.*
4820 ändern, *to change, alter.*
4821 schwermüthig, *a.*, *melancholy.*
4822 heiter, *a.*, *cheerful.*
4823 Muth, *m.*, *mood, frame of mind, courage.*
4824 guten Muthes sein, *to be of good cheer.*
4825 fassen, *to lay hold of.*
4826 Muth fassen, *to take courage.*
4827 sich fassen, *to compose one's self.*
4828 sich gehen lassen, *to indulge one's inclinations.*
4829 Sache, *f.*, *thing, matter.*
4830 verzagen, *to despond.*
4831 glauben, *to believe.*
4832 **die** Gefahr, *danger.*
4833 Herr, *master, lord;* (*before a proper name*) *Mr.;* (*in address*) *sir.*
4834 arg, *a. and adv., bad, evil.*
4835 wirklich, *a. and adv., real.*
4836 versichern, *to assure, assert.*
4837 sagen, *to say, tell.*
4838 Meinung, *f.*, *opinion.*
4839 unverhohlen, *unconcealed, candidly, frank.*
4840 gegenwärtig, *a.*, *present.*
4841 die Acht (2514), *heed, care.*
in Acht nehmen, *to take care, be careful.*
4842 ausarten (70), *to degenerate, deteriorate.*
4843 vorhanden, *a.*, *at hand, in existence.*
4844 beunruhigen (2346), *to alarm, make uneasy.*
4845 Arzeneitrank, *m.*, *potion.*
4846 ausmachen, *to make much or little, care much or little.*
wenn es nichts macht (*abbreviation for ausmacht*), *if you don't mind it.*
4847 erleichtern, *to make lighter, easier.*
4848 **reiben, rieb, gerieben,** *to rub.*
einreiben, *to rub in.*
4849 sich einstellen, *to present one's self, appear.*
4850 ähnlich, *a.*, *similar.*
4851 besuchen, *to visit, come to see.*
4852 ermangeln, *to fail, be wanting, deficient.*
4853 **bringen, brachte, gebracht,** *to bring.*
zubringen, *to pass (one's time).*
4854 zum Theil (704), *partly.*

4855 nichts desto weniger, *nevertheless*.
4856 Unterbrechung, *f.*, *interruption*.
4857 durch'schlafen, *to pass sleeping, to sleep all night long*.
4858 nachlassen, *to abate, cease*.
4859 jedenfalls, *at all events, in any case*.
4860 sich verhalten, *to behave, demean or comport one's self*.
4861 Schluck, *draught, gulp, swallow*.
4862 bekommen (*w. dat.*), *to suit, agree with, do good*.
4863 verspüren, *to perceive, be aware of, feel*.
4864 Linderung, *f.*, *alleviation, mitigation, relief*.
4865 um vieles, *by much, far*.
4866 sagen, *to say, tell*.
4867 ja, *you know, why*.
4868 bedeuten, *to signify, mean*.
4869 es hat nichts zu bedeuten, *it does not signify, it matters not*.
4870 versprechen, *to promise*.
4871 Folge, *f.*, *consequence, effect*.
4872 aufstehen, *to get up*.
4873 ein paar, *a few, some few*.
4874 morgen, *adv.*, *to-morrow*.
4875 verlassen, *to leave*.
4876 Hunger, Durst bekommen, *to get hungry, thirsty*.
4877 es schmeckt (1433) mir, *I like it* (lit. *it tastes well to me*).
4878 keineswegs, *by no means* (**noways**).
4879 **geniessen, genoss, genossen,** *to enjoy, to take, to taste (food or drink)*.
4880 oft or öfters, *often, frequently*.
4881 sich einschränken, *or* }
4882 sich beschränken, }
    *to confine one's self, restrict one's self* (auf- *to*).
4883 möglich, *a. and adv.*, *possible*.
    so viel als möglich, *as much as possible*.
4884 Mehlspeise, *f.*, *farinaceous food*.
4885 gern, *adv.*, *willingly, fain*.
    gern haben, *to like*.
4886 sich enthalten (*w. gen.*), *to abstain from*.
4887 sich halten, *to take care of one's health*.
4888 so Gott will *or* will's Gott, (*if*) *God please* (*if it pleases God*).
4889 wieder auf den Beinen sein, *to be restored to health*.
4890 **rathen, rieth, gerathen,** *to advise*.
4891 das Land, *country*.
4892 Monat, *month*.

4893 verändern, *to change, alter.*
4894 Veränderung, *f., change, alteration.*
4895 Luftveränderung, *change of air.*
4896 vornehmen, *to undertake, to go about.*

## Conversation 5.

4897 **heissen, hiess, geheissen,** *to be called.*
4898 buchstabieren, *to spell.*
4899 **der** Name (*gen.* -ns), *name.*
4900 **schreiben, schrieb, geschrieben,** *to write.*
4901 alt, *a., old.*
4902 wohnen, *to dwell, reside.*
4903 wo—her? *or* woher? *where from.*
4904 lange, *long.*
4905 Staat (*gen.* -es, *plur.* -en), *state.*
4906 die Vereinigten Staaten, *United States (of America).*
4907 Geschäft, *business.*
4908 Profession (ōn'), *profession, trade.*
4909 verheirathen, *to marry.*
sich verheirathen, *to get married.*
4910 unverheirathet, *unmarried, single.*
4911 Wittwer, *widower.*
4912 Wittwe, *widow.*
4913 leben, *to live.*
4914 Eltern (*plur. only*), *parents.*
4915 **sterben, starb, gestorben,** *to die.*
4916 Geschwister (*pl.*), *brothers and sisters.*
4917 ander, *a., other.*
4918 das Leben, *life.*
4919 früher, *adv., formerly, previously.*
4920 einmal, *once.*
4921 Weise, *f., wise, manner.*
4922 sich erfreuen (*w. gen.*), *to be glad, rejoice.*
4923 **das** Kind, *child.*
4924 Familie, *f., family.*
4925 wann? *when?*
4926 diesmal, *this time, for the present.*
4927 **beginnen, began, begonnen,** *to begin.*
4928 wie? *how?*
4929 worüber, *of what, about what.*
4930 weshalb? *wherefore?*
4931 klagen (über), *to complain (of).*
4932 zuerst, *adv., at first; for the first time.*
4933 sich erinnern, *to remember, recollect.*
4934 wodurch? *whereby? by what means?*

4935 aufmerksam, *a. and adv.*, *attentive.*
4936 beschreiben, *to describe, give a description.*
4937 veranlassen, *to give cause or occasion for; to suggest (a thought).*
4938 irgend ein, *some, any.*
4939 wahrnehmen, *to observe, notice, to perceive.*
4940 bemerken, *to notice; to remark, mention.*
4941 anhalten, *to continue, last.*
4942 das Übel, *evil, illness, hurt.*
4943 sich zuziehen (174), *to bring upon one's self, incur, occasion.*
4944 Fall (1583), *m.*, *fall.*
4945 einen Fall thun, *to get or have a fall.*
4946 Schlag, *m.*, *blow, stroke.*
4947 **schlagen, schlug, geschlagen,** *to strike, beat.*
4948 sich überarbeiten, *to overwork one's self.*
4949 Lebensweise (4921), *manner of life.*
4950 sich anstrengen, *to exert or fatigue one's self.*
4951 Mangel, *m.*, *want, need.*
Mangel leiden, *to suffer want, be in distress.*
4952 Kummer, *m.*, *grief, trouble.*
4953 Sorge, *f.*, *sorrow, care.*
4954 Schnapps (3906), *(common) brandy.*
Schnäppschen *(dim.)*
4955 Spirituosen (*pl.*), *spirits, alcoholic liquors.*
4956 rauchen, *to smoke.*
4957 sich erkälten, *to catch cold.*
4958 vielleicht, *adv., perhaps.*
4959 zu Rathe (3838) ziehen, *to consult, ask counsel of.*
4960 bisher, *adv., hitherto, till now.*
4961 anwenden, *to apply, make use of.*
4962 merklich (4940), *perceptibly, discernibly.*
4963 verbessern, *to improve.*
4964 sogar (*or abbreviated* gar), *even.*
4965 genau, *a. and adv., accurate, exact.*
4966 ungefähr, *adv., nearly, by a rough guess.*
4967 Ort, *m.*, *place.*
4968 angeben, *to state, declare, specify.*
4969 vorziehen, *to prefer.*
4970 Stellung, *f., position, posture.*
4971 einmal (*after an imperative*), *just, pray, please, as* husten Sie einmal, *cough please! just cough!*
4972 jetzig, *a., present.*
4973 offener Leib, *open bowels.*
4974 Halsschmerzen (*pl.*), *sore throat.*
4975 geneigt zu, *disposed, liable to.*
4976 ausbleiben, *to stay out; be wanting.*

## Conversation 6.

4977 vornehmlich, *adv.*, *principally, mainly.*
4978 innerhalb, *within.*
4979 das Knochenleiden, *bone affection, disease of the bones.*
4980 anhaltend, *a. and adv.*, *persistent, continual.*
4981 sich niederlegen, *to lie down.*
4982 **die Nacht** (*pl.* **Nächte**), *night.*
bei Nacht, *at night.*
4983 Tag, *m.*, *day.*
bei Tag, *by day.*
Tagesstunde (4809), *hour of the day.*
4984 bestimmt, *fixed, definite, particular.*
4985 Anstrengung, *f.*, *straining, effort, exertion.*
4986 mindern, *to diminish, lessen.*
4987 Morgen, *m.*, *morning.*
4988 des Morgens, *in the morning.*
4989 ehe, *adv.*, *before.*
4990 vorübergehend, *a.*, *passing, transient.*
4991 Gegenstand, *m.*, *object.*
4992 begleiten, *to accompany.*
begleitet sein, *to be associated.*
4993 Diätfehler, *m.*, *indiscretion in diet.*
4994 begehen (2350), *to commit.*
4995 theilweise, *partial.*
4996 Taubheit, *f.*, *deafness.*
4997 **entstehen, entstand, entstanden,** *to arise, rise from or out of.*
4998 folgen, *to follow.*
4999 afficieren, *to affect.*
5000 unbestimmt, *a.*, *uncertain.*
5001 zurückkehren, *to return, recur.*
5002 thätig, *a.*, *active.*
5003 Überanstrengung (4950), *f.*, *over-fatigue.*
5004 begrenzen, *to limit, confine, circumscribe.*
5005 Intensität, *f.*, *intensity.*
5006 gleichmässig, *a. and adv.*, *uniform.*
5007 worin, *wherein.*
5008 gleichzeitig, *a. and adv.*, *at the same time.*

## Conversation 7.

5009 vorangehen, *to precede.*
5010 weder—noch, *neither—nor.*
5011 zurückhalten, *to hold back, retain.*
5012 geschlechtlich, *a.*, *sexual.*
5013 Excess, *m.*, *excess.*
5014 **die Schuld,** *fault, offence.*
5015 sich etwas zu Schulden kommen lassen, *to commit, to be guilty of.*
5016 Übermüdung, *overweariness, fatigue.*

5017 Durchnässung, *f.*, *getting wet all over.*
5018 Nachtruhe, *f.*, *a night's rest.*
5019 überhaupt, *adv.*, *generally.*
5020 Strick, *m.*, *rope.*

## Conversation 8.

5021 niederdrücken, *to depress.*
5022 verfallen, *to fall, sink.*
5023 comatös, *comatous.*
5024 Anzeichen, *sign, indication; foreboding.*
5025 vorausgehen, *to precede.*
5026 Schädelverletzung, *f.*, *injury of the skull.*
5027 stattfinden, *to take place.*
5028 erleiden, *to suffer, to sustain.*
5029 existieren, *to exist.*
5030 Eiterungsprocess, *m.*, *suppuration.*
5031 bestehen, *to exist, subsist.*
lange bestehen, *to continue, to last.*

## Conversation 9.

5032 Veranlassung, *f.*, *occasion, cause.*
5033 Ursache, *f.*, *cause.*
5034 pflegen, *to be accustomed, used, wont, to use.*
5035 täglich, *a. and adv.*, *daily.*
5036 Zwischenraum, *m.*, *interval, interstice.*
5037 beschränken, *to limit.*
5038 beschränkt sein, *to be limited.*
5039 verbreiten, } sich verbreiten, } *to spread, to be distributed.*
5040 Verlauf, *m.*, *course.*
5041 hervorrufen, *to call forth, to evoke, bring on.*
5042 unbedeutend, *trivial, unimportant.*
5043 Luftzug, *m.*, *current of air.*
5044 knarren, *to creak, jar.*
5045 **die** Thür, *door.*
5046 unmässig, *a. and adv.*, *intemperate, unreasonable.*
5047 **der** Genuss, *use (of food, drink, etc.), taking of food.*
5048 das Leberleiden, *disease of the liver.*
5049 Nierenleiden, *n.*, *disease of the kidneys.*
5050 Herzkrankheit, *heart disease*
5051 Blutmangel, *m.*, *anæmia.*
5052 **gerathen, gerieth, gerathen,** *to fall into, get into.*
5053 Unordnung, *f.*, *disorder.*
5054 **heben, hob, gehoben,** *to remove (diseases).*
5055 Anwendung, *f.*, *applying.*
5056 befeuchten, *to moisten, wet.*
5057 **die** Leinwand, *linen, linen cloth.*
5058 Verdünstung, *f.*, *evaporation.*

## Conversations 10, 11, 12.

5059 verhindern, *to prevent, hinder.*
5060 einnehmen, *to take up, occupy*
5061 Speichelfluss, *m., salivation.*
5062 Schwellung, *f., swelling.*
5063 die Geduld, *patience.*
5064 siphilitish, *a., syphilitic.*
5065 Athembeschwerde, *f., difficulty in breathing.*
5066 Bedeutung, *f., importance, consequence.*
5067 Ruhe, *f., rest.*
5068 lästig, *a. and adv., troublesome.*
5069 Untersuchung, *f., examination.*
5070 Gewächs, *n., growth, excrescence.*
5071 das Stimmband, *vocal cord.*
5072 schlechterdings, *adv., by all means.*

## Conversation 13.

5073 herauslaufen, *to run out, return.*
5074 Flüssigkeit, *f., fluid, liquid.*
5075 vorstehen, *to stand out, project.*
5076 einführen, *to introduce.*
5077 erfahren, *to learn, find out, experience.*
5078 Kieferwinkel, *m., angle of the jaw.*
5079 hinaufschiessen, *to shoot up.*
5080 seitdem, *adv., since that time, ever since.*
5081 Verbesserung, *f., improvement.*
5082 sich freuen, *to be glad.*
5083 **bersten, barst, geborsten,** *to burst.*

## Conversation 14.

5084 Versuch, *m., experiment, attempt.*
5085 verschieden, *a., various, different.*
5086 Verschwärung, *f., ulceration.*
5087 Stelle, *f., place.*
5088 Quecksilberdämpfe (*pl.*), *vapors of mercury.*
5089 eigenthümlich, *a., peculiar.*
5090 vermuthen, *to conjecture, presume, guess.*
5091 Zäpfchen, *uvula.*
5092 Gehörsinn, *sense of hearing.*

## Conversation 15.

5093 Unterkieferdrüse, *f., submaxillary gland.*
5094 Wirkung, *f., action, effect.*
5095 entgegenarbeiten, *to counteract.*

5096 reichen, *to reach.*
5097 anfangs, *in the beginning.*
5098 gleich=sogleich, *immediately, directly.*
5099 erst, *adv., not before, not till.*
5100 spät, *a. and adv., late.*
5101 erst, (*num.*) *first.*
5102 Krankheitserscheinung, *f., appearance or manifestation of a disease.*

## Conversation 16.

5103 Kitzeln, *n., tickling.*
5104 räuspern, *to hawk, clear the throat.*
5105 das Räuspern, *hawking.*
5106 gefälligst, *if you please, please.*
5107 sich geben, *to yield, give way.*
5108 Hauptsache, *f., main point, principal thing.*
5109 ja, *adv.* (*to enforce or modify the sense*), ja nicht, *on no account, by no means.*
5110 suchen, *to seek, look for.*
5111 reizlos, *a., not irritating.*

## Conversation 17.

5112 zufällig, *a. and adv., accidental.*
5113 Hinderniss, *n., impediment.*
5114 selbst, *adv., even.*
5115 Schwierigkeit, *f., difficulty.*
5116 durchbringen, *to bring through, swallow.*
5117 **das** Stück (*pl.* -e), *piece.*
5118 **gerinnen, gerann, geronnen,** *to coagulate, clot, curdle.*

## Conversations 18, 19.

5119 hindurch, *adv., through, throughout.*
5120 verbinden (98), *to connect.*
5121 nun, *adv., now; well.*
5122 **vermeiden, vermied, vermieden,** *to avoid.*
5123 Staub, *m., dust.*
5124 lüften, *to air, ventilate.*
5125 mischen, *to mix.*
5126 aushusten, *to cough up.*
5127 Wärmegefühl, *n., sense or perception of warmth.*
5128 vorhergehen, *to precede.*
5129 sich ängstigen, *to be alarmed.*
5130 und zwar, *and that.*
5131 Erhitzung, *f., heating, excitement.*
5132 Gemüthsaufregung, *f., emotion, agitation of mind.*
5133 beengen, *to straiten, narrow.*

5134 Kleidungsstücke (*pl.*), *articles of wearing apparel.*
5135 säuerlich, *a.*, *acidulated.*
5136 sorgen für, *to take care of, to look or see to.*
5137 Leibesöffnung, *f.*, *laxness, openness of the bowels.*

## Conversations 20, 21.

5138 Blutbrechen, Bluterbrechen, } *vomiting of blood.*
5139 Beklommenheit, *f.*, *oppression, anxiety.*
5140 Brechgefühl, *n.*, *nausea.*
5141 halbverdaut, *half digested.*
5142 herauspringen, *to spurt out.*
5143 schaumig, *a.*, *frothy.*
5144 Bedrückung, *f.*, *oppression.*
5145 **verschwinden, verschwand, verschwunden,** *to disappear.*
5146 wiederkehren, *to return, reappear.*
5147 Heftigkeit, *f.*, *violence, severity.*
5148 halten auf, *to take care of, be particular about.*
5149 staubig, *a.*, *dusty.*
5150 verschliessen (405), *to close.*

## Conversation 22.

5151 unmittelbar, *a.* and *adv.*, *immediate.*
5152 erwachen, *to wake up.*
5153 währen, *to last, continue.*
5154 unangenehm, *a.*, *disagreeable.*
5155 entleeren, *to empty, evacuate.*
5156 Fall, *case;* im Falle, *in case, if.*
5157 **giessen, goss, gegossen,** *to pour.*
5158 Taschentuch, *n.*, *pockethandkerchief.*
5159 einhauchen, *to inhale.*

## Conversation 23.

5160 schleimig, *a.*, *mucous, slimy.*
5161 reiswasserähnlich, *a.* reiswasserartig, *a.* } *resembling rice-water.*
5162 ausschweifen, *to be dissolute, licentious.*
5162 ausschweifend, *dissolute, licentious.*
5163 gehörig, *a.*, *proper, requisite.*
5164 Tabakqualm, *m.*, *tobacco fumes.*

5165 shädlich, *a.*, *pernicious, noxious.*
5166 Wohnung, *f.*, *dwelling.*
5167 empfehlenswerth, *a.*, *commendable.*
5168 Aufenthalt, *m.*, *sojourn.*
5169 Waldluft, *f.*, *woodland air.*
5170 Jahreszeit, *f.*, *season.*
5171 gleichförmig, *a.*, *uniform.*
5172 sich bedienen (*w. gen.*), *to make use of.*
5173 Oberkörper, } *upper part of*
5174 Oberleib, } *the body.*
5175 sich entschlagen (*w. gen.*), *to rid one's self of.*

## Conversation 24.

5176 haften, *to hold (fast), adhere, stick.*
5177 lungenkrank, *a.*, *lung-sick.*
5178 Beziehung, *f.*, *relation, respect, regard.*
5179 abendlich, *a.*, *vespertine, evening . . .*
5180 Verschlimmerung, *f.*, *deterioration, exacerbation.*
5181 aufheben, *to lay up, keep, preserve.*

## Conversation 25.

5182 äussern, *to show, manifest.*
5183 sich ———, *to appear, break out.*
5184 Schüttelfrost, *m.*, *rigor, severe chill.*
5185 **beklemmen, beklomm, beklommen,** *to pinch, oppress.*
5186 spärlich, *a.*, *scanty.*
5187 **der** Gebrauch, *use.*
5188 nothwendig, *a.*, *necessary.*
5189 hinlänglich, *adv.*, *sufficiently*
5190 einschliessen, *to inclose, encompass.*
5191 hinzufügen, *to add.*
5192 tauchen, *to dive, dip.*
5193 **auswinden, wand, gewunden,** *to wring out.*
5194 erneuern, } *to renew, re-*
     erneuen, } *peat.*
5195 Erleichterung, *f.*, *relief.*

## Conversation 26.

5196 erträglich, *a.*, *bearable.*
5197 Rippenbruch, *m.*, *fracture of a rib.*
5198 Ausleerung, *f.*, *evacuation.*
5199 erzielen, *to aim at, attain.*
5200 behalten, *to retain.*
5201 wiederholen, *to repeat.*
5202 ertragen, *to bear, sustain, endure.*

5203 flanell, *a.*, *flannel*.
5204 Wickelband, *n.*, *swathing band, roller*.
5205 einweichen, *to soak, drench*.
5206 grob, *a.*, *coarse, gross*.
5207 falten, *to fold*.

5208 anliegen, *to lie close or near to, to fit well*.
5209 verrücken, *to dislocate, displace*.
5210 eindringen, *to penetrate, reach into, enter*.
5211 unberührt, *untouched*.

## Conversation 27.

5212 vorkommen, *to come before one's eyes, to occur*.
5213 Wasseransammlung, *f.*, *accumulation of water, dropsical swelling*.
5214 Schnelligkeit, *f.*, *quickness*.
5215 meiden (5122), *to avoid*.
5216 das Mahl, *meal*.
5217 gewähren, *to afford, bestow upon*.
5218 genügend, *a.*, *sufficient*.
5219 während, *during*.

## Conversations 28, 29, 30, 31.

5220 Wiederwillen, *m.*, *aversion*.
5221 färben, *to color, stain*.
5222 Stand, *m.*, *state, condition*.
5223 im Stande sein, *to be in a condition, to be able to*.
5224 herbeiführen, *to bring on*.
5225 erhitzen, *to heat, make hot*.
5226 Einfluss, *m.*, *influence*.
5227 Körpertheil, *m.*, *part of the body*.
5228 nüchtern, *a.*, *jejune, on an empty stomach*.
5229 Einnahme, *f.*, *taking*.
5230 unlöschbar, *a.*, *unquenchable*

## Conversation 32.

5231 Missbehagen, *n.*, *discomfort*.
5232 plagen, *to torment*.
5233 geruchlos, *a.*, *without smell*.
5234 zubereiten, *to prepare*.
5235 ergeben, *a.*, *given to, devoted; abandoned* (*to vice*).
5236 schlafen gehen, *to go to bed*. das Schlafen gehen, *going to bed*.
5237 Eisstücke (*pl.*), *pieces of ice*.

## Conversation 33.

5238 aufhören, *to cease*.
5239 in der Regel, *as a general rule, generally*.
5240 enthalten, *to contain*.
5241 schwärzen, *to blacken*.

5242 vollkommen, *a. and adv.*, *perfect, complete.*
5243 verbleiben (4812), *to remain.*
5244 vorzüglich, *a. and adv.*, *chief, special, principal.*
5245 wochenlang, *for weeks.*
5246 lauwarm, *m.*, *lukewarm, tepid.*
5247 **empfehlen, empfahl, empfohlen**, *to recommend.*

## Conversations 34, 35, 36, 37.

5248 martervoll, *a.*, *full of torments, excruciating.*
5249 zugleich, *adv.*, *along with, at the same time.*
5250 verschaffen, *to procure, to help to.*
5251 sofort, *adv.*, *forthwith, immediately.*
5252 halbstündig, *of half an hour*
5253 schütteln, *to shake.*
5254 ausschliesslich, *exclusively.*

## Conversation 38.

5255 Gallenstein, *m.*, *gall-stone, biliary calculus.*
5256 unüberwindlich, *insuperable, invincible.*
5257 **aufweisen, wies, gewiesen,** *to show forth, present.*
5258 besonder, *a.*, *particular, peculiar.*
5259 hauptsächlich, *a. and adv.*, *chiefly, particularly.*
5260 schlürfen, *to sip; sup.*
5261 vermittelst, *by means of.*
5262 Strohhalm, *m.*, *straw halm.*
5263 die Vorsicht, *precaution.*
5264 je, *at a time*, je drei, vier, *three, four at a time (abbreviation for* jedesmal).

## Conversation 39.

5265 führen, *to lead.*
5266 Forderung, *f.*, *demand.*
5267 unbeachtet, *unnoticed.*
5268 unbeachtet lassen, *to take no notice of, disregard.*
5269 Nachlässigkeit, *f.*, *negligence, carelessness.*
5270 sofortig, *a.*, *instantaneous.*
5271 leisten, *to do, perform, execute*
5272 Folgeleistung, *obedience.*
5273 bloss, *barely, simply, only.*
5274 aufweichen, *to mollify, soften.*
5275 ferner, *adv.*, *furthermore.*
5276 breiig, *a.*, *pappy.*
5277 lieber (*comp. of* lieb), *rather, sooner than, in preference.*

## Conversations 40, 41.

5278 **kneifen, kniff, gekniffen,** *to pinch.*
5279 Eingeweidewurm, *m.*, *helminth.*
5280 gütlich, *a. and adv.*, *amicable.*
 sich gütlich thun, *to indulge one's self.*
5281 Stuhlzwang, *m.*, *tenesmus.*
5282 Kothstücke (*pl.*), *pieces of fæcal matter.*
5283 unzulänglich, *insufficient.*
5284 überfüllen, *to fill to excess, to overcrowd.*
5285 Gerstenmehl, *n.*, *barley meal*
5286 zuträglich, *a.*, *beneficial, useful.*

## Conversation 42.

5287 Hemorrhoidalknoten (*pl.*), *hemorrhoidal ulcers, piles.*
5288 schwierig, *a.*, *difficult.*
5289 gemächlich, *a. and adv.*, *easy, comfortable.*
5290 üppig, *a. and adv.*, *luxurious.*
 üppig leben, *to live luxuriously.*
5291 stammen, *to originate.*
5292 örtlich, *a.*, *local.*
5293 radical (káhl), *a. and adv.*, *radical.*
5294 curieren, *to cure.*
5295 entarten, *to degenerate.*
5296 äusserst, *a. and adv.*, *exceedingly.*
5297 hemmen, *to stop, arrest, stay* (*progress*).
5298 sich legen, *to lie down,* fig. *to fall, subside, abate.*
5299 schmal, *small, narrow; scanty*
5300 schmale Kost, *low diet.*
5301 sich unterziehen (*w. dat.*), *to take upon one's self, to undergo.*

## Conversation 43.

5302 zu Muthe sein (wohl, übel), *to feel* (*at ease, uneasy*).
5303 sich hervordrängen, *to press forward, protrude.*
5304 zurückbringen, *to bring back, replace.*
5305 beölen, *to oil.*
5306 vorsichtig, *a. and adv.*, *carefully.*
5307 sonstig, *a.*, *other.*
5308 auswischen, *to wipe out, to sponge.*

## Conversation 44.

5309 Strahl (*gen.* -es, *pl.* -en), *m.*, *flash (of water)*, *jet, spurt.*
5310 Absatz, *m., intermission.* in Absätzen, *intermittingly.*
5311 beifügen, *to add.*
5312 **geschehen, geschah, geschehen,** *to be done; to happen.*
5313 ablassen, *to draw off.*
5314 ununterbrochen (2002), *uninterrupted, continuous.*
5315 müssen, *must* (log. *necessity, i. e., it must be according to my judgment*), *I suppose.*
5316 lehren, *to teach.*
5317 einbringen, *to bring into, introduce.*
5318 helfen, *to help.*
5319 Bodensatz, *m., sediment.*
5320 Hefen (*pl.*), *dregs.*
5321 sich zeigen (443), *to appear, to show one's self.*
gleich (5098).
5322 so wie, *as soon as.*
5323 fertig, *a., done, finished.*
5324 fertig sein, *to have done or finished.*

## Conversations 45, 46, 47, 48.

5325 unvermischt, *unmixed.*
5326 Stockung, *f., stoppage.*
5327 Erschütterung, *concussion, shaking, violent shock.*
5328 **schweigen, schwieg, geschwiegen,** *to be silent,* fig. id., *to be hushed.*
5329 Umstrickung, *f., constriction*
5330 Ausgang, *m., outlet.*
5331 ungegohren, *unfermented.*
5332 Stein, *m., stone, calculus.*
5333 Hüftbad, *n., hip-bath.*
5334 abwärts, *adv., down, downward.*
5335 sich erstrecken, *to stretch, extend.*
5336 aufschwellen (2398),*to swell.*

## Conversation 49.

5337 einwirken, *to operate on, to influence.*
5338 Abnahme, *f., reduction, abatement.*
5339 Gefühlsstumpfheit, *f., loss of feeling.*
5340 entwickeln, *to develop.*
5341 **steigen, stieg, gestiegen,** *to raise, increase.*
5342 bloss, *a., bare, naked, nude.*
5343 wählen, *to select.*
5344 Boden, *m., soil, floor, bottom.*
5345 kalkig, *a., chalky.*
5346 die freie Luft, *open air.*

## Conversations 50, 51.

5347 Leibwäsche, *f.*, *linen, body-linen.*
5348 Leistenbeule, *f.*, *bubo.*
5349 erlauben, *to allow, permit.*
5350 Glied (392), *penis.*
5351 fleischlich, *a. and adv., fleshly, carnal.*
5352 Umgang, *m.*, *intercourse.*
5353 Drüsenbeule, *bubo.*
5354 **das** Werg, *oakum.*
5355 Eiterausfluss, *m.* }
  Eiterabfluss, *m.* }
  *discharge of pus or matter.*
5356 aufsaugen, *to absorb.*
5357 bedürfen (*w. gen*), *to need, want, require.*
5358 befreien, *to free, rid of.*
5359 ausfindig (*used exclusively with* machen).
  ausfindig machen, *to find (out), to discover, search out.*
5360 wofern, *conj., unless, provided that.*
5361 genügen, *to be sufficient.*

## Conversations 52, 53, 54.

5362 reichen (5096), *to reach, hand, give.*
5363 verdünnen, *to make thin, attenuate, dilute.*
5364 anweisen, *to point out, show, direct.*
5365 Anweisung, *f.*, *direction, instruction.*
5366 Schambug, *m.*, *groin.*
5367 unterwerfen (2344), *to subject.*
5368 unterworfen sein, *to be subjected.*
5369 ergreifen (2122), *to seize, lay hold of.*
5370 ergriffen sein, *to be seized.*
5371 unbedingt, *a. and adv., absolute.*
5372 Abwaschung, *f.*, *lotion.*
5373 zudecken, *to cover.*
5374 einschliessen (5190), *to encase.*

## Conversation 55.

5375 Sumpf, *m.*, *marsh, swamp.*
5376 sumpfig, *a.*, *marshy, swampy*
5377 miasmatisch, *a.* }
  Miasmen erzeugend, }
  *miasmatic, miasmal.*
  Miasma (*pl.* -en), *miasma.*
5378 sich aufhalten, *to sojourn.*
5379 sich übermüden, *to over-tire one's self, over-fatigue.*
5380 Sonnenhitze, *f.*, *heat of the sun.*

5381 fieberfrei, *a.*, *free from fever.*
5382 harntreibend, *diuretic, ischuretic.*
5383 völlig, *a. and adv., full, complete.*
5384 wesentlich, *essential.*

## *Conversations 56, 57.*

5385 unbewusst, *a. and adv., unconscious of.*
5386 stets, *adv., always, constantly.*
5387 meistens, *adv., mostly, for the most part.*
5388 umherwandern, *to ramble.*
5389 Erschlaffung, *f., relaxation, lassitude.*
5390 abstumpfen, *to blunt, dull, deaden.*
5391 abgestumpft, *dull.*
5392 fortbestehen (2302), *to continue, subsist.*
5393 **scheren, schor, geschoren,** *to shear.*
abscheren, *to shave off, shear.*
5394 Gummibeutel, *m., rubber bag.*
5395 stossen (736), *to break small, crush.*
5396 Begiessung, *f., affusion.*
5397 wasserdicht, *water-proof.*
5398 das Zeug (*pl.* -e), *stuff, cloth.*
5399 murmeln, *to mutter, mumble.*
5400 betragen, *to amount to or come to.*
Betrag, *m., amount.*
5401 Verhältniss, *n., proportion.*
nach Verhältniss, *in proportion.*
5402 entziehen (174), *to withdraw*
5403 hingegen, *adv. and conj., on the contrary, whereas.*
5404 festsetzen, *to establish, fix.*
5405 ermuntern, *to awake, rouse.*

## *Conversation 58.*

5406 ausbrechen, *to come out.*
5407 übersteigen (675), *to exceed.*
5408 hinfahren über, *to pass, skim (over), to touch upon.*
**fahren, fuhr, gefahren,** *to move (to move in a vehicle), go, pass.*
5409 abtrocknen, *to dry.*
5410 sorgfältig, *a. and adv., careful.*
5411 dämpfen, *to stew.*
5412 abhelfen (5318), *to remedy.*
5413 Unze, *f., ounce.*
5414 zerschmelzen, *to melt (away).*
**schmelzen, schmolz, geschmolzen,** *to melt.*
5415 belieben, *to like, to be pleased.*
5416 das Belieben, *liking, pleasure.*
nach Belieben, *at pleasure, as you please.*
5417 fortschaffen, *to remove.*

## Conversation 59.

5418 sich ausbreiten, *to extend.*
5419 **lesen, las, gelesen,** *to read.*
5420 verdunkeln (710), *to darken.*
sich verdunkeln, *to darken, grow dim.*
5421 deutlich, *a. and adv., distinct.*
5422 vor Kurzem (1991), *lately.*
5423 matt (1565), *a., dull, dim, weak.*
5424 Zwielicht, *n., twilight.*
5425 Ferne, *f., distance.*
5426 in der Ferne, *in or at a distance.*
5427 Buchstabe (*gen.* -ns, *pl.* -n), *letter.*
5428 doppelt, *a. and adv., double.*
5429 Ring, *m., ring.*
5430 Nebel, *m., mist, fog, nebula.*
einen Nebel vor den Augen haben, *to be dim-sighted.*
5431 grell, *a., dazzling, glaring.*
5432 einwärtskehren, *to introvert, turn inwards.*
5433 Zoll, *m., inch.*
5434 das Gevierte (*from* vier, *four*), *square.*
ins Gevierte, *square.*
5435 Blutegel (Blutigel *is incorrect*), *leech.*
5436 einstellen, *to discontinue, to stop.*
5437 träufeln, *to drop, trickle.*
5438 auflegen, *to apply.*
5439 Stecknadelkopf, *m., pinhead*
5440 Fleck *or* Flecken, *speck.*

5441 Regenbogenhautentzündung, *f., iritis.*
5442 ernstlich, *earnestly.*
5443 zerstören, *to destroy.*
5444 Augennerv, *optic nerve.*
5445 wahrscheinlich, *a. and adv., probable.*
5446 beseitigen, *to remove, take away.*
5447 schwimmen, *to swim, float.*
5448 aushöhlen, *to excavate.*
5449 passen, *to fit, suit.*
5450 passend, *fitting, proper.*
5451 sich eignen, *or* geeignet sein, } *to be proper.*
5452 geeignet, *p. p., proper, suitable.*
5453 lähmen, *to lame, paralyze.*
5454 durchführen, *to pass through*
5455 öffnen, } *to open.*
5456 sich öffnen,
5457 körnig, } *granular.*
gekörnt,
5458 unternehmen, *to undertake.*
5459 einwärts wenden, *to turn in or inwards.*
5460 das Innere, *interior.*
5461 Ausnahme, *f., exception.*
5462 aus Mangel an, } wegen Mangels an, *for want of.*
5463 gerade, *straight.*
5464 Brille, *f., spectacles.*
5465 **verbergen, verbarg, verborgen,** *to hide from.*

## Conversation 60.

5466 erkältet (4957) sein, *to have a cold.*
5467 durchbohren, *to perforate.*
5468 das Mittelohr, *middle ear.*
5469 Ansammlung, *f.*, *collection.*
5470 aufblähen, *to inflate.*
5471 weg sein, *to be gone.*
5472 zu Wege bringen, *to cause.*
5473 beträchtlich, *considerable.*
5474 absterben (4915), *to die.*

## Conversation 61.

5475 gelegentlich, *a. and adv.*, *occasional.*
5476 Blutfluss, *bloody flux, hemorrhage.*
5477 Schreck, *m.* } *fright,*
    Schrechen, *m.* } *terror.*
5478 Zusammenhang, *connection.*
5479 Zeitdauer, *f.*, *duration of time.*
5480 ungeheuer, *a. and adv.*, *enormous, immense.*
5481 neulich, *adv.*, *lately, newly.*

## Conversations 62, 63, 64.

5482 verzögern, *to retard, delay.*
5483 wegbleiben, *to stay or keep away.*
5484 **betreffen, betraf, betroffen,** *to concern.*
    was betrifft, *as for, as to, in point of.*
5485 Rücksicht, *f.*, *respect, regard, reference.*
    mit Rücksicht auf, *with regard to.*
5486 rücksichtlich, *adv.*, *regarding.*
5487 abgehen, *to come off, to be discharged.*
5488 mit, *adv.*, *together with, likewise.*
5489 ordentlich, *a. and adv.*, *orderly, verily, in fact.*
5490 **das** Übermass, *excess.*
5491 Verminderung, *f.*, *diminution.*
5492 Erguss, *m.*, *effusion.*
5493 gleichartig (71), *of the same kind and nature, analogous.*
5494 zur Zeit, *at the time.*
5495 unrichtige Wochen haben (1275), *to miscarry.*
5496 eiterartig, *purulent, puriform.*
5497 sich herschreiben (4900), *to date.*
5498 unerlässlich, *a. and adv.*, *indispensable.*
5499 hinreichend, *sufficient.*
5500 Stütze, *f.*, *support.*
5501 **bieten, bot, geboten,** *to offer.*

### Conversation 65.

5502 segnen, *to bless.*
5503 in gesegneten Umständen (1250), *in the family way.*
5504 gefasst (4827), *p. p., composed, calm.*
    gefasst auf, *ready, prepared for.*
5505 anordnen, *to arrange.*
5506 auf einander, *upon one another.*
5507 nicht besonders (4799), (*supply* gut), *so-so.*
5508 hübsch, *a., fine, pretty.*
5509 anlegen, *to apply to the breast.*
5510 anfassen, *to take hold of.*
5511 Warzenschützer, *artificial nipple* (lit. *protector of the nipple*).

### Conversation 66.

5512 aufdecken, *to uncover.*
5513 verhungern, *to starve, famish.*
5514 Zweck, *m., purpose, aim.*
5515 zweckmässig, *a. and adv., answering a purpose, suitable.*
5516 Zufall, *m., casualty.*
5517 krampfhafte Zufälle (1710), *fits, attacks.*
5518 bekannt, *known.*
5519 aufschreien (2482), *to (give a) shriek, raise an outcry.*
5520 die Windel, *swaddling.*
5521 die Windeln nassmachen, *to wet the swaddling.*

### Conversations 67, 68.

5522 ärgerlich, *a., angry, peevish.*
5523 wiederwärtig, *a., adverse, odious.*
5524 scheuen, *to shun, avoid.*
5525 **wenden, wandte, gewandt,** *to turn.*
    sich abwenden, *to turn away.*
5526 verdriesslich, *a. and adv., peevish.*
5527 belustigen, *to amuse.*
5528 unterhalten, *to entertain, amuse.*
5529 fesseln, *to fetter;* fig. *to catch, rivet.*
5530 Aufmerksamkeit, *f., attention.*
5531 allenthalben, *everywhere.*
5532 **waschen, wusch, gewaschen,** *to wash.*
    auswaschen, *to wash out.*

### Conversations 69, 70.

5533 sich hin und her werfen, *to toss about.*
5534 zurückwerfen, *to throw back.*
5535 erhalten, *to maintain.*
5536 zulassen, *to admit.*
5537 Wasserkessel, *m., kettle of water.*

## Conversations 71, 72, 73.

5538 bewachen, *to watch, guard.*
5539 regeln, *to regulate.*
5540 Strömung, *f., current.*
5541 durchaus, *by all means, absolutely.*
5542 stochern, *to pick, poke.*
5543 verbrauchen, *to consume, use up.*

## Conversation 74.

5544 Zahnreihe, *f., row of teeth.*
5545 befallen (1583), *to befall.* befallen sein *or* werden, *to be attacked, seized (with a disease).*
5546 die Schwitzcur, *sweating cure.*
5547 sich setzen, *to seat one's self.*
5548 daneben, *near it.*
5549 herrühren, *to arise, come from, emanate.*
5550 übertragen, *to transfer.*
5551 Zweifel, *m., doubt.*
5552 Dienst, *m., service.*
5553 einen Dienst leisten (5271), *to render a service.*
5554 plombieren, *to plug.*
5555 ausfüllen, *to fill up.*
5556 Goldblatt (*dim.* blättchen), *gold leaf.*
5557 befürchten, *to fear.*
5558 shadhaft, *bad, rotten, injured.*
5559 halten für (gut, schlecht), *to think, consider as . . .*
5560 tagelang, *adv., for days, whole days.*
5561 Geisteskräfte (*pl.*, 17), *mental faculties.*
5562 ansehen, *to look at.*
5563 ausspülen, *to rinse out.*
5564 zusammendrücken, *to press together.*
5565 **die** Sorgfalt, *care, solicitude.*
5566 verwenden auf, *to bestow upon, devote to.*
5567 Zahnpilz (3522, *dim.* -chen), *fungus of the teeth.*
5568 zusetzen, *to set to, add.*
5569 Essigäther, *acetic ether.*
5570 abbürsten, *to brush.*
5571 tüchtig, *a. and adv., thoroughly.*
5572 Zahnstein, *m., tartar.*
5573 entgegenwirken, *to counteract, prevent.*
5574 zeitweilig, *from time to time, occasional.*
5575 abscheuern, *to scour.*
5576 feingepulvert, *finely powdered.*
5577 Zahnpulver, *n., tooth powder.*
5578 entrichten, *to pay, discharge*
5579 schuldig sein, *to owe.*
5580 **das** Honorar, *fee.* was beträgt Ihr Honorar, *what does your fee amount to?* (5400).

5581 honorieren, *to pay a fee.*
5582 verzeihen, *to pardon.*
5583 entstellen, *to disfigure.*
5584 kunstvoll, *a., artificial.*
5585 ersetzen, *to supply the place of one.*
5586 kosten, *to cost.*

## Conversation 75.

5587 besorgen, *to take care of, to attend to; to apprehend, be afraid.*
5588 leid thun, *to be sorry for.*
5589 anfertigen, *to make, compose.*
Anfertigung, *preparation.*
5590 Auftrag, *m., errand, commission.*
5591 verbinden, *to oblige one.*
5592 bereiten, *to prepare.*
5593 ausdrücklich, *a. and adv., explicit.*
5594 verfälschen, *to adulterate.*
5595 unverfälscht, *unadulterated.*
5596 daher, *adv. and conj., therefore; for that reason.*
5597 Nachbarschaft, *f., neighborhood.*
5598 Nähe, *f., nearness, neighborhood.*
5599 lieben, *to love; like, be fond of.*
5600 verabreichen, *to administer, offer.*
5601 zählen, *to count.*
5602 Tropfenzähler, *drop-counter.*
5603 anschaffen, *to procure.*
sich ——, *to furnish one's self with.*
5604 aufkochen, *to boil up.*
5605 das Quart (*pl. -e*), *quart.*
5606 schnupfen, *to snuff.*
5607 trocknen, *to dry.*
5608 Tropfglas, *n., medicine dropper.*
5609 aufbewahren, *to preserve.*
sich aufbewahren lassen, *to keep well.*
5610 luftdicht, *a. and adv., airtight.*
5611 Stöpsel, *m., stopper, cork.*
5612 Gebrauchsanweisung, *f., direction for use.*
5613 aufschreiben, *to write down.*
5614 Aufschrift, *f., direction.*
5615 die Schachtel, *box, bandbox.*
5616 das Hausmittel, *family medicine, old woman's remedy.*
5617 probat (ba'ht), *approved, tried.*
5618 sich bewähren, *to prove, hold good or true.*
5619 doctern, *to doctor.*
5620 erfahren (5077), *a., experienced.*
5621 schon, *adv., certainly, truly, no doubt.*
schon das Vertrauen, *the very confidence, the bare confidence.*

5622 das Vertrauen, *confidence*.
5623 einflössen, *to infuse*, fig. *to impress with, inspire with*.
5624 beitragen, *to contribute, be instrumental*.
5625 Recht (426) haben, *to be right*.
5626 theilen, *to divide; to share (one's opinion)*.
5627 quacksalbern, *to act or play the quack*.
5628 das Quacksalbern, } *quack-*
die Quacksalberei, } *ery*.
5629 überlassen, *to leave, give up to*.
5630 vertrauen (*w. dat.*), *to confide, trust*.
5631 gewissenhaft, *a. and adv., conscientious*.
5632 wissenschaftlich, *a. and adv., scientific*.
5633 gebildet, *educated*.

# CONVERSATIONS.

### 1. At a private Call.

Good morning, Sir; what can I do for you?

I am afraid I need your assistance, and therefore I have taken the liberty to send for you.

What's the matter?

I am not well at all.

I feel very ill.

I feel very uncomfortable to-day.

How long have you been ill?

How long have you felt unwell?

Since yesterday.

How do you find yourself at present?

I am uncommonly weak; I can hardly stand on my legs.

How did your sickness commence? How were you taken ill?

It began day before yesterday, with a shivering (chill).

Did you perspire then?

Yes, I perspired profusely.

### 1. Bei einem Privatbesuch.

D. Guten Morgen, mein Herr, womit kann ich Ihnen dienen?

P. Ich fürchte, Ihren Beistand nöthig zu haben, und darum war ich so frei, Sie rufen zu lassen.

D. Was fehlt Ihnen?

P. Ich bin gar nicht wohl.

Ich fühle mich recht (4792) (or sehr) unwohl.

Ich befinde mich heute sehr unwohl.

D. Seit wann fühlen Sie sich unwohl?

Seit wann sind Sie krank?

P. Seit gestern.

D. Wie befinden Sie sich in diesem Augenblicke?

P. Ich bin ungewöhnlich matt; ich habe Mühe, mich auf den Beinen zu halten.

D. Wie hat Ihr Leiden angefangen?

P. Es fing vorgestern mit einem Frösteln (mit einem Fieberschauer) an.

D. Haben Sie sodann geschwitzt?

P. Ja, sehr.

| | |
|---|---|
| *Where do you feel pain now ?* | D. Wo haben (*or* empfinden) Sie jetzt Schmerz ? |
| *I have no pain, but I feel sick and sometimes am ready to vomit.* | P. Ich habe keinen Schmerz, aber ich fühle Übelkeit und zuweilen Neigung zum Erbrechen. |
| *My head aches terribly.* | Ich habe heftiges Kopfweh. |
| *My head swims. I feel giddy.* | Es ist mir schwindelig, *or* Ich habe Schwindel. |
| *Have you a pain in your side ?* | D. Haben Sie einen Schmerz (*or* Schmerzen) in der Seite ? |
| *Yes, Sir; especially when I attempt to take a long breath.* | P. Ja, besonders wenn ich tief zu athmen versuche. |
| *How is your appetite ?* | D. Wie ist Ihr Appetit ? |
| *Have you any appetite ?* | Haben Sie etwas Appetit ? |
| *I have no appetite at all.* | P. Ich habe gar keinen Appetit. |
| *I scarcely feel any inclination to eat.* | Ich habe fast gar keine Lust zum Essen. |
| *I dislike all food.* | Es schmeckt mir nichts. |
| *Are you thirsty ?* | D. Haben Sie Durst ? |
| *I feel an intolerable thirst.* | P. Ich habe unerträglichen Durst. |
| *Do you sleep soundly ?* | D. Schlafen Sie gut ? |
| *My sleep is much disturbed, and I dream a great deal.* | P. Mein Schlaf ist sehr unruhig, und ich träume viel. |
| *Let me feel your pulse.* | D. Lassen Sie mich Ihren Puls fühlen. |
| *You are feverish.* | Sie sind fieberisch. |
| *You have a little fever.* | Sie haben etwas Fieber. |
| *Pray ( please to), draw your breath* | Bitte, holen Sie einmal Athem. |
| *Do you not feel any pain ?* | Fühlen Sie keinen Schmerz ? |
| *I breathe with difficulty.* | P. Ich athme nur schwer. |
| *Show me your tongue.* | D. Lassen Sie Ihre Zunge sehen. |
| *Your tongue is not quite clean.* | Ihre Zunge ist nicht ganz rein. |
| *Your tongue is very foul.* | Ihre Zunge ist sehr belegt. |
| *Have you a bad taste in your mouth ?* | Haben Sie einen üblen Geschmack im Munde ? |

| | |
|---|---|
| Yes, especially when I awake, I have a quite bitter, sour, brassy taste in my mouth. | P. Ja, besonders beim Aufwachen habe ich einen ganz bittern, sauren, metallischen Geschmack im Munde. |
| You have a disordered stomach. | D. Sie haben einen verdorbenen (1824) Magen. |
| Your stomach is loaded. | Ihr Magen ist unrein. |
| You must take a little medicine. | Sie müssen etwas Arzenei nehmen (*or* etwas einnehmen). |
| I'll write you a prescription. | Ich will Ihnen etwas verschreiben. |
| Get this medicine made up, and take a table-spoonful every two hours. | Lassen Sie diese Arzenei machen, und nehmen Sie alle zwei Stunden einen Esslöffel voll davon. |
| Must I do anything besides? (*else?*) | P. Habe ich sonst noch etwas zu beobachten? |
| No, only take care to keep yourself warm. | D. Nein, halten Sie sich nur recht warm. |
| You must remain in bed; your room. | Sie müssen im Bette bleiben, das Zimmer hüten. |
| Take care not to catch cold. | Hüten Sie sich vor Erkältung. |
| What may I eat? | P. Was darf ich essen? |
| You ought not to eat anything to-day. | D. Sie sollten heute nichts essen. |
| I'll call again to-morrow morning, and hope to find you better. | Ich werde morgen früh wiederkommen und hoffentlich wird es dann mit Ihnen besser gehen. |

### 2. *At a private Call.*

### 2. Bei einem Privatbesuch.

| | |
|---|---|
| How do you find yourself since yesterday? | D. Wie befinden Sie sich seit gestern? |
| How are you to-day? | Wie geht's heute? |
| How is your health do-day? | Wie geht's mit Ihrer Gesundheit heute? |

| | |
|---|---|
| *How do you feel to-day?* | Wie fühlen Sie sich heute? |
| *Do you feel stronger, weaker?* | Fühlen Sie sich stärker, schwächer? |
| *How are your spirits, good or bad?* | Wie sind Sie aufgelegt; gut oder schlecht? |
| *I am not better.* | P. Es geht nicht besser. |
| *My condition is still the same.* | Mein Zustand ist noch immer derselbe. |
| *My state of health has not changed a particle.* | Es hat sich in meinem Zustande noch nichts geändert. |
| *Why are you so low-spirited?* | D. Warum sind Sie so verstimmt (2624)? |
| *Why are your spirits so depressed?* | Warum sind Sie so schlecht aufgelegt? |
| *What makes you so melancholy?* | Was macht Sie so schwermüthig? |
| *Why are you so uneasy?* | Weshalb sind Sie so unruhig? |
| *How can I be in good humor (or cheerful, or in a kindly mood), when I grow worse from day to day? (or when I don't see the slightest improvement in my condition?)* | P. Wie kann ich gut aufgelegt sein (or heiter or guten Muthes sein), da ich ja von Tag zu Tag schlechter werde? (or da ich ja nicht die geringste Besserung in meinem Zustande sehe?) |
| *Be of good cheer!* | D. Fassen Sie Muth! |
| *Compose yourself!* | Fassen Sie sich! |
| *Don't yield to your humors!* | Sie müssen sich nicht gehen lassen! |
| *Do not lose heart for so small a matter!* | D. Lassen Sie den Muth nicht sinken (2464) wegen einer so geringen Sache. |
| *Do not despond!* | Verzagen Sie nicht! |
| *Take my word for it, you are not in danger.* | Glauben Sie mir, Sie sind in keiner Gefahr. |
| *Oh, Doctor! you don't know how sick I am. I grow weaker and weaker every day.* | P. Ach, Herr Doctor! Sie wissen gar nicht, wie krank ich bin. Ich werde alle Tage schwächer. |

| | |
|---|---|
| *You make your disease worse than it really is. I dare say, you will recover.* | D. Sie machen Ihre Krankheit ärger, als sie wirklich ist. Ich kann versichern, Sie werden wieder genesen. |
| *It is true, you are improving slowly, but you improve.* | Es geht zwar langsam mit Ihrer Genesung; aber es geht besser. |
| *Dear Doctor, tell me your opinion without reserve.* | P. Bester Herr Doctor! sagen Sie mir Ihre Meinung ganz unverhohlen. |
| *My opinion is, that your illness is not dangerous in its present stage; but you must take care lest it should become so. It might otherwise turn to a malignant fever. There is nothing at present to make you uneasy.* | D. Meine Meinung ist, dass Ihre Krankheit in dem gegenwärtigen Zustande nicht gefährlich ist; Sie müssen sich aber in Acht nehmen, dass sie es nicht wird. Sie könnte sonst in ein bösartiges Fieber ausarten. Es ist gegenwärtig nichts vorhanden, was Sie beunruhigen könnte. |
| *I shall prescribe you a potion.* | Ich will Ihnen einen Arzeneitrank verschreiben. |
| *If it is all the same, I had rather take the remedy in pills, as it is very hard for me to swallow either potion or powder.* | P. Wenn es nichts macht, so würde ich das Mittel lieber in Pillen nehmen, da ich nur sehr schwer Arzeneitränke oder Pulver hinunter schlucken kann. |
| *Yes, you can have your medicine in pills.* | D. Ja, Sie können die Arzenei in Pillen bekommen. |
| *If the pain in your side should reappear, you can have a poultice put on your side, or you may apply mustard poultices.* | Wenn der Schmerz in der Seite wiederkommen sollte, so lassen Sie sich einen Umschlag auf die Seite legen, oder Sie können auch Senfumschläge machen. |
| *I have so much difficulty in breathing, can there nothing be done to ease it?* | P. Das Athmen fällt mir so schwer; kann man nichts thun, um es zu erleichtern? |

| | |
|---|---|
| I also prescribed an ointment to rub your chest (breast) with. That will ease your breathing. | D. Ich habe Ihnen auch eine Salbe verschrieben, um sich die Brust damit einzureiben. Das wird Ihnen das Athmen erleichtern. |
| What diet must I keep? | P. Welche Diät muss ich halten (or beobachten)? |
| If your appetite should come again, you can take some light nourishment that digests easily, such as broth, chicken, gruel, and the like. | D. Wenn sich der Appetit (die Esslust) wieder einstellen sollte, so können Sie etwas leicht verdauliche Nahrung zu sich nehmen, wie Fleischbrühe, Huhn, Grütze und Ähnliches. |
| Pray, Doctor, come again tomorrow to see me. | P. Bitte, Herr Doctor, besuchen Sie mich morgen wieder. |
| I will not fail. | D. Ich werde nicht ermangeln. |

## 3. At a private Call. — 3. Bei einem Privatbesuch.

| | |
|---|---|
| How are you to-day? | D. Wie befinden Sie sich heute? |
| I feel myself much better. | P. Ich fühle mich viel besser. |
| How have you passed the night? | D. Wie haben Sie die Nacht zugebracht? |
| I have not been so much agitated, and I slept a little. | P. Ich war nicht so aufgeregt und habe ein wenig geschlafen. |
| Did you sleep soundly? | D. Haben Sie gut geschlafen? |
| My sleep was somewhat disturbed, but nevertheless refreshing. | P. Mein Schlaf war zum Theil unruhig, aber nichts desto weniger erfrischend. |
| I slept all night with few interruptions. | Ich habe mit wenig Unterbrechungen die Nacht durchgeschlafen. |
| Did your physic work well? | D. Hat das Abführmittel gut gewirkt? |
| Very well. | P. Sehr gut. |
| How many stools have you had? | D. Wie oft sind Sie zu Stuhl gegangen? |

Five or six.
Have you still a bitter taste in your mouth when you awake?

A little, but not so much as yesterday.
Let me feel your pulse!

The fever is much abated; but it is not yet entirely off.

At any rate you are a great deal better to-day.
Yes, I feel greatly relieved.

Get this medicine made up, and take twenty drops every two hours.
Whenever you feel feverish, you must keep yourself as tranquil as possible.

I often feel thirst.
A little raspberry vinegar in water would relieve your thirst.

P. Fünf bis sechsmal.
D. Haben Sie noch beim Aufwachen einen bittern Geschmack im Munde?

P. Ein wenig, aber nicht so viel als gestern.
D. Lassen Sie mich Ihren Puls fühlen.

Das Fieber hat sehr nachgelassen; aber es ist noch nicht ganz weg.

Jedenfalls finde ich Sie heute bedeutend besser.
P. Ja, ich fühle mich sehr erleichtert.

D. Lassen sie diese Arzenei machen, und nehmen Sie alle zwei Stunden 20 Tropfen.
So oft Sie sich fieberhaft (or fieberisch) fühlen, müssen Sie sich so ruhig als möglich verhalten.

P. Ich habe oft Durst.
D. Etwas Himbeeressig mit Wasser würde Ihren Durst stillen.

### 4. At a private Call.

How are you to-day?
I feel a great deal better.
Did you take the medicine I prescribed?
I took it at one draught.

Did it do you good?

### 4. Bei einem Privatbesuch.

D. Wie geht es heute?
P. Es ist mir weit besser.
D. Haben Sie die Arznei genommen, die ich Ihnen verschrieb?
P. Ich nahm sie auf einen Schluck.

D. Ist sie Ihnen bekommen?

| | |
|---|---|
| *I experienced some relief.* | P. Ich verspürte einige Linderung. |
| *Do you feel any headache ?* | D. Haben Sie noch Kopfweh ? |
| *Much less than I did; I am getting much better.* | P. Viel weniger als früher, ich fühle mich um vieles leichter. |
| *I told you, you know, it would be nothing serious.* | D. Ich sagte Ihnen ja, es habe nichts zu bedeuten. |
| *I can promise you that it will have no bad consequences.* | Ich kann Ihnen versprechen, dass es keine üblen Folgen haben wird. |
| *May I get up a little ?* | P. Kann ich ein wenig aufstehen ? |
| *Yes, for an hour or two.* | D. Ja, ein paar Stunden. |
| *May I leave my bed to-morrow ?* | P. Darf ich morgen das Bett verlassen ? |
| *We shall see about that to-morrow.* | D. Das werden wir morgen sehen. |
| *What must I drink when I am thirsty ?* | P. Was darf ich trinken, wenn ich Durst bekomme ? |
| *Some lemonade.* | D. Etwas Limonade. |
| *May I drink cold water (iced water) ?* | P. Darf ich kaltes Wasser (Wasser mit Eis) trinken ? |
| *But very moderately.* | D. Nur sehr mässig (1994). |
| *May I eat anything I like ?* | P. Darf ich alles essen, was mir schmeckt ? |
| *By no means, you must still live very low.* | D. Keineswegs! Sie müssen noch eine strenge Diät beobachten. |
| *You must eat no fat meat, nothing sour, and sparingly of vegetables.* | D. Sie dürfen keine fetten Speisen essen, nichts Saures und nur wenig Gemüse. |
| *You may often take some broth.* | Geniessen Sie öfters etwas Fleischbrühe. |
| *Confine yourself to farinaceous substances as much as possible.* | Beschränken Sie sich so viel als möglich auf Mehlspeisen. |
| *What would you like to have for breakfast (dinner, supper) ?* | Was möchten Sie gern zum Frühstück (Mittagbrod, Abendbrod) haben ? |
| *May I have some cold meat ?* | P. Darf ich etwas kaltes Fleisch essen ? |

*No, it would disagree with you. You must abstain from meat entirely at present.*

D. Nein, das würde Ihnen schlecht bekommen. Sie müssen sich für jetzt des Fleisches gänzlich enthalten.

*May I drink some beef-tea?*

P. Darf ich etwas Bouillon geniessen?

*Yes, it would do you good.*

D. Ja, das würde Ihnen bekommen.

*Keep a strict diet a little longer, and in a very short time you will recover your health and strength.*

Halten Sie sich nur etwas länger, und in sehr kurzer Zeit, werden Sie wieder auf den Beinen sein.

*I should advise you to go to the country for a month.*

Ich würde Ihnen rathen, einen Monat auf's Land zu gehen.

*You must try a change of air.*

Sie müssen eine Luftveränderung vornehmen.

*You must take moderate exertion.*

Sie müssen sich mässige Bewegung machen.

### 5. *At a Hospital.*  5. In einem Hospitale.

*What is your name?*  Wie heissen Sie?

*How do you spell it?*  Wie buchstabieren Sie Ihren Namen?
Wie schreiben Sie sich?

*How old are you?*  Wie alt sind Sie?

*Where do you live?*  Wo wohnen Sie?

*Where do you come from?*  Wo kommen Sie her?

*What is your native country?*  In welchem Lande sind Sie geboren?

*How long have you been in the United States?*  Wie lange sind Sie in den Vereinigten Staaten?

*What is your business?*  Was ist Ihr Geschäft?
Was für Geschäft haben Sie?
Was sind Sie von Profession?

*Are you married, single, widowed (a widower, a widow)?*  Sind Sie verheirathet, unverheirathet, verwittwet (ein Wittwer, eine Wittwe)?

| | |
|---|---|
| *Are your parents living ?* | Leben noch Ihre Eltern ? |
| *How old are they ?* | Wie alt sind dieselben ? |
| *At what age did they die ?* | In welchem Alter sind sie gestorben ? |
| *Of what sickness ?* | An welcher Krankheit ? |
| *Have you brothers and sisters ?* | Haben Sie Geschwister ? |
| *How many of them are still living ?* | Wie viele von ihnen leben noch ? |
| *Of what sickness and at what age did the others die ?* | An welcher Krankheit und in welchem Alter sind die andern gestorben ? |
| *What sicknesses have you had in your life ?* | Welche Krankheiten haben Sie schon in Ihrem Leben gehabt ? |
| *Have you suffered from the same or a similar sickness before ?* | Waren Sie früher einmal auf dieselbe oder ähnliche Weise krank ? |
| *Did you enjoy good health as a child ?* | Haben Sie sich als Kind einer guten Gesundheit erfreut ? |
| *Were you in good health formerly?* | Haben Sie sich früher in guter Gesundheit befunden ? |
| *Did you suffer from anything formerly ?* | Haben sie früher an irgend etwas gelitten ? |
| *Do you know of any hereditary affection in your family ?* | Wissen Sie von irgend einem Leiden, das in Ihrer Familie erblich vorhanden ist ? |
| *When were you taken sick ?* | Wann wurden Sie diesmal krank ? |
| *When did your suffering commence ?* | Wann hat Ihr Leiden begonnen ? |
| *How were you taken sick ?* | Wie sind Sie krank geworden ? |
| *Of what did you first complain ?* | Worüber klagten Sie zuerst ? |
| *Do you remember where you first felt the pain ?* | Erinnern Sie sich, wo Sie zuerst den Schmerz verspürt ? |
| *Do you remember what first called your attention to your suffering?* | Erinnern Sie sich, wodurch Sie zuerst auf Ihr Leiden aufmerksam geworden ? |
| *Describe your first symptoms, or what first made you think you were sick.* | Beschreiben Sie die ersten Symptome, oder was Sie zuerst veranlasst hat, sich krank zu denken. |

| | |
|---|---|
| *Did you have a chill?* | Empfanden Sie ein Frösteln? |
| *Did you have any fever?* | Hatten Sie irgend ein Fieber? |
| *Did you have any pain?* | Hatten Sie irgend einen Schmerz? |
| *What did you next observe?* | Was nahmen Sie sodann wahr (*or* bemerkten Sie sonst)? |
| *And then what?* | Und dann, was sonst? |
| *How long did that continue?* | Wie lange hielt das an (*or* dauerte (48) das)? |
| *What brought on your illness?* | Wodurch haben Sie sich wohl das Übel zugezogen? |
| *Have you had a fall?* | Haben Sie einen Fall gethan (*or* Sind Sie gefallen)? |
| *Has anything struck you or fallen on you?* | Haben Sie einen Stoss oder Schlag bekommen (*or* Sind Sie gestossen oder geschlagen worden)? |
| *Have you overworked yourself?* | Haben Sie sich überarbeitet? |
| *What is your mode of life?* | Wie ist Ihre Lebensweise? |
| *Are you obliged to overexert yourself?* | Müssen Sie über Ihre Kräfte arbeiten (*or* sich anstrengen)? |
| *Do you suffer want?* | Sind Sie dem Mangel ausgesetzt (*or* Leiden Sie Mangel) (Noth)? |
| *Have you troubles, cares?* | Haben Sie viel Kummer, Sorgen? |
| *Are you fond of spirits, as wine, liquor, beer?* | Sind Sie ein Freund von Spirituosen, wie Wein, Schnapps, Bier? |
| *Do you like to drink often?* | Lieben Sie öfters ein Schnäppschen? |
| *Do you sometimes take one more glass than you should?* | Trinken Sie manchmal eins mehr, als Sie sollten? |
| *Do you smoke much?* | Rauchen Sie stark? |
| *Do you chew?* | Kauen Sie? |
| *Have you taken cold?* | Haben Sie sich erkältet? |
| *Have you perhaps slept in a damp room?* | Haben Sie vielleicht in einem feuchten Zimmer geschlafen? |
| *Have you consulted any other physician?* | Haben Sie irgend einen Arzt schon zu Rathe gezogen? |
| *What remedies have you made use of heretofore?* | Welche Mittel haben Sie bisher angewandt? |
| *What is your present condition?* | Wie ist Ihr Zustand jetzt? |

| | |
|---|---|
| *Has it visibly grown worse, better, or has it remained the same?* | Hat er sich merklich verschlimmert, verbessert, oder ist er unverändert derselbe geblieben? |
| *Have you a sensation of weight in your head and limbs, of lassitude, uneasiness, or even pain somewhere?* | Haben Sie ein Gefühl der Schwere im Kopf und in den Gliedern, der Mattigkeit, Unbehaglichkeit oder gar Schmerz irgend wo? |
| *Can you tell me exactly or nearly, where you have the sensation (or the pain)?* | Können Sie mir genau oder ungefähr den Ort sagen (*or* angeben), wo Sie das Gefühl (*or* den Schmerz) haben? |
| *Can you move in all directions freely, without pain?* | Können Sie sich frei und ohne allen Schmerz nach allen Seiten hin bewegen? |
| *Do you feel equally uncomfortable in every position of the body?* | Fühlen Sie sich in jeder Lage des Körpers gleich unbehaglich? |
| *Do you prefer any one posture, standing, sitting, lying, or lying on the right or left side or on the back?* | Ziehen Sie irgend eine Stellung, Stehen, Sitzen, Liegen, oder die Lage auf der rechten Seite, der linken, oder dem Rücken vor? |
| *Is your breathing in any way impeded?* | Ist Ihr Athmen in irgend einer Weise beschwert? |
| *Draw a very long and deep breath.* | Athmen Sie einmal recht tief und langsam ein. |
| | Holen Sie einmal recht tief und langsam Athem. |
| | Schöpfen Sie einmal recht tief und langsam Athem. |
| *Keep your breath as long as you can.* | Halten Sie Ihren Athem an (995), so lange Sie können. |
| *Cough!* | Husten Sie einmal! |
| *How is your digestion?* | Wie ist Ihre Verdauung? |
| *How were your appetite and digestion prior to your present sickness?* | Wie war Ihr Appetit und Ihre Verdauung vor Ihrer jetzigen Krankheit? |
| *Are your bowels in order?* | Ist Ihre Darmthätigkeit in Ordnung? *or* |

| | |
|---|---|
| *Have you open bowels regularly?* | Haben Sie regelmässig offenen Leib? |
| *Do you know how your stools (alvine evacuations) look?* | Wissen Sie, wie Ihre Darmentleerungen (or Stuhlgänge) aussehen? |
| *Had you a chill, fever, vomiting, diarrhœa (loose bowels), constipation (costiveness), trembling, convulsions?* | Haben Sie Frost, Fieber, Erbrechen, Durchfall, Verstopfung, Zittern, Krämpfe gehabt? |
| *Have you any urinary trouble?* | Haben Sie irgend Beschwerden beim Harnen? |
| *Do you often suffer from palpitation, soar throat, hoarseness?* | Leiden Sie oft an Herzklopfen, Halsschmerzen, Heiserkeit? |
| *Do you easily catch cold?* | Erkälten Sie sich leicht? |
| *Are you liable to fainting spells?* | Sind Sie zu Ohnmachten geneigt? |
| *Do you have your regular monthlies?* | Ist Ihre Periode in Ordnung? Haben Sie Ihr Monatliches (3258) regelmässig? |
| *Have your monthly courses stopped?* | Ist Ihr Monatliches ausgeblieben? |

## 6. Headache, Vertigo. — 6. Kopfschmerz, Schwindel.

| | |
|---|---|
| *Where is the seat of the pain?* | Wo ist der Sitz des Schmerzes? |
| *Where is the particular locality of the pain?* | Wo verspüren Sie vornehmlich den Schmerz? |
| *Do you feel the pain within the head (skull) or externally?* | Verspüren Sie den Schmerz innerhalb des Schädels oder äusserlich? |
| *Is the pain aggravated upon my pressing here?* | Wird der Schmerz vermehrt, wenn ich hier drücke? |
| *Have you any disease of the bones?* | Haben Sie irgend ein Knochenleiden? |
| *Do you suffer from any affection of the teeth, the ears, the eyes?* | Leiden Sie an den Zähnen, den Ohren, den Augen? |
| *Is the pain persistent or paroxysmal?* | Ist der Schmerz anhaltend derselbe, oder tritt er in schwerern oder leichtern Anfällen (Paroxysmen) auf. |

| | |
|---|---|
| *Is the pain increased suddenly by rising, stooping, lying down, or by moving your head from one side to the other?* | Empfinden Sie mehr Schmerz, wenn Sie plötzlich sich erheben, sich bücken, sich niederlegen, oder wenn Sie den Kopf von einer Seite zur andern bewegen? |
| *Is the pain worse at night or by day, or at any particular hour of the day?* | Ist der Schmerz grösser bei Nacht oder bei Tage, oder zu irgend einer bestimmten Tagesstunde? |
| *Is the pain increased by long sleep, by bodily or mental fatigue?* | Wird der Schmerz durch langen Schlaf, oder durch körperlich und geistig ermüdende Anstrengungen vermehrt? |
| *Is your headache sometimes very sharp (piercing) and constant?* | Ist der Kopfschmerz zuweilen sehr stechend und andauernd? |
| *Is the pain relieved by stimulants or food?* | Wird der Schmerz durch Stimulantien oder Nahrung gemindert? |
| *Is the pain worse in the morning, before food has been taken?* | Ist der Schmerz heftiger des Morgens, ehe Sie Nahrung zu sich genommen haben? |
| *Have you sometimes a transitory sensation of falling, of dizziness, a feeling of illusory movements of external objects?* | Haben Sie zuweilen ein vorübergehendes Gefühl von Fallen, von Schwindel, oder eine Empfindung, als wenn die äussern Gegenstände sich bewegten? |
| *Is this feeling relieved by closing your eyes?* | Wird dieses Gefühl durch Schliessen der Augen erleichtert? |
| *Is this sensation associated with a dull headache, and with gastric disturbances?* | Ist dieses Gefühl von dumpfem Kopfschmerz, von Störungen des Magens begleitet? |
| *Have you perhaps been careless regarding your diet?* | Sollten Sie vielleicht einen Diätfehler begangen haben? |
| *Are you free from suffering between the attacks?* | Sind Sie in der Zeit zwischen solchen Anfällen frei von Leiden? |
| *Is the vertigo associated with partial deafness and ringing in the ears.* | Ist das Schwindelgefühl von theilweiser Taubheit oder Sausen in den Ohren begleitet? |

| | |
|---|---|
| *Did the sickness begin with vomiting, followed by suddenly developed buzzing in the ear?* | Hat die Krankheit etwa mit Erbrechen begonnen, dem plötzlich entstehendes Sausen in den Ohren folgte? |
| *Is pain produced in the affected ear by noise of any kind?* | Wird Schmerz durch irgend ein Geräusch in dem afficierten Ohre hervorgerufen? |
| *Is the abnormal sensation of short or longer duration?* | Ist das abnorme Gefühl von kurzer oder längerer Dauer? |
| *Is the sensation more or less severe?* | Ist es mehr oder weniger heftig? |
| *Do you momentarily lose all consciousness?* | Verlieren Sie dabei für den Augenblick das Bewusstsein oder nicht? |
| *Does the vertigo recur at uncertain times, even while actively engaged, or in bed and half asleep?* | Kehrt das Schwindelgefühl zu unbestimmten Zeiten zurück, vielleicht sogar wenn Sie thätig, oder im Bette und halb im Schlafe sind? |
| *Is your health good otherwise?* | Ist sonst Ihre Gesundheit gut? |
| *Did your disease arise from overtaxing the brain?* | Ist Ihr Leiden vielleicht in Folge einer Überanstrengung des Gehirns entstanden? |
| *Is the vertigo relieved in the recumbent position?* | Wird das Schwindelgefühl in liegender Stellung verbessert? |
| *Is the headache dull or of steady intensity, or more violent and paroxysmal?* | Ist der Kopfschmerz dumpf und von statig (or immer) gleicher Intensität, oder ist er heftig und tritt in Paroxysmen auf? |
| *Was the headache sudden in its development; is it uniform and general instead of neuralgic and limited?* | Ist der Kopfschmerz plötzlich aufgetreten, ist er gleichmässig und allgemein, statt neuralgisch und begrenzt? |
| *How long have you been suffering from it?* | Seit wie lange schon leiden Sie daran? |
| *Is your sense of vision impaired, and wherein does this impairment consist?* | Ist Ihre Sehkraft beeinträchtigt, und worin besteht diese Beeinträchtigung? |

| | |
|---|---|
| Do you see only half an object, do you see it double? | Sehen Sie nur die Hälfte eines Gegenstandes, sehen Sie denselben doppelt? |
| Does your head ache at the same time? | Leiden Sie gleichzeitig auch an Kopfschmerz? |
| Has the patient perhaps been struck on his head? | Ist der Patient vielleicht auf den Kopf geschlagen worden? |

## 7. Paralysis. — 7. Lähmung.

| | |
|---|---|
| Did the paralysis occur suddenly? | Ist die Lähmung plötzlich erschienen? |
| Was it preceded by a sudden pain in the back? | Ging ihr ein plötzlicher Schmerz im Rücken voran? |
| Is the patient unable to retain his urine and his fæces? | Kann der Kranke weder seinen Urin noch seine Excremente zurückhalten? |
| Did the patient indulge in sexual excesses? | Hat der Patient sich geschlechtliche Excesse zu Schulden kommen lassen? |
| Did the paralysis occur after fatigue or violent exertion, or in consequence of exposure to cold or wet? | Ist die Lähmung nach Übermüdung oder heftigen Anstrengungen aufgetreten, oder in Folge von Erkältung, Durchnässung? |
| Is the difficulty in motion much greater on arising after a night's rest, or indeed whenever the patient has been for any length of time in a recumbent posture? | Ist die Bewegung mehr erschwert, wenn der Patient nach der Nachtruhe Morgens aufsteht, oder überhaupt längere Zeit in liegender Stellung zugebracht? |
| Is the pain more severe as I press upon the spine? | Ist der Schmerz grösser, wenn ich auf das Rückgrat drücke? |
| Is the pain aggravated by movements of any kind? | Wird der Schmerz durch jede Bewegung verschlimmert? |
| Has the patient sometimes a sensation, as if a cord had been drawn around his body? | Hat der Patient zuweilen ein Gefühl, als ob ein Strick um seinen Leib gezogen wäre? |

Attempt to walk or stand with the eyes closed.

Versuchen Sie mit geschlossenen Augen zu gehen oder zu stehen.

## 8. Coma.

Has the patient fallen or been struck upon his head?
Has he applied himself too intensely to study?
Had he perhaps great trouble or depressing cares?

Was the attack very sudden?
Had the patient, before sinking into the comatose sleep, more or less headache, nausea, confusion of thought, or even convulsions?

Did signs of deranged intellect precede the attack?

Has the patient sustained an injury of the skull?
Is there a discharge of the ears?

Was there a protracted suppuration in any part of the body?

## 8. Schlafsucht.

D. Ist der Patient auf den Kopf gefallen oder geschlagen worden?
Hat er sich geistig zu sehr angestrengt?
Hatte er vielleicht grossen Kummer, oder niederdrückende Sorgen?

Kam der Anfall plötzlich?
Hatte der Patient, ehe er in den comatösen Schlaf verfiel, mehr oder weniger Kopfschmerz, Brechneigung, Confusion der Gedanken, oder gar Convulsionen?

Sind dem Anfalle Anzeichen von Geistesstörung vorausgegangen?

Hat der Kranke eine Schädelverletzung erlitten?
Hat der Kranke einen Ausfluss aus den Ohren?

Existirt ein schon lange bestehender Eiterungsprozess?

## 9. Neuralgia.

Did the pain arise without external cause?
Do you know any cause for it?

How often does it usually take place?

## 9. Neuralgie (Nervenschmerz).

D. Ist der Schmerz ohne äussere Veranlassung aufgetreten?
Wissen Sie irgend eine Ursache für denselben?

Wie oft pflegt er stattzufinden?

| | |
|---|---|
| *Daily; every other day.* | P. Täglich; jeden andern Tag. |
| *Often the paroxysms are separated by intervals of longer or shorter duration.* | Oft sind die Anfälle durch Zwischenräume von kürzerer oder längerer Dauer von einander geschieden. |
| *Is the pain (aching) always limited to the same organ, or is it irregularly distributed over the whole body?* | D. Ist der Schmerz immer auf dasselbe Organ beschränkt, oder verbreitet er sich unregelmässig über den ganzen Körper? |
| *Is it associated with spasms?* | Ist er von Krämpfen begleitet? |
| *Can you describe what course the pain generally follows?* | Können Sie mir angeben, welchen Verlauf der Schmerz nimmt? |
| *Where have you the pain?* | Wo haben Sie den Schmerz? |
| *Here, in my face.* | P. Hier, im Gesicht. |
| *What brings it on?* | D. Was ruft ihn hervor? |
| *Often the most trivial cause; a current of air, a slight touch, a jar of my bed, a knock at the door.* | P. Oft der unbedeutendste Umstand, ein Luftzug, eine leise Berührung, das Knarren der Bettstelle, ein Klopfen an die Thür. |
| *Do you make use of too much coffee, tea, tobacco, and alcoholic drinks?* | D. Sind Sie vielleicht unmässig im Genusse von Kaffee, Thee, Taback, oder alkoholischen Getränken? |
| *You feel the pain in your shoulder, have you a disease of the liver; in your thigh, have you a disease of the kidneys; in your left arm, do you suffer from heart-disease?* | Sie verspüren den Schmerz in der Schulter, haben Sie ein Leberleiden? in der Hüfte, haben Sie ein Nierenleiden? im linken Arm, leiden Sie an einer Herzkrankheit? |
| *Do you suffer from debility, anæmia, gout, rheumatism?* | Leiden Sie an Schwäche, Blutmangel, Gicht, Rheumatismus? |
| *Are your digestive organs out of order?* | Sind Ihre Verdauungsorgane in Unordnung gerathen? |
| *You may remove the malady by applying magnetism, especially a constant galvanic stream.* | Sie werden das Übel dadurch heben, dass Sie Magnetismus, besonders einen constanten galvanischen Strom in Anwendung bringen. |

*Inhale chloroform, whenever the pain is very severe.*
*Moisten a piece of linen with chloroform and apply it to the painful part. Bind a piece of oiled silk around it, to prevent evaporation.*

Athmen Sie, wenn der Schmerz sehr heftig ist, Chloroform ein.
Befeuchten Sie ein Stück Leinwand mit Chloroform, und legen Sie es auf die schmerzende Stelle. Binden Sie ein Stück Wachstaffet darüber, um die Verdünstung zu verhindern.

## 10. *Hemicrania.*    10. Migräne (halbseitiger Kopfschmerz).

*Does the pain occupy one side or both sides of the head?*
*How long does the attack last?*
*Is the pain augmented by noise?*

*Is the pain accompanied by a disorder of sight?*
*Is there any vomiting or nausea, a sense of weight, numbness and tingling in the limbs?*

*Is the pain relieved by warmth?*

D. Nimmt der Schmerz eine oder beide Seiten des Kopfes ein?
Wie lange hält der Anfall an?
Wird der Schmerz durch Geräusch vermehrt?
Ist der Schmerz von Störung im Sehen begleitet?
Haben Sie Erbrechen oder Brechneigung, ein Gefühl von Schwere, von Taubsein und Kribbeln in den Gliedern?
Wird der Schmerz durch Anwendung von Wärme gehoben?

## 11. *Glossitis.*    11. Zungenentzündung.

*Where is the pain?*
*In my tongue; it is burning.*
*Is your mouth very dry?*
*Yes, I also have an intense thirst and constant salivation.*

*Your tongue is swollen very much. Where did the swelling commence?*
*In front.*

D. Wo ist der Schmerz?
P. In der Zunge; er ist brennend.
D. Ist Ihr Mund sehr trocken?
P. Ja, auch habe ich einen sehr grossen Durst und fortwährenden Speichelfluss.
D. Die Zunge ist sehr geschwollen. Wo hat die Schwellung begonnen?
P. An dem vorderen Theile.

| | |
|---|---|
| Your tongue is inflamed; you must have patience, until the inflammation subsides. | D. Ihre Zunge ist entzündet; Sie müssen Geduld haben, bis die Entzündung nachlässt. |
| Have you injured it or were you stung by a venomous insect? | Haben Sie dieselbe verletzt, oder sind Sie von einem giftigen Insect gestochen worden? |
| You have ulcers on your tongue; have you had syphilis? | Sie haben Geschwüre an der Zunge; sind Sie siphilitisch? |

## 12 a. Sore Throat. — 12 a. Halsweh.

| | |
|---|---|
| You are very hoarse, have you become so suddenly? Yes. | D. Sie sind sehr heiser. Sind Sie es plötzlich geworden? P. Ja. |
| Do you cough and expectorate? | D. Husten Sie, und werfen Sie dabei aus? |
| Neither the one nor the other. | P. Weder das Eine noch das Andere. |
| Do you feel pain or difficulty in breathing? No, Sir. | D. Empfinden Sie Schmerz oder Athembeschwerden? P. Nein. |
| It is of no consequence, and a result of over-exerting your voice. You have used your voice too much in singing or speaking. A short period of rest will remove your trouble. | D. Es ist von keiner Bedeutung, und eine Folge von Überanstrengung Ihrer Stimme. Sie haben Ihre Stimme zu stark beim Singen oder Sprechen gebraucht. Eine kurze Zeit der Ruhe wird das Übel heben. |

## 12 b. — 12 b.

| | |
|---|---|
| I am very hoarse and have a troublesome cough. | P. Ich bin sehr heiser und habe einen lästigen Husten. |
| Is your cough dry? | D. Ist der Husten trocken? |
| No, I also expectorate. | P. Nein, ich werfe auch aus? |
| Can you breathe without difficulty? | D. Können Sie ohne Beschwerde athmen? |
| No, Sir, I cannot breathe very freely. | P. Nein. Ich kann nicht ganz frei athmen. |

*You caught a severe cold and thereby an inflammation of the vocal cords.*

D. Sie haben sich eine schwere Erkältung und dadurch eine Entzündung der Stimmbänder zugezogen.

### 12 c.

*I have considerable difficulty in breathing.*

P. Ich habe bedeutende Beschwerde beim Athmen.

*You have, I find by examination, a growth on one of your vocal cords, which by all means should be removed.*

D. Sie haben, wie die Untersuchung zeigt, ein Gewächs (eine Geschwulst) auf einem der Stimmbänder. Dasselbe muss schlechterdings entfernt werden.

### 13. *Tonsilitis.*

### 13. Acute Entzündung der Mandeln.

*I have a very sore throat.*

P. Ich habe einen sehr schlimmen (wehen) Hals.

*Can you swallow without pain ?*

D. Können Sie ohne Schmerzen schlucken?

*No, the difficulty in swallowing is very severe.*

P. Nein, die Schluckbeschwerden sind sehr heftig.

*Do liquids return through your nose ?*

D. Laufen Ihnen genossene Flüssigkeiten wieder aus der Nase heraus?

*Is it painful to you to talk ?*

Macht Ihnen das Sprechen Schmerzen?

*Let me inspect your throat.*

Lassen Sie mich Ihnen in den Hals sehen.

*Your tonsils are swollen and prominent.*

Ihre Mandeln sind angeschwollen und stehen vor.

*As we cannot separate your jaws sufficiently, let me introduce my finger, to find out the condition of the affected parts.*

Da wir die Kiefer nicht weit genug auseinander bekommen, so lassen Sie mich den Finger einführen, um den Zustand der afficierten Theile zu erfahren.

| | |
|---|---|
| *You have an acute inflammation of the tonsils.* | D. Sie haben eine acute Entzündung der Mandeln. |
| *How long have you had the pain?* | Wie lange schon haben Sie den Schmerz? |
| *Have you much pain at the angles of the jaws, shooting to the ear?* | Haben Sie grossen Schmerz an den Kieferwinkeln, der bis zum Ohre hinaufschiesst? |
| *You have an unpleasant taste in your mouth, do you experience relief and an improvement in deglutition? Yes.* | Sie haben einen schlechten Geschmack im Munde, fühlen Sie seitdem Erleichterung, eine Verbesserung im Schlucken? P. Ja. |
| *I am glad of it. The abscess has burst.* | D. Das freut mich. Der Abscess (das Geschwür) ist geborsten. |

## 14. Diseases of the Mouth.

## 14. Krankheiten des Mundes.

| | |
|---|---|
| *You feel pain in the cavity of your mouth. Let me inspect it.* | D. Sie empfinden Schmerz in der Mundhöhle. Lassen Sie mich dieselbe inspiciren. |
| *Is any attempt at chewing painful?* | Ist jeder Versuch zum Kauen schmerzhaft? |
| *Is your taste impaired?* | Ist Ihr Geschmack beeinträchtigt? |
| *Does saliva flow from your mouth?* | Fliesst Ihnen Speichel aus dem Munde? |
| *I also see ulcerations at various parts. You have an inflammation of the cavity of the mouth.* | Ich sehe auch an verschiedenen Stellen Verschwärungen (or Geschwüre). Sie haben eine Mundhöhlenentzündung. |
| *Have you swallowed hot or corrosive substances?* | Haben Sie heisse oder ätzende Flüssigkeiten verschluckt? |
| *Do you suffer from gastric disorder?* | Leiden Sie am Magen? |
| *I have also pain in the jaws and my teeth are becoming loose.* | P. Ich habe auch Schmerz in den Kiefern, und die Zähne werden mir lose. |

*Have you inhaled vapors of mercury or used mercurial remedies?*

*The peculiar nauseous breath, the enlarged tongue and the copious discharge of saliva make me apprehend it.*

*Your uvula is swollen.*

*Have you coryza (cold in the head) and fever?*

*Is your sense of hearing also impaired?*

D. Haben Sie Quecksilberdämpfe eingeathmet oder innerlich Quecksilber gebraucht?

Der eigenthümliche Geruch Ihres Athems, die vergrösserte Zunge und der so reichliche Speichelfluss lassen mich das vermuthen.

Ihr Zäpfchen ist angeschwollen.

Haben Sie Schnupfen und Fieber?

- Ist Ihr Gehörsinn auch beeinträchtigt?

### 15. Diphtheria.

*How did the disease begin?*

*With a sore throat.*

*Is your neck somewhat stiff?*

*Does it hurt you as I press here (on the glands at the angles of the jaw)?*

*Have you much pain in the head?*

*Have you a sense of weakness and prostration?*

*We must counteract the depressing effect of the malady by nourishment and stimulants.*

*Let me examine the cavity of your mouth and see how far the swelling extends, whether the gums and larynx are free.*

*Have you paroxysms of cough and dyspnœa?*

### 15. Diphtheritis.

D. Wie hat die Krankheit begonnen?

P. Mit Halsschmerzen.

D. Ist Ihnen der Hals etwas steif?

Schmerzt es, wenn ich hier (auf die Unterkieferdrüsen) drücke?

Haben Sie starkes Kopfweh?

Haben Sie ein Gefühl von Schwäche und Erschöpfung?

Wir müssen durch Nahrung und Reizmittel der erschöpfenden Wirkung der Krankheit entgegen arbeiten.

Lassen Sie mich Ihre Mundhöhle untersuchen und sehen, wie weit die Schwellung reicht, ob Gaumen und Luftröhre frei sind.

Haben Sie Anfälle (Paroxysmen) von Husten und Athemnoth?

| | |
|---|---|
| *Have you any difficulty in deglutition (swallowing)?* | *D.* Haben Sie Schluckbeschwerden? |
| *Has the patient's voice changed? No, Sir.* | Ist die Stimme des Kranken verändert? Nein. |
| *Did difficulty in deglutition occur from the onset or later?* | Traten die Schluckbeschwerden gleich Anfangs oder erst später auf? |
| *Where were the first manifestations of the disease, in the throat or in the larynx?* | Wo waren die ersten Krankheitserscheinungen, im Rachen oder im Kehlkopf? |

## *16. Laryngitis.*   16. Entzündung des Kehlkopfes.

| | |
|---|---|
| *Where is the seat of your affliction?* | *D.* Wo ist der Sitz Ihres Leidens? |
| *Here (pointing to the larynx) in my throat.* | *P.* Hier, im Halse. |
| *What is your sensation there?* | *D.* Was für Gefühl haben (*or* verspüren) Sie da? |
| *A feeling of tickling and irritation.* | *P.* Ein Gefühl von Kitzeln und Reizung. |
| *Do you often cough?* | *D.* Husten Sie oft? |
| *Yes, the cough is but trifling, though annoying; it is rather a constant disposition to clear the throat.* | *P.* Ja, der Husten ist unbedeutend, aber quälend; er besteht mehr in der Neigung zum Räuspern. |
| *Have you any difficulty in swallowing?* | *D.* Haben Sie Beschwerden beim Schlucken? |
| *Yes, but not much.* | *P.* Ja, aber nicht viel. |
| *Please, have a spoon brought for me.* | *D.* Bitte, lassen Sie mir einen Löffel bringen. |
| *Please to open your mouth, a little wider, as wide as you can.* | Machen Sie gefälligst den Mund auf; bitte, etwas weiter! so weit Sie können. |
| *You have a slight attack of laryngitis.* | Sie haben einen leichten Anfall von einer Kehlkopfentzündung. |

*That will pass away soon.*

*The principal thing is, speak as little as possible, and be careful not to speak with effort, but very low. The affected organ requires the greatest rest.*

*And then try to have uniformly warm and pure air for breathing, and not irritating food and drink.*

### 17. *Stricture of the Œsophagus.*

*Have you any difficulty and pain in deglutition?*

*Are these associated with hiccough?*

*Have you a burning sensation between the shoulders, in the course of the tube?*

*Have you swallowed hot water or corrosive substances?*

*Yes, sir, I have accidentally swallowed nitric acid (sulphuric acid).*

*Have you an impediment in swallowing?*

*Cannot the patient swallow even liquid food without great difficulty?*

*Do you sometimes expectorate clots of blood without cough or vomiting?*

D. Das wird sich bald geben.

Die Hauptsache ist, sprechen Sie so wenig als nur möglich, und ja nicht mit Anstrengung, sondern ganz leise. Das afficierte Organ verlangt die grösste Ruhe.

Suchen Sie sodann gleichmässig warme und reine Luft zum Athmen, und reizlose Speisen und Getränke.

### 17. Verengerung der Speiseröhre.

D. Haben Sie Beschwerden oder Schmerzen beim Schlucken?

Sind dieselben von Rülpsen (Aufstossen) begleitet?

Haben Sie eine brennende Empfindung zwischen den Schultern, im Verlaufe der Speiseröhre?

Haben Sie heisses Wasser oder ätzende (corrodierende) Flüssigkeiten verschluckt?

P. Ja, ich habe zufällig Saltpetersäure (Schwefelsäure) verschluckt.

D. Haben Sie ein Hinderniss beim Schlucken?

Kann der Kranke selbst flüssige Nahrung nur mit grosser Schwierigkeit durchbringen?

Speien Sie zuweilen ohne Husten oder Erbrechen Stücke geronnenen Blutes aus?

| | |
|---|---|
| Please to open your mouth as wide as you can, and let me pass the instrument. | D. Öffnen Sie gefälligst, so weit Sie können, den Mund, und lassen Sie mich das Instrument einführen. |

### 18. Cough. — 18. Husten.

| | |
|---|---|
| Do you cough much? | D. Husten Sie viel? |
| Very much; constantly; by day and all night. | P. Sehr viel; beständig; am Tage und die Nacht hindurch. |
| How long have you had this cough? | D. Haben Sie diesen Husten schon lange? |
| I have been suffering from it for some time. | P. Ich leide schon längere Zeit daran. |
| Is your cough accompanied by expectoration? | D. Ist der Husten mit Auswurf verbunden? |
| No, it is a dry cough. | P. Nein, es ist ein trockner Husten. |
| Well, you must take care of yourself, and avoid everything that provokes cough, especially dust, smoke, and cold air. Your sleeping room must be moderately warmed by night, and ventilated (aired) by day. | D. Nun, Sie müssen sich wohl in Acht nehmen (4841), und Alles vermeiden, was zum Husten reizt, ganz besonders Staub, Rauch und kalte Luft. Ihr Schlafzimmer muss bei Nacht mässig erwärmt und am Tage gelüftet sein. |

### 19. Hæmoptysis. — 19. Bluthusten.

| | |
|---|---|
| What was the nature of the sputum? | D. Wie war der Auswurf beschaffen? |
| It was purulent mixed with blood. | P. Er war eitrig mit Blut gemischt. |
| Was the blood fluid or coagulated? | D. War das Blut flüssig oder in geronnenem Zustande? |
| Was it florid or dark? | War es hellroth oder dunkel? |
| Was it in a small or large quantity? | War es in geringer oder grösserer Menge? |
| Since when have you perceived blood in your expectoration? | Seit wann bemerken Sie Blut in Ihrem Auswurf? |

*How long have you been coughing up blood?*

D. Wie lange mag es her sein, dass Sie Blut aushusten?

*Was the bleeding preceded by pain in the chest, tickling and perception of warmth, palpitation, difficulty in breathing, etc.?*

Gingen der Blutung Brustschmerz, Kitzeln und Wärmegefühl, Herzklopfen, Athembeschwerden u. d. gl. voran?

*You need not be alarmed, but you must keep your bed for some time in a pure, cool (not cold) air, and in a sitting rather than recumbent posture.*

Sie brauchen sich nicht zu ängstigen, doch müssen Sie für einige Zeit im Bette bleiben, und zwar in einer kühlen (nicht kalten) Luft und in einer mehr sitzenden als liegenden Stellung.

*Avoid speaking as much as you possibly can, all over-excitements, emotions of mind, and especially all inhalations provoking cough.*

Vermeiden Sie, so viel Sie irgend können, das Sprechen, alle Erhitzungen, Gemüthsaufregungen, so wie ganz besonders alle Einathmungen, die zum Husten reizen.

*Lay off all clothing that tends to compress the chest.*

Legen Sie alle beengenden Kleidungsstücke ab.

*Then I recommend you cold acidulated drinks, as also mild, not hot food.*

Sodann empfehle ich Ihnen säuerliches, kaltes Getränk, so wie milde, nicht heisse Speisen.

*See that the bowels are kept open, and avoid everything that might cause palpitation of the heart.*

Sorgen Sie auch für Leibesöffnung, und vermeiden Sie Alles, was Herzklopfen veranlassen könnte.

## 20. Hæmatemesis.

## 20. Blutsturz.

*Was the vomiting of blood preceded by a feeling of weight and uneasiness in the epigastric region?*

D. Ging dem Bluterbrechen ein Gefühl der Schwere und Beklommenheit in der Magengegend voraus?

*Yes, I had decided nausea.*

P. Ja, ich hatte ein starkes Brechgefühl.

*How did the ejected matter look?*

D. Wie hat das Erbrochene ausgesehen?

*It was a dark-colored blood mixed with half-digested food.*

P. Es war dunkelgefärbtes, mit halbverdauter Nahrung vermischtes Blut.

*Did the vomiting take place once or several times?*

D. Fand das Erbrechen einmal oder mehrere Male statt?

*The vomited blood was florid and it spurted out in jets.*

P. Das erbrochene Blut war hellroth, und es sprang in Stössen heraus.

*What taste did you perceive in your mouth?*

D. Was für Geschmack verspürten Sie im Munde?

*A saltish taste.*

P. Einen salzigen Geschmack.

*Was your mouth filled with blood without any effort, or did the patient expectorate a quantity of light-red and frothy blood after a slight cough?*

D. Hat der Kranke ohne welche Anstrengung oder nach unbedeutendem Husten eine Quantität hellrothen und schaumigen Blutes ausgeworfen?

### 21. Bronchitis.

### 21. Bronchialkatarrh.

*Where do you feel pain?*

D Wo verspüren Sie Schmerz?

*I have soreness behind the breast-bone, and pain in my limbs.*

P. Ich habe ein Wehgefühl hinter dem Brustbein und auch Schmerz in den Gliedern.

*Can you breathe freely?*

D. Können Sie frei athmen?

*No, I feel a slight oppression in breathing and a tickling in the throat.*

P. Nein, ich fühle eine leichte Bedrückung beim Athmen und ein Kitzeln im Halse.

*What is the nature of the cough?*

D. Wie tritt der Husten auf?

*Paroxysmal.*

P. In Hustenanfällen (Paroxysmen).

*Does the cough become loose easily or with difficulty?*

D. Löst sich der Husten leicht oder schwer?

*With difficulty, and the sputum is clear and frothy.*

P. Schwer, und der Auswurf ist klar und schaumig.

*Now it is looser than at the beginning, and less fatiguing. The sputum looks yellowish or greenish, sometimes streaked with blood.*

Er ist jetzt loser als anfangs und weniger anstrengend. Der Auswurf sieht jetzt gelblichgrün, zuweilen auch blutgestreift aus.

| | |
|---|---|
| *Do you eject small or large quantities?* | D. Werfen Sie kleine oder grössere Quantitäten aus? |
| *Is the cough persistent?* | Ist der Husten andauernd? |
| *No, sometimes it disappears entirely, and then it reappears with more than its previous severity.* | P. Nein, zuweilen verschwindet er gänzlich und kehrt mit um so grösserer Heftigkeit wieder. |
| *Be particular about pure, uniformly warm air by day and by night, in summer and winter. Therefore avoid dusty air, and whenever you happen to be exposed to it, I advise you to close your mouth and nose by a respirator.* | D. Halten Sie stets auf reine, gleichmässig warme Luft bei Tag und bei Nacht, im Sommer und im Winter. Vermeiden Sie daher staubige Luft, und sollten Sie solcher ausgesetzt sein, so rathe ich Ihnen, Mund und Nase durch einen Respirator zu verschliessen. |
| *Also avoid going out into the cold from a warm place.* | Meiden Sie auch aus der Wärme in die Kälte zu gehen. |
| *Let your fare be mild, consisting principally of milk.* | Ihre Kost sei mild, und bestehe besonders in Milch. |
| *Avoid continuous speaking.* | Vermeiden Sie anhaltendes Sprechen. |

## 22. *Asthma*.     22. Brustkrampf.

| | |
|---|---|
| *What do you complain of particularly?* | D. Worüber klagen Sie besonders? |
| *Of great distress in breathing.* | P. Über grosse Athemnoth. |
| *How does it occur?* | D. Wie tritt diese auf? |
| *In paroxysms, and is attended with wheezing.* | P. In Krampfanfällen und ist von einem pfeifenden Tone begleitet. |
| *Are these attacks preceded by any feeling?* | D. Geht diesen Anfällen irgend ein Gefühl voraus? |
| *Yes, I sometimes have a feeling of suffocation immediately before.* | P. Ja, ich habe manchmal unmittelbar vor denselben ein Gefühl des Erstickens. |

| | |
|---|---|
| Not always, they often occur suddenly. I wake up out of my sleep and hear the wheezing sound. | P. Nicht immer, oft treten sie plötzlich auf. Ich erwache aus dem Schlafe und höre den pfeifenden Ton beim Athmen. |
| How long does the attack generally last? | D. Wie lange dauert gewöhnlich der Anfall? |
| For the most part it passes off with copious expectoration after a few hours. But sometimes it lasts for days; by day it is better and by night worse. | P. Er geht meist nach einigen Stunden unter reichlichem Auswurf vorüber. Oft währt er aber auch Tage lang. Am Tage geht's dann besser, und während der Nacht schlimmer. |
| Do you know any cause for these spasms? | D. Wissen Sie irgend eine Ursache für diese Anfälle? |
| Have you at any time inhaled irritating fumes or disagreeable vapors? | Haben Sie zu irgend einer Zeit reizende Dämpfe oder unangenehm riechende Stoffe eingeathmet? |
| Did digestive troubles, bronchitis, or a cold in the head precede the paroxysm? | Sind dem Anfalle Verdauungsbeschwerden, Bronchitis oder Schnupfen vorangegangen? |
| Did you for some time prior to the asthmatic paroxysm pass a dark-colored, heavy urine? | Haben Sie längere oder kürzere Zeit vor dem Anfalle einen dunkelgefärbten, schweren Urin entleert? |
| Are you intemperate in your habits? | Leben Sie zu flott? |
| Have you any heart trouble? | Haben Sie irgend ein Herzleiden? |
| Let the patient take this medicine, a teaspoonful every two hours; and if an attack occurs, pour a few drops from this vial upon a pocket handkerchief or a napkin and let him inhale the vapor of it. | Lassen Sie den Patienten von dieser Medicin alle zwei Stunden einen Theelöffel voll nehmen. Bei eintretendem Hustenanfalle giessen Sie einige Tropfen aus diesem Fläschchen auf ein Taschentuch oder eine Serviette und lassen ihn die Dünste einhauchen. |

## 23. Disease of the Chest.

*How did the affection set in?*

With a cough.

*What was the nature of the cough?*

It was dry at the beginning, then it was followed by mucilaginous rice-water sputa, which now are thick and purulent.

*Do you perspire freely?*

Yes, especially at night, and these night-sweats weaken me very much.

*Have you lost in flesh?*

A great deal; I have fallen away very much since the beginning of the disease, and I feel an increasing weakness.

*Have you been obliged to work much in dusty, bad air?*

*Have you indulged in sexual excesses?*

You have a disease of the chest, and must take good care, lest it become incurable. Your condition requires bodily, sexual, and mental rest, nourishing (particularly animal food, with the proper amount of water, fat, and salt. The air to be inhaled must be **pure** (free from dust, smoke, tobacco fumes, injurious gases) and **warm**, and this by night as well as by day. Let your dwelling, particularly the sleep-

## 23. Brustkrankheit.

D. Wie hat die Krankheit angefangen?

P. Mit Husten.

D. Wie war der Husten?

P. Er war anfangs trocken, dann folgte ihm ein schleimiger reiswasserähnlicher Auswurf, der jetzt dick und eitrig ist.

D. Schwitzen Sie viel?

P. Ja, besonders des Nachts, und diese Nachtschweisse schwächen mich sehr.

D. Haben Sie an Gewicht verloren?

P. Sehr, ich bin seit dem Beginn der Krankheit sehr abgemagert und fühle eine zunehmende Schwäche.

D. Haben Sie viel in staubiger, schlechter Luft arbeiten müssen?

Haben Sie ein ausschweifendes Leben geführt?

Sie sind brustkrank, und müssen sich wohl in Acht nehmen, dass die Krankheit nicht unheilbar werde. Ihr Zustand verlangt körperliche, geschlechtliche und geistige Ruhe, nahrhafte (besonders thierische) Kost, mit der gehörigen Menge von Wasser, Fett und Salz. Die einzuathmende Luft muss **rein** (frei von Staub, Rauch, Tabakqualm, schädlichen Gasen) und **warm** sein, und dies ebenso

ing room, be dry, sunny, and well ventilated. It is also commendable to sojourn in an open but warm and pure woodland air. During the cold and stormy season you ought to remain within the uniform temperature of your room, or on going out always make use of a respirator. Also avoid everything that compresses the lungs, especially continuous sitting with the body bent forward. You must also dispense with all beverages that produce palpitation of the heart and heat.

wohl bei Nacht wie bei Tag. Die Wohnung, besonders das Schlafzimmer, sei trocken, sonnig und wohlgelüftet; besonders empfehlenswerth ist der Aufenthalt in freier, aber warmer, reiner Waldluft. Während der kalten und stürmischen Jahreszeit müssten Sie in der gleichförmigen Temperatur des Zimmers verbleiben, oder beim Ausgehen sich stets des Respirators bedienen. Vermeiden Sie auch Alles, was die Lunge beengt, besonders anhaltendes Sitzen mit vorgebeugtem Oberkörper. Sie müssen sich auch aller Getränke entschlagen, die Herzklopfen und Hitze erzeugen.

## 24. Phthisis.

Have you spit blood?
Have you perceived blood in your expectoration?
Did a hemorrhage occur?

Had you a hemorrhage?
Does coughing hurt you?

Where do you feel the pain?
Is the pain acute, is it stationary, or do you feel it in different places?

Has any member of your family (your father, mother, brother, etc.) suffered from weak lungs?

## 24. Schwindsucht.

D. Haben Sie Blut gespieen?
Haben Sie in Ihrem Auswurf Blut bemerkt?
Hat ein Blutsturz stattgefunden?
Hatten Sie einen Blutsturz?
Fühlen Sie Schmerz beim Husten?
Wo fühlen Sie den Schmerz?
Ist der Schmerz heftig? haftet er stets an derselben Stelle, oder fühlen Sie ihn an verschiedenen Stellen?
War Jemand in Ihrer Familie (Ihr Vater, Ihre Mutter, Ihr Bruder, etc.) lungenkrank?

| | |
|---|---|
| *Had any member of your family cough, spitting of blood?* | D. Hat Jemand in Ihrer Familie an Husten, Blutspeien gelitten? |
| *No, they had good health.* | P. Nein, sie waren ganz gesund. |
| *Yes, my father (brother) died of lung disease (of breast complaint).* | Ja, mein Vater (Bruder) ist an einem Lungenleiden (an der Brustkrankheit) gestorben. |
| *Did you enjoy good health in every other respect?* | D. Waren Sie in jeder Beziehung sonst gesund? |
| *Do you catch cold easily?* | Erkälten Sie sich leicht? |
| *Do you get out of breath easily?* | Kommen Sie leicht ausser Athem? |
| *Do evening exacerbations take place in your malady?* | Finden abendliche Verschlimmerungen in Ihrem Leiden statt? |
| *Please to loosen the clothing over your breast.* | Bitte, entblössen Sie Ihre Brust. |
| *Take a deep, long breath!* | Holen Sie tief und langsam Athem. |
| *Do you feel pain in doing so?* | Fühlen Sie Schmerz dabei? |
| *It hurts me to take a long breath.* | P. Ich kann vor Schmerz nicht tief athmen. |
| *Can you cough without pain?* | D. Können Sie ohne Schmerz husten? |
| *No, Sir, it hurts me even to move at all.* | P. Nein, jede Bewegung schmerzt mich. |
| *Be so good as to fold your arms across the chest and bend forward.* | D. Seien Sie so gut, kreuzen Sie die Arme über die Brust, und beugen Sie sich vorwärts. |
| *Place your arms over your head, thus!* | Bringen Sie die Arme über den Kopf; so! |
| *Sit up!* | Setzen Sie sich auf! |
| *Stand up!* | Stehen Sie auf! |
| *Lean forward, backward a little!* | Beugen Sie sich ein wenig nach vorn (*or* vorwärts). Beugen Sie sich nach hinten (*or* rückwärts). |
| *Please to lie still.* | Bitte, liegen Sie still! |

| | |
|---|---|
| *How is the sputum? How does it look?* | D. Wie ist der Auswurf? Wie sieht er aus? |
| *Is it fetid?* | Ist derselbe sehr übelriechend? |
| *Keep it for me; I must examine it.* | Heben Sie mir denselben auf; ich muss ihn untersuchen. |

## 25. Pneumonia.    25. Lungenentzündung.

| | |
|---|---|
| *How did the sickness come on?* | D. Wie hat sich die Krankheit anfänglich (*or* im Anfange) geäussert? |
| *It began with a severe chill, followed by a fever.* | P. Mit einem Schüttelfrost, worauf Fieber folgte. |
| *Where is the pain?* | D. Wo ist der Schmerz? |
| *How is it? More oppressive and annoying than pungent.* | D. Wie ist er? P. Mehr drückend und beklemmend, als stechend. |
| *How is your breathing?* | D. Wie ist das Athmen? |
| *Short and restricted.* | P. Kurz und beklommen. |
| *Try to draw a deeper breath.* | D. Versuchen Sie tiefer einzuathmen. |
| *It is very painful for me to breathe deeply.* | P. Tiefes Einathmen ist für mich sehr schmerzhaft. |
| *How is the patient's cough?* | D. Wie ist der Husten des Patienten? |
| *Troublesome and painful, with dark-colored blood expectoration. At the beginning the cough was dry with but little expectoration.* | P. Lästig und schmerzhaft mit blutigem Auswurf. Anfangs war der Husten trocken, mit nur wenig Auswurf. |
| *How is his urine? Scanty.* | D. Wie ist der Harn? P. Spärlich. |
| *Has the patient been delirious?* | D. Hat der Patient phantasiert? |
| *Since when has his speaking been more laborious?* | Seit wann ist sein Sprechen erschwert? |
| *Has the patient been healthy and vigorous hitherto?* | War der Patient bisher gesund und kräftig? |
| *Has the patient been sickly?* | War der Patient kränklich? |
| *Was his constitution shattered by excesses?* | War seine Constitution durch Excesse erschüttert? |

| | |
|---|---|
| Has the patient been unconscious? | D. War der Patient bewusstlos? |
| Has he temporarily lost the use of his limbs? | Hat er auf einige Zeit den Gebrauch seiner Glieder verloren? |
| The patient must be bled. | Es muss dem Patienten zur Ader gelassen werden. |
| Absolute rest in the horizontal position is necessary for the patient. | Es ist dem Patienten absolute Ruhe in horizontaler Lage nothwendig. |
| The patient must remain in bed in a well ventilated room, the temperature of which is kept as near 60° Fahrenheit as possible. | Der Patient muss im Bette bleiben in einem gut gelüfteten Zimmer, dessen Temperatur möglichst nahe 60° Fahrenheit bleibt. |
| Give a tablespoonful (teaspoon) of this medicine every two hours till the blood disappears from the sputa. | Geben Sie ihm (ihr) alle zwei Stunden einen Esslöffel (Kaffeelöffel, Theelöffel) voll von dieser Medicin, bis das Blut aus dem Auswurf verschwindet. |
| Give him (her) small quantities of nutritious fluid food in the form of milk, milk and eggs, oatmeal gruel or meat-broth, frequently repeated. | Geben Sie ihm (ihr) öfters, aber immer nur kleine Quantitäten nahrhafter, flüssiger Speise in der Form von Milch, Milch und Eier, Hafergrütze, oder auch Fleischbrühe. |
| Give him (her) milk punch or brandy and beef-tea in small quantities and frequently. | Geben Sie ihm (ihr) Milchpunsch oder Branntwein und Beefthee in kleinen Quantitäten und öfters. |
| Relieve his thirst by acidulated, or alkaline drinks. | Stillen (löschen) Sie seinen (ihren) Durst durch säuerliche oder alkalische Getränke: |
| Make warm linseed-meal poultices, sufficiently large to thoroughly inclose the affected side, adding a small quantity of mustard, and change them as often as they become cool. | Machen Sie ihm (ihr) warme Leinsamenumschläge, hinlänglich gross, um die leidende Seite ganz einzuschliessen, und wechseln Sie dieselben, sobald sie kalt werden. Sie können auch etwas Senf hinzufügen. |

| | |
|---|---|
| When his fever is very high, put cold water compresses on his breast (especially the affected side). Dip cloths in cold water, wring them out well. Repeat them every five minutes, until relief is obtained. | D. Wenn das Fieber hoch ist, machen Sie ihm (ihr) Kaltwasserumschläge auf die Brust (besonders die leidende Stelle). Tauchen Sie ein Tuch in kaltes Wasser, und winden Sie es gut aus. Alle fünf Minuten müssen die Umschläge erneuert werden, bis Erleichterung kommt. |

## 26. Peritonitis.    26. Bauchfellentzündung.

| | |
|---|---|
| Where do you feel the pain? | D. Wo fühlen Sie den Schmerz? |
| It is worst here. | P. Er ist am schlimmsten hier. |
| Is it bearable? Yes. | D. Ist er erträglich? P. Ja. |
| No, Sir; it is almost unbearable, even the softest touch hurts. | P. Nein, er ist fast unerträglich; die leiseste Berührung thut weh. |
| Since when has your belly been so distended? | D. Seit wann ist der Leib so aufgetrieben? |
| Does the patient ordinarily lie on the affected side? | Liegt der Patient gewöhnlich auf der leidenden Seite? |
| Do you know of any cause for your suffering? | Können Sie irgend eine Ursache für Ihr Leiden angeben? |
| Have you sustained a fracture of a rib? | Haben Sie einen Rippenbruch erfahren? |
| Have you been injured by a stab or a shot-wound in this region? | Sind Sie durch eine Stich- oder Schusswunde in dieser Gegend verletzt worden? |
| The patient needs absolute rest. | Der Patient bedarf absoluter Ruhe. |
| Put warm poultices on the painful place. | Legen Sie ihm warme Breiumschläge auf die schmerzende Stelle. |
| Try to obtain an evacuation by clysters. | Suchen Sie durch Klystiere eine Ausleerung zu erzielen. |
| The food must be mild, easily digestible. | Die Kost sei mild, leicht verdaulich und so wenig reizbar als möglich. |

| | |
|---|---|
| Is the stomach of the patient able to retain food, or does he vomit immediately? | D. Kann der Magen des Patienten Speise behalten, oder bricht er dieselbe sofort wieder aus? |
| Give him small quantities of meat-broth and wine every two hours. | Geben Sie ihm (ihr) alle zwei Stunden kleine Quantitäten von Fleischbrühe und Wein. |
| Should the stomach be unable to retain food, inject a third of a pint of strong meat soup slowly through the rectum. When relief has been obtained in consequence thereof, repeat the same injection after three hours. | Sollte der Magen nichts bei sich behalten können, so spritzen Sie ihm (ihr) durch den After eine drittel Pinte ($\frac{1}{6}$ Quart) von starker Fleischbrühe langsam ein. Wenn der Schmerz in Folge davon nachgelassen, so wiederhohlen Sie nach drei Stunden dieselbe Einspritzung. |
| If the abdomen can bear pressure, a flannel roller should be firmly applied around the body. | Wenn der Unterleib Druck ertragen kann, so legen Sie ihm (ihr) ein flanellenes Wickelband fest um den Leib. |
| Soak a piece of coarse flannel four to eight double, in warm (but not hot) water; wring it well, and apply it so as to cover the whole abdomen. Lay over it a cover of oiled silk or india rubber stuff. This compress should fit as closely as possible and not be displaced, lest air enter between the skin and the compress. Change the compress two or three times a day. It may remain untouched as long as it is moist and warm. | Tauchen Sie ein grobes, vier- bis achtfach gefaltetes Stück Flanell in warmes (nicht heisses) Wasser ein; winden Sie dasselbe gut aus, und legen Sie es auf den ganzen Unterleib (Bauch). Legen Sie darüber einen Umschlag von Wachstaffet oder Kautschuckstoff. Diese Compresse muss möglichst genau anliegen und nicht verrückt werden, damit keine Luft zwischen dieselbe und die Haut eindringe. Wechseln Sie die Compresse zwei oder dreimal des Tages. So lange sie warm und feucht ist, lassen Sie sie unberührt liegen. |

## 27. Diseases of the Heart.

| | |
|---|---|
| Have you pain in the region of the heart? | D. Haben Sie Schmerz in der Herzgegend? |
| Have you severe headaches, vertigo, or apoplectic attacks? | Haben Sie heftigen Kopfschmerz, Schwindel und apoplectische Anfälle? |
| Have you uneasy dreams? | Haben Sie unruhige Träume? |
| Do you often suddenly awake from your sleep? | Erwachen Sie plötzlich aus dem Schlafe? |
| Has a hemorrhage occurred? | Ist ein Blutsturz vorgekommen? |
| Where did the dropsical swelling begin? | Wo hat die Wasseransammlung begonnen? |
| About the ankles and the feet. | P. Um die Fussgelenke herum und an den Füssen. |
| Are the feet more puffy in the evening than in the morning? | D. Sind die Füsse mehr des Abends geschwollen als des Morgens? |
| Had a persistent pain in the heart, or in the left side of the neck and arm preceded? | War ein andauernder Schmerz am Herzen, oder an der linken Seite des Halses und Armes vorangegangen? |
| Do you suffer from palpitation of the heart? | Leiden Sie an Herzklopfen? |
| Does your heart often pulsate with increased quickness and violence? | Schlägt das Herz oft mit erhöhter Schnelligkeit und Heftigkeit? |
| Does it beat with regularity, or is the beat irregular, sometimes slow, at times fast, or occasionally intermitting? | Schlägt das Herz regelmässig oder schlägt es bald langsam, bald schnell oder setzt es gar zuweilen aus? |
| Have you noticed that the palpitation returned after breakfast, or whenever hot tea or hot coffee had been taken? | Bemerkten Sie, dass das Herzklopfen nach dem Frühstück, oder nachdem Sie heissen Thee oder heissen Kaffee getrunken, wiederkehrte? |

| | |
|---|---|
| In this case you must avoid all hot drinks, and drink cold water at breakfast and at each meal. | D. Sie müssen in dem Falle alle heissen Getränke meiden, und kaltes Wasser beim Frühstück und jedem Mahle trinken. |
| Do you smoke inordinately? | Rauchen Sie unmässig? |
| Do you study too much, and take insufficient rest and exercise? | Studieren Sie zu viel und gewähren sich nicht genügende Ruhe und körperliche Bewegung? |
| Are you subject to hysteria? | Leiden Sie an Hysterie? |
| Are you regular in your monthlies? | Ist Ihr Monathliches in Ordnung? |
| Are you gouty, rheumatic? | Leiden Sie an der Gicht, an Rheumatismus? |
| Take the medicine every quarter or every half hour, according to the severity of the attack. | Nehmen Sie die Medicin jede Viertel- oder Halbestunde, je nach der Heftigkeit des Anfalls. |
| During the paroxysm, lie flat on your back, and loosen the clothing about the neck and chest. | Während des Anfalls legen Sie sich flach auf den Rücken, und machen sich die Kleider um Hals und Brust auf. |

## 28. Diseases of the Stomach.

## 28. Magenkrankheiten.

| | |
|---|---|
| Have you an absolute aversion to taking any kind of food? | D. Haben Sie einen absoluten Widerwillen gegen jede Nahrung? |
| Are you not able to partake of certain articles? | Können Sie gewisse Nahrungsmittel nicht vertragen? |
| Have you nausea, vomiting, acidity, flatulency, pain? | Haben Sie Übelkeit, Erbrechen, Säure, Blähungen, Schmerz? |
| Is your digestion disturbed? | Ist Ihre Verdauung gestört? |
| Your tongue is coated. | Ihre Zunge ist stark belegt. |
| How is your appetite? | Wie ist Ihr Appetit? |
| Ravenous; I have a ravenous appetite. | P. Äusserst gierig; ich habe einen Heisshunger. |

| | |
|---|---|
| *Have you a sensation of burning in the pit of the stomach accompanied with eructation of acid substances?* | D. Haben Sie ein Gefühl von Brennen in der Magengegend, mit Aufstossen saurer Massen? |
| *Make use of a little magnesia or bi-carbonate of soda.* | Gebrauchen Sie Magnesia oder doppelt-kohlensaures Natrum. |
| *Have you a sensation of constriction in the breast, short breathing and palpitation of the heart?* | Haben Sie ein Gefühl der Beklemmung in der Brust, Kurzathmigkeit und Herzklopfen? |
| *Is your sleep broken by uneasy dreams?* | Wird Ihr Schlaf durch unruhige Träume gestört? |
| *Have you belching which has the taste of rotten eggs?* | Haben Sie Aufstossen, das den Geschmack von faulen Eiern hat? |

## 29. *Nausea, Vomiting.* — 29. Brechneigung, Erbrechen.

| | |
|---|---|
| *Were you previously in good health?* | D. Waren Sie sonst gesund? |
| *Is everything that you swallow immediately rejected?* | Wird was Sie verschlucken, sofort wieder ausgebrochen? |
| *At what period does the vomiting happen, before meals or after meals, and how long afterward?* | Um welche Zeit tritt das Erbrechen auf, vor oder nach den Mahlzeiten, und wie lange danach? |
| *What is ejected?* | Was wird ausgebrochen? |
| *Food. Liquid mixed with saliva and some mucus. Half digested food.* | P. Nahrung. Flüssigkeit mit Speichel und etwas Schleim vermischt. Halb-verdaute Nahrung. |
| *Is there much retching, or is the act of vomiting often repeated, and is the vomited material stained with bile?* | D. Findet starkes Würgen statt, oder wird der Brechakt oft wiederholt, und ist das Ausgebrochene mit Galle gefärbt? |
| *Have you noticed pus or blood in the ejected matter?* | Haben Sie Eiter oder Blut in der erbrochenen Masse wahrgenommen? |

Has the ejected matter the appearance of coffee ground ?

D. Hat das Erbrochene das Aussehen von Kaffeesatz?

### 30. Stomach-ache.

Is your pain slight or violent?

Does your pain consist more of a feeling of soreness than actual pain?

Is your pain increased by pressure or not?

Is your pain augmented or relieved by the taking of food?

Is your pain more severe soon after meals, or when the stomach is full?

Is it more severe after a heavy meal of animal food, than after a light one of farinaceous substances and milk?

Is the pain brought on by certain articles of food which your stomach does not tolerate or is unable to digest?

Was the attack brought on by exposure to cold and damp?

Have you been in the habit of drinking cold water when heated?

Had you a sudden and violent emotion?

### 30. Magenschmerz.

D. Ist der Schmerz gering, oder heftig?

Besteht Ihr Schmerz mehr in einem Gefühl von Wundsein, als von wirklichem Schmerze?

Wird Ihr Schmerz durch Druck vermehrt oder nicht?

Wird Ihr Schmerz durch Zunahme von Nahrung verschlimmert oder gelindert?

Ist Ihr Schmerz heftiger bald nach den Mahlzeiten, oder wenn der Magen voll ist?

Ist er heftiger nach einer schweren Mahlzeit animalischer Kost, als nach einer leichten aus Mehlspeisen und Milch bestehenden?

Wird der Schmerz bei Ihnen von gewissen Nahrungsmitteln hervorgerufen, die Ihr Magen nicht verträgt, oder nicht zu verdauen im Stande ist?

Ist der Anfall durch Erkältung und Durchnässung herbeigeführt?

Pflegten Sie kaltes Wasser bei erhitztem Körper zu trinken?

Haben Sie eine plötzliche und heftige Gemüthsaufregung erfahren?

| | |
|---|---|
| *Have you been exposed to exhausting influences?* | D. Waren Sie erschöpfenden Einflüssen ausgesetzt? |
| *How is the pain?* | Wie ist der Schmerz? |
| *It is violent and agonizing, but not constant.* | P. Er ist heftig und qualvoll, aber nicht andauernd. |
| *Is the pain relieved by the recumbent position and by external pressure?* | D. Wird der Schmerz durch die horizontale Lage und durch äussern Druck gemildert? |
| *Do you complain of neuralgic pains in other parts of the body?* | Klagen Sie über Neuralgie an andern Körpertheilen? |
| *Have you a constant craving for food?* | Besteht oft ein Heisshunger für Nahrung bei Ihnen? |
| *How do you feel in the morning, when your stomach is empty?* | Wie befinden Sie sich des Morgens in nüchternem Zustande? |
| *Is the pain eased by a hearty breakfast?* | Wird der Schmerz nach Einnahme eines reichlichen Frühstücks gelindert? |

## *31. Acute Gastritis.* — 31. Akute Magenentzündung.

| | |
|---|---|
| *Have you swallowed any irritating or corroding substance?* | D. Haben Sie irgend eine reizende oder ätzende Substanz verschluckt? |
| *Did you ever before suffer from indigestion?* | Haben Sie früher je an Unverdaulichkeit gelitten? |
| *How long prior to the appearance of the pain had your digestion been disordered?* | Wie lange vor dem Erscheinen des Schmerzes war Ihre Verdauung gestört? |
| *Which preceded, the pain, or the indigestion?* | Was ging voran, der Schmerz oder die Verdauungsstörung? |
| *Is your pain increased by swallowing liquids as well as solids?* | Wird Ihr Schmerz beim Schlucken von flüssiger Nahrung sowohl als von fester vermehrt? |
| *Does it hurt when I press you here?* | Thut's Ihnen weh, wenn ich hier drücke? |

| | |
|---|---|
| *When do you feel nausea?* | D. Wann fühlen Sie Brechneigung? |
| *Constantly.* | P. Fortwährend. |
| *Do you vomit frequently?* | D. Brechen Sie oft? |
| *What do you vomit?* | Was brechen Sie aus? |
| *Almost all that I swallow, and large quantities of a greenish fluid.* | P. Fast Alles, was ich verschlucke, und grosse Quantitäten einer grünen Flüssigkeit. |
| *Are your bowels free?* | D. Haben Sie offnen Leib? |
| *No, Sir, I suffer from costiveness.* | P. Nein, ich leide sehr an Verstopfung. |
| *Do you feel any distress in your head? Yes.* | D. Haben Sie Beschwerden im Kopfe? P. Ja. |
| *Do you feel thirst? Yes, unquenchable.* | D. Empfinden Sie Durst? P. Ja, unlöschbaren. |

### 32. Chronic Gastritis.

### 32. Chronische Magenentzündung.

| | |
|---|---|
| *Have you a sensation of discomfort, of pressure, and of soreness at the pit of the stomach, aggravated by food?* | D. Haben Sie ein Gefühl von Missbehagen, von Druck und Schmerzen in der Magengrube, die durch Nahrungsaufnahme verschlimmert werden? |
| *Do you feel a burning there, and an inward heat, even when the stomach is empty?* | Empfinden Sie ein Brennen daselbst und innerliche Hitze sogar bei leerem Magen? |
| *Is your appetite capricious or impaired?* | Ist Ihr Appetit launisch oder verringert? |
| *Do you complain of fermentation, heart-burn, or flatulency?* | Empfinden Sie Gährung, Sodbrennen, Flatulenz? |
| *Are your bowels constipated?* | Ist Ihr Stuhlgang verstopft? |
| *Do you suffer from chilliness?* | Empfinden Sie Frösteln? |
| *How are your spirits?* | Wie ist Ihre Stimmung? |
| *Depressed.* | P. Niedergedrückt. |
| *Have you thirst?* | D. Haben Sie Durst? |
| *Yes, an annoying thirst torments me, and vomiting after meals.* | P. Ja, ein quälender Durst plagt mich und Erbrechen, welches nach dem Essen auftritt. |

| | |
|---|---|
| *Do you also vomit when your stomach is empty?* | D. Brechen Sie auch bei leerem Magen? |
| *Yes, the ejected matter is fluid, clear and without smell.* | P. Ja, die erbrochene Masse ist flüssig, hell und geruchlos. |
| *Is the ejected matter colorless?* | D. Ist die Masse, die Sie erbrechen, farblos? |
| *Do you live on coarse (badly prepared) food?* | Geniessen Sie schlecht zubereitete Speisen? |
| *Do you frequently drink to excèss?* | Sind Sie dem Trunke ergeben? |
| *Do you eat greedily?* | Essen Sie übermässig? |
| *Take three of the pills I prescribed fasting, and three more before bedtime.* | Nehmen Sie drei von den Pillen, die ich verschrieben, nüchtern, und drei weitere vor dem Schlafengehen. |
| *Put hot water fomentations continually over the region of the stomach, as hot as you can bear them.* | Legen Sie beständig heisse Wasserumschläge über die Magengegend, so heiss Sie dieselben vertragen können. |
| *Drink iced water, but only in small quantities, or take small pieces of ice in your mouth* | Trinken Sie Eiswasser, aber nur in kleinen Quantitäten, oder nehmen Sie kleine Eisstücke in den Mund. |

### 33. *Gastric Ulcer.*    33. Magengeschwür.

| | |
|---|---|
| *What is the character of the pain?* | D. Wie ist der Schmerz? |
| *Generally, a continuous dull feeling; sometimes it is burning or gnawing.* | P. Gewöhnlich andauernd dumpf, andere Male wieder brennend oder nagend. |
| *Where is its situation?* | D. Wo ist der Sitz des Schmerzes? |
| *Commonly in the middle of the epigastric region?* | P. Gewöhnlich in der Mitte der Magengegend. |
| *Is the same place very sensitive to the touch?* | D. Wird an derselben Stelle der Schmerz durch Druck erhöht? |
| *In addition to this continued feeling, do paroxysms of more violent pains occur?* | Treten neben diesem beständigen Schmerzgefühle auch noch Anfälle von heftigern Schmerzen auf? |

| | |
|---|---|
| *How long do these last?* | *D.* Wie lange dauern diese? |
| *For several hours; sometimes with trifling intermissions, for days.* | *P.* Mehrere Stunden; manchmal mit geringen Unterbrechungen mehrere Tage. |
| *By what are these paroxysms induced?* | *D.* Wodurch werden diese Paroxysmen hervorgerufen? |
| *By pressure or by food; sometimes they come on suddenly, even when the stomach is empty.* | *P.* Durch Druck oder Nahrung; manchmal kommen sie auch plötzlich, selbst bei leerem Magen. |
| *Do the symptoms remit from time to time?* | *D.* Lassen die Symptome von Zeit zu Zeit nach? |
| *Yes, there are sometimes long intervals during which all pain ceases, and I can take food without the least inconvenience.* | *P.* Ja, zuweilen finden lange Intervalle statt, während welcher alle Schmerzen aufhören, und ich ohne die geringste Beschwerde Nahrung nehmen kann. |
| *Have you suffered in this way very long?* | *D.* Leiden Sie schon lange so? |
| *Is the vomiting excessive?* | Brechen Sie viel? |
| *When is the pain regularly rendered more acute?* | Wann wird der Schmerz in der Regel heftiger? |
| *Within about a quarter of an hour after eating, and remains so as long as food occupies the stomach.* | *P.* Etwa eine Viertelstunde nach dem Essen, und bleibt so, so lange der Magen Nahrung enthält. |
| *Does the recumbent position afford you any relief?* | *D.* Gewährt es Ihnen Linderung, wenn Sie liegen? |
| *Do you also vomit blood?* | Brechen Sie auch Blut? |
| *Yes, it is pure and red, or sometimes blackened.* | *P.* Ja, helles und rothes, oder auch geschwärztes Blut. |
| *When does the vomiting occur?* | *D.* Wann findet das Erbrechen statt? |
| *Immediately or some time after the food has been swallowed.* | *P.* Sofort oder einige Zeit nach Verschluckung der Speisen. |

| | |
|---|---|
| Usually it happens so speedily that it seems to be a regurgitation. | P. Gewöhnlich bald, so dass es mehr ein Regurgitieren ist. |
| The most complete rest possible ought to be given the stomach for a few days. You will therefore have to remain in the recumbent posture, and get your nourishment by nutritive enemata. | D. Es muss dem Magen die möglichst vollkommenste Ruhe für einige Tage gegeben werden. Sie werden daher in vollkommen ruhiger horizontaler Lage verbleiben, und Ihre Ernährung durch Klystiere erhalten müssen. |
| I am glad to find you better to-day. | Ich freue mich, Sie heute besser zu finden. |
| Your principal food should for the present be milk and lime-water, and as the pain diminishes we shall add some other articles. | Ihre vorzüglichste Nahrung sollte für jetzt Milch und Kalkwasser sein, und je nachdem der Schmerz sich vermindert, werden wir andere Dinge hinzufügen. |
| You must take lukewarm pappy food for weeks. Take it frequently, but in small quantities. I recommend you particularly milk-diet. Drink Karlsbad water. Avoid fat and sour food, and drinking cold wine, beer, water. | Sie müssen wochenlang lauwarme, breiige Speisen geniessen, und zwar oft, aber in kleinen Quantitäten. Ich empfehle Ihnen besonders eine Milchdiät. Trinken Sie Karlsbader Brunnen. Vermeiden Sie fette und saure Speisen, so wie kalten Wein, Bier, Wasser zu trinken. |

## 34. Gastric Cancer. — 34. Magenkrebs.

| | |
|---|---|
| Is the pain influenced by the taking of food? | D. Wird der Schmerz grösser oder geringer, wenn Sie Nahrung zu sich nehmen? |
| Do you invariably feel the pain in the same place? | Fühlen Sie den Schmerz immer nur an einer Stelle? |

| | |
|---|---|
| *No, Sir, it is here and there, in different places.* | P. Nein; er ist bald hier, bald da, an verschiedenen Stellen. |
| *What is the character of the pain?* | D. Wie ist der Charakter des Schmerzes? |
| *Its character is varying; now it is dull, now gnawing or lancinating. Sometimes it is slight; at other times it evokes excruciating agonies.* | P. Verschieden; bald ist er dumpf, bald nagend oder lancinierend. Zuweilen ist er unbedeutend, dann wiederum ruft er die martervollsten Qualen hervor. |
| *Have you noticed a swelling in the neighborhood of the stomach?* | D. Haben Sie eine Geschwulst in der Magengegend wahrgenommen? |
| *Have you sour eructations?* | Haben Sie saures Aufstossen? |
| *Do you suffer from flatulency?* | Leiden Sie an Blähungen? |
| *Have you gradually lost flesh?* | Haben Sie nach und nach an Gewicht verloren? |
| *Do you frequently vomit? When?* | Brechen Sie häufig? Wann? |
| *Within a brief time after meals.* | P. Gleich nachdem ich Nahrung zu mir genommen. |
| *What is the appearance of the ejected matter?* | D. Wie sieht das Erbrochene aus? |
| *It has a black or reddish-brown appearance (as coffee ground).* | P. Es sieht schwarz oder rothbraun (wie Kaffeesatz) aus. |
| *Do red spots appear sometimes in the afternoon on the cheeks of the patient?* | D. Zeigen sich zuweilen des Nachmittags rothe Flecke auf den Wangen des Patienten? |
| *Is cancer hereditary in your family?* | Ist in Ihrer Familie der Krebs erblich? |
| *Use very nourishing fluid food (eggs, meat-broth).* | Geniessen Sie sehr nahrhafte, flüssige Kost (Eier, Fleischbrühe). |
| *Whatever you drink must be warm.* | Was Sie trinken, muss warm sein. |
| *I also recommend you warm poultices upon the abdomen (belly).* | Auch empfehle ich Ihnen warme Umschläge auf den Unterleib (Bauch). |

## 35. Colic.

*Where have you the pain?*
*Here, in the neighborhood of the navel, and at the same time in other parts of the belly.*
*Are your bowels free?*
*No, they are constipated?*
*Is the pain increased as I press here?*
*No, rather diminished.*
*Do you suffer from indigestion?*

*Have you been enfeebled by any exhausting malady?*

*Have you eaten any indigestible food?*
*Have you overloaded your stomach (or eaten too much)?*

*I will prescribe an opiate (a purgative) for you.*

*Enemata may give you relief.*

*Apply warm linseed or mustard poultices over the abdomen, or bags filled with hot chamomile, or heated sand, or heated salt.*

*Your diet should be sago and arrow-root, with a little brandy. Also for some time after your recovery, it should be light.*

## 35. Kolik.

D. Wo haben Sie den Schmerz?
P. Hier, in der Nähe des Nabels und zugleich an andern Stellen des Bauches.
D. Haben Sie offenen Leib?
P. Nein, ich bin verstopft.
D. Wird der Schmerz verstärkt, wenn ich hier drücke?
P. Nein, eher gemindert.
D. Leiden Sie an Verdauungsschwäche?
Sind Sie durch irgend eine erschöpfende Krankheit geschwächt worden?
Haben Sie etwas Unverdauliches genossen?
Haben Sie sich Ihren Magen zu sehr überladen (or zu viel gegessen)?
Ich will Ihnen ein Opiat (ein Abführungsmittel) verschreiben.
Klystiere werden Ihnen Erleichterung verschaffen.
Machen Sie sich warme Umschläge von Leinsamen oder Senf auf den Leib (Bauch), oder Säckchen gefüllt mit heissen Kamillen oder mit heissem Sand, heissem Salz.
Ihre Diät muss aus Sago und Arrow-Wurzel mit etwas Branntwein bestehen. Auch einige Zeit nach Ihrer Herstellung muss sie leicht sein.

## 36. Bilious Colic.

Are the violent pains accompanied by vomiting?

Have you severe retching and vomiting?

Does the attack recur after certain intervals?

Keep the passages (the stools), I must look at them to-morrow.

Take a tablespoonful of the medicine every four hours.

Have a warm bath prepared immediately, as hot as the patient can bear it.

Use Vichy and Karlsbad waters.

## 37. Enteritis.

How did the disorder begin?

With great colicky pains.

With a chilly sensation and fever, and extreme thirst.

Is there nausea and vomiting?

Yes, sometimes the most distressing fits of retching.

Do the bowels move?

No, the bowels are constipated; sometimes constipation alternates with diarrhœa.

Have you noticed any blood in your stools?

## 36. Biliöse Kolik.

D. Treten die heftigen Schmerzen zugleich mit Erbrechen auf?

Haben Sie heftiges Würgen und Erbrechen?

Wiederholt sich der Anfall nach gewissen Zwischenräumen?

Heben Sie mir die Excremente (den Stuhlgang) auf; ich muss sie (ihn) morgen besichtigen.

Nehmen Sie von der Medicin alle vier Stunden einen Esslöffel voll.

Lassen Sie sofort ein warmes Bad machen, so heiss es der Patient vertragen kann.

Trinken Sie auch Vichy und Karlsbad Brunnen.

## 37. Entzündung der Eingeweide.

D. Wie hat die Krankheit begonnen?

P. Mit grossen, kolikartigen Schmerzen.

Mit einem Frostanfall, Fieber und ausserordentlichem Durst.

D. Haben Sie Brechneigung und Erbrechen?

P. Ja, zuweilen die qualvollsten Anfälle von Würgen.

D. Haben Sie Stuhlgang?

P. Nein, ich leide an Verstopfung; zuweilen wechselt dieselbe mit Diarrhöe ab.

D. Haben Sie Blut in Ihrem Stuhlgang bemerkt?

| | |
|---|---|
| Yes, several times a small quantity. | P. Ja, einige Male ein wenig. |
| How is your appetite? | D. Wie ist's mit Ihrem Appetit? |
| It is completely lost, but there is extreme thirst. | P. Der ist gänzlich geschwunden; aber ausserordentlichen Durst habe ich. |
| Where is the pain? | D. Wo ist der Schmerz? |
| Mostly here, near the navel, but thence it shifts to various parts of the abdomen? | P. Meist hier, nahe dem Nabel, doch zieht er sich von da nach verschiedenen Theilen des Bauches. |
| Is it increased by my pressure here? | D. Wird er vermehrt, wenn ich hier drücke? |
| O, very much! | P. O sehr! |
| Do you find any relief by lying on your back and drawing up your thighs? Yes. | D. Finden Sie Linderung, wenn Sie auf dem Rücken liegen und die Schenkel einziehen? P. Ja. |
| Take the medicine I prescribed, in four portions, with intervals of half an hour. Shake the bottle well. | D. Nehmen Sie die Medicin, die ich verschrieben, auf viermal in halbstündigen Zwischenräumen. Schütteln Sie die Flasche gut. |
| Apply a flannel roller firmly round the patient's abdomen. | Binden Sie dem Patienten ein flanellenes Wickelband fest um den Leib (Bauch). |
| Dip a towel in warm water, fold it up, and lay it, covered with a dry one, over the patient's abdomen. Let it lie all night, and change it several times during the day. | Falten Sie ein in warmes Wasser getauchtes Handtuch zusammen, und legen Sie es mit einem trockenen darüber dem Patienten auf den Leib (Bauch). Lassen Sie es die ganze Nacht liegen, und am Tage wechseln Sie es mehrere Male. |
| The diet must consist exclusively of warm, fluid, mild and mucilaginous, but nourishing food (also meat soups, but no meat). | Die Diät muss ausschliesslich aus warmen, flüssigen, milden und schleimigen, aber nahrhaften Nahrungsmitteln bestehen (auch Fleischsuppen, aber kein Fleisch). |

### 38. *Ileus Volvulus.*

*For how many days have you noticed that your bowels have not moved?*

*Did you take a purgative?*

*Yes, I have taken my usual and then a more active one, but without any effect.*

*Your abdomen is somewhat distended. Does it hurt you as I press here or here?*

*No, or but slightly.*

*Had you any fever in the beginning?*

*How many days were you constipated, when the stercoraceous vomiting made its appearance?*

*Have you a hernia?*

*Have you previously suffered from the passage of gall-stones?*

*Have you ever before had attacks of constipation overcome with difficulty?*

*Do your fæces present peculiarities in shape and size and are they sometimes mixed with blood?*

*Is the quantity of the urine passed by you the same as formerly, or smaller?*

*Did hiccough appear early and violently?*

*Did vomiting occur early, and was no fæcal matter noticeable in the ejected matter?*

### 38. **Darmverschliessung.**

D. Seit wie vielen Tagen bemerkten Sie ein Ausbleiben des Stuhlgangs?

Haben Sie ein Abführmittel genommen?

P. Ja, ich habe mein gewöhnliches und dann ein stärkeres genommen, doch immer ohne Erfolg.

D. Ihr Leib ist etwas aufgetrieben, thut's weh, wenn ich hier oder hier drücke?

P. Nein, oder nur wenig.

D. Hatten Sie anfangs Fieber?

Wie viele Tage waren Sie schon ohne Stuhl, als das Kotherbrechen sich gezeigt?

Haben Sie einen Bruch?

Haben Sie schon an Gallensteinen gelitten?

Hatten Sie schon früher Anfälle von fast unüberwindlicher Verstopfung?

Weisen Ihre Excremente eine besondere Gestalt und Grösse auf, und sind sie zuweilen mit Blut untermischt?

Ist die Quantität des von Ihnen gelassenen Harnes dieselbe wie früher, oder ist sie kleiner?

Ist das Schlucken früh und heftig aufgetreten?

Ist Erbrechen früh aufgetreten, aber kein Koth in dem Erbrochenen wahrgenommen worden?

| | |
|---|---|
| *Is the pain chiefly or entirely in the neighborhood of the umbilicus?* | D. Ist der Schmerz hauptsächlich oder allein in der Nähe des Nabels? |
| *You must reduce the quantity of food and drink, in every possible way. Restrict the latter to small but frequent sips (through a long straw or tube) of cool iced liquids. Food is to be given, as strong beef tea, soup, or milk, with equal precaution. I have prescribed pills for you; take three of them four times a day.* | Sie müssen für jetzt auf jede mögliche Weise Ihr Essen und Trinken verringern. Beschränken Sie dieses auf weniges aber öfteres Schlürfen von eisigkalten Getränken vermittelst eines längern Strohhalms oder Röhrchens. Speisen, wie starker Beefthee, Suppen, müssen mit gleicher Vorsicht genossen werden. Ich habe Ihnen Pillen verschrieben; nehmen Sie je drei viermal des Tages. |

## 39. Constipation.     39. Verstopfung.

| | |
|---|---|
| *Do you lead a sedentary life?* | D. Führen Sie eine sitzende Lebensweise? |
| *How are your evacuations usually? Are they frequent and easy?* | Wie sind Ihre Darmentleerungen gewöhnlich? Sind dieselben oft und leicht? |
| *No, Sir, they are generally infrequent and difficult.* | P. Nein, gewöhnlich sind sie selten und beschwerlich. |
| *Do you suffer from giddiness, headache?* | D. Leiden Sie an Schwindel, Kopfweh? |
| *Do you suffer from indigestion, palpitation of the heart, from neuralgic pains?* | Leiden Sie an Verdauungsbeschwerden, Herzklopfen, an neuralgischen Schmerzen? |
| *Are your hands and feet often cold?* | Sind Ihnen die Hände und Füsse oft kalt? |
| *What is the nature of your diet?* | Wie ist die Art Ihrer Diät? |
| *Are you dyspeptic or suffering from piles?* | Leiden Sie an Dyspepsie oder an Hemorrhoiden? |

*Do you sometimes neglect the calls of nature from carelessness or because circumstances prevent their being obeyed at the proper time?*

D. Lassen Sie zuweilen die natürliche Forderung zum Stuhlgang unbeachtet, sei's aus Nachlässigkeit oder weil äussere Umstände die sofortige Folgeleistung verhindern?

*Have you too little bodily exercise?*

Machen Sie sich zu wenig körperliche Bewegung?

*Have you noticed any change in the size or form of what you pass?*

Haben Sie an Ihren Excrementen eine Veränderung in der Grösse und Gestalt derselben bemerkt?

*Have you any trouble with your stomach or liver?*

Leiden Sie am Magen oder an der Leber?

*Take one of these pills after each meal. If the bowels don't move by these means, use clysters, either emollient of warm water, or stimulating with soap, salt, oil or turpentine.*

Nehmen Sie nach jeder Mahlzeit eine von diesen Pillen. Sollten Sie aber hierdurch keinen offenen Leib bekommen, dann gebrauchen Sie Klystiere, entweder bloss aufweichende von warmem Wasser, oder reizende mit Seife, Salz, Öl oder Terpentin.

*Furthermore, mild laxatives, as cider, prunes, fruits, vegetables, castor-oil.*

Ferner, milde Abführmittel, wie Äpfelwein, Pflaumen, Obst, Gemüsen, Ricinusöl.

*Take easily digestible food, mostly fluid and pappy, and more vegetable than animal, in small quantities and rather frequently. Furthermore I recommend large amounts of fluids (water, beer), and physical exercise.*

Geniessen Sie leicht verdauliche, meist flüssige und breiige, und mehr pflanzliche als thierische Kost, in geringer Menge und lieber oft. Ferner empfehle ich Ihnen reichlichen Genuss von Getränken (Wasser, Bier) und körperliche Bewegung.

### 40. *Diarrhœa.*
### 40. Diarrhöe.

*Of what do you complain?*
*Of frequent fluid stools.*

D. Worüber klagen Sie?
P. Über häufige, flüssige Darmentleerungen.

| | |
|---|---|
| How do they look? | D. Wie sehen dieselben aus? |
| They are now colorless. | P. Sie sind jetzt farblos. |
| Have you any pain? | D. Haben Sie Schmerz? |
| Sometimes griping and colicky pains in the bowels, commonly a sensation of uneasiness in the abdomen. | P. Zuweilen Bauchgrimmen und Kneifen, gewöhnlich ein Gefühl von Unbehagen im Leibe. |
| Let me see your tongue. | D. Lassen Sie mich Ihre Zunge sehen. |
| Has your drinking water been impure? | War das Wasser unrein, das Sie getrunken? |
| Have you been much exposed to moisture, to contaminated air? | Waren Sie der Feuchtigkeit oder verdorbener Luft ausgesetzt? |
| Do you suffer from worms? | Leiden Sie an Eingeweidewürmern? |
| Have you indulged in too much eating and drinking? | Haben Sie sich im Essen oder Trinken zu gütlich gethan? |
| Have you eaten any indigestible food? | Haben Sie unverdauliche Nahrung zu sich genommen? |
| Have the stools a bilious appearance and are they very fetid? | Haben die Diarrhöen bei Ihnen ein biliöses (galliges) Aussehen, und sind dieselben sehr übelriechend? |
| Is the diarrhœa accompanied by vomiting? | Haben Sie auch Erbrechen? |

## 41. Dysentery.  41. Ruhr.

| | |
|---|---|
| What is your complaint? | D. Worüber klagen Sie? |
| Frequent and painful passages of slime mixed with blood. | P. Über häufige und schmerzliche Entleerungen von Schleim und Blut. |
| Have you pain also at other times, and where? | D. Ist auch sonst Schmerz vorhanden und wo? |
| Yes, more or less pain is always present; it is here. Sometimes it intermits, or changes its position. | P. Ja, mehr oder weniger Schmerz ist immer anwesend; er ist hier. Zuweilen setzt er aus oder wechselt den Ort. |

| | |
|---|---|
| *Have you at times a disagreeable, heavy feeling near the anus?* | D. Haben Sie zuweilen ein Gefühl des Unbehagens und der Schwere am After? |
| *Yes, and I have consequently a continual desire to go to stool, accompanied by tenesmus, which is very distressing.* | P. Ja, und dadurch empfinde ich fortwährend Drang und Stuhlzwang, der sehr quälend ist. |
| *Do the stools contain much blood?* | D. Enthält der Stuhl viel Blut? |
| *No, Sir, but little; it is rather bloody mucus.* | P. Nein, nur wenig; es ist mehr blutiger Schleim. |
| *Have you noticed any lumps of fæcal matter?* | D. Bemerkten Sie Kothstücke darunter? |
| *Sometimes a few small ones.* | P. Zuweilen einige (kleine). |
| *How do the discharges look now?* | D. Wie sehen die Entleerungen jetzt aus? |
| *They resemble the washings of meat; like jelly; greenish in color; very dark and mucous.* | P. Fleischwasserähnlich; geléeartig; grünlich; sehr dunkel und schleimig. |
| *Is there also vomiting, hiccough?* | D. Haben Sie auch Erbrechen, Schlucken? |
| *What is your diet? Has it an excess of animal or vegetable food?* | Wie ist Ihre Kost? Besteht sie mehr aus Fleisch oder Pflanzenspeisen? |
| *Is it sufficient or insufficient?* | Ist dieselbe reichlich oder unzulänglich? |
| *Is it prepared well or badly?* | Ist sie gut oder schlecht zubereitet? |
| *Have you eaten any unripe fruits?* | Haben Sie unreifes Obst gegessen? |
| *Do you drink to excess?* | Sind Sie unmässig im Trinken? |
| *Is your house overcrowded?* | Ist Ihre Wohnung mit Menschen überfüllt? |
| *Is it well ventilated?* | Ist sie gut gelüftet? |
| *Have you slept on damp ground or have you been exposed to cold?* | Haben Sie auf feuchtem Boden geschlafen oder sonst wie sich erkältet? |

| | |
|---|---|
| *Take the powder prescribed, remain perfectly still in bed, and abstain from fluids for at least three hours. If thirst is urgent, appease it by small bits of ice or a teaspoonful of iced water.* | D. Nehmen Sie das Pulver, das ich Ihnen verschrieben; bleiben Sie sodann vollkommen ruhig im Bette, und enthalten sich wenigstens 3 Stunden lang jedes Getränks. Sollte der Durst zu gross sein, dann stillen Sie ihn durch kleine Eisstücke oder einen Theelöffel voll Eiswasser. |
| *Take milk boiled with flour, arrow-root, or barley-meal frequently, night and day. It should be taken cold, even with ice, and in small quantities at a time. Strong beef-tea, or Liebig's extract of flesh, is also very useful.* | Geniessen Sie mit Arrow-wurzel oder Gerstenmehl gekochte Milch recht oft, bei Tag und Nacht. Sie müssen dieselbe aber kalt nehmen, sogar mit Eis, und immer nur in kleinen Quantitäten. Starker Beefthee und Liebig's Fleischextract sind auch sehr zuträglich. |

## 42. *Hemorrhoids, Piles.*    42. Hämorrhoiden (goldene Ader, Mastdarmblutfluss).

*Are the piles to be seen outside of the anus?* — D. Sind die Hämorrhoidalknoten äusserlich zu sehen?

*Are they external or internal?* — Sind sie äusserlich oder innerlich?

*Are they bleeding?* — Sind sie fliessend?

*Yes, the evacuation of my bowels is accompanied by hemorrhage.* — P. Ja, beim Stuhlgange kommen Blutungen vor.

*No, they are blind piles.* — Nein, sie sind blind.

*Is the evacuation difficult and painful?* — D. Ist der Stuhlgang schwierig und schmerzhaft?

*Yes, I have a burning and stinging sensation in the anus.* — P. Ja, ich fühle ein Brennen und Stechen im After.

| | |
|---|---|
| Can you stand or sit with comfort? | D. Können Sie gemächlich stehen oder sitzen? |
| Do you find relief in the horizontal position? | Finden Sie Erleichterung, wenn Sie liegen? |
| Are your habits of life luxurious and sedentary? | Führen Sie eine üppige und sitzende Lebensweise? |
| Do you suffer from any disease of the liver? | Haben Sie irgend ein Leberleiden? |
| Are you suffering from habitual constipation? | Leiden Sie an habitueller Verstopfung? |
| I must make an examination to see whether the bleeding does not originate from the mucous membrane of the rectum, which could be cured radically by means of local treatment. | Ich muss Sie untersuchen, ob die Blutungen nicht aus der Mastdarmschleimhaut stammen und durch eine örtliche Behandlung radical curiert werden können. |
| Please to leave us alone for a few moments! | Bitte, lassen Sie uns einen Augenblick allein! |
| Please to unbutton your clothes! | Knöpfen Sie sich gefälligst auf! |
| Kneel on this chair! So, thus! | Knieen Sie hier auf diesen Stuhl! So, so! |
| The mucous membrane is degenerated. | Die Schleimhaut ist entartet. |
| The piles are exceedingly inflamed. | Die Hämorrhoidalknoten sind äusserst entzündet. |
| Try injections of cold water into the anus, and see whether the bleedings will be stopped thereby. | Versuchen Sie Einspritzungen von kaltem Wasser in den After, und sehen Sie, ob die Blutungen dadurch gehemmt werden. |
| The inflammation will subside by retaining the recumbent position for some time, and by applying poultices, fomentations, and by low diet. | Die Entzündung wird sich legen, wenn Sie sich für einige Zeit in horizontaler Lage verhalten und Überschläge, Bähungen anwenden und einer schmalen (magern) Kost sich unterziehen. |

## 43. The Fissure of the Anus. Prolapsus recti.

*Is the pain during the passage of the evacuations intense?*

Yes, and for some hours afterwards I have great discomfort, smarting and itching.

*You have a fissure of the anus.*

*Take the pills three times daily; they will procure regular and somewhat soft evacuations.*

*All the tumors protruding through the anus, or the lower portion of the rectum itself, must be replaced by soft pressure immediately after the evacuation; they could otherwise easily get inflamed.*

*If they do not easily return, the forefinger should be oiled and pushed up carefully into the anus, and it will convey the protruded intestine with it.*

*You should afterwards retain the recumbent position for some hours.*

*Sponge the parts after every evacuation with cold water, or soap and water.*

## 43. Die Fissur des Afters, der Mastdarmvorfall.

D. Ist der Schmerz während des Stuhlgangs heftig?

P. Ja, und auch einige Stunden nachher ist es mir sehr unbehaglich zu Muthe, es schmerzt und juckt mich.

D. Sie haben eine Fissur des Afters.

Nehmen Sie dreimal des Tages die Pillen; sie werden Ihnen regelmässige und ziemlich weiche Stuhlgänge verschaffen.

Alle durch den After sich hervordrängenden Geschwülste oder das untere Theil des Mastdarms selbst müssen sofort nach der Entleerung durch sanften Druck wieder zurückgebracht werden, weil sie sonst leicht sich entzünden könnten.

Sollten dieselben nicht leicht zurückzubringen sein, dann beölen Sie sich den Zeigefinger und stossen vorsichtig in den After, wodurch der Mastdarm wie die sonstigen sich hervordrängenden Theile zurückkehren werden.

Sie sollten nachher einige Stunden in liegender Stellung verbleiben.

Wischen Sie nach jeder Entleerung die Theile mit einem in kaltes Wasser getauchten Schwamme oder mit Seife und Wasser aus.

## 44. Diseases of the Urinary Organs.

## 44. Krankheiten der Harnorgane.

Can you make water as usual, and does it flow in a constant stream?
D. Können Sie wie gewöhnlich Wasser lassen, und geht dasselbe in vollem Strahle ab?

No, Sir, it comes slowly and scantily.
P. Nein, es geht nur langsam und spärlich ab.

I do not pass as much as usual.
Es kommt nicht so viel wie gewöhnlich.

The urine only dribbles out, or comes in drops.
Der Urin kommt nur tropfenweise.

It comes out but in jets, and often I can only make water in certain positions.
Er kommt in Absätzen und oft kann ich nur in gewissen Stellungen urinieren.

Do you feel a straining?
D. Empfinden Sie ein Drängen?

Yes, Sir, continually; I am obliged to pass water every few minutes.
P. Ja, fortwährend; mich drängt das Wasser alle paar Minuten.

Take the powders every half hour, till they act.
D. Nehmen Sie die Pulver jede halbe Stunde, bis zur Wirkung.

Put hot linseed poultices upon the abdomen, adding to the linseed two tablespoonfuls of the prescription.
Machen Sie sich auch warme Leinsamenumschläge auf den Unterleib, und fügen Sie zwei Esslöffel voll von der verschriebenen Tinktur dem Leinsamen bei.

Have you any scalding or lancinating sensation in voiding urine?
Haben Sie ein Brennen oder Schneiden beim Urinieren?

Yes, it burns like fire sometimes.
P. Ja, es brennt manchmal wie Feuer.

I am afraid there is some obstruction, either in the urethra or in the bladder, which causes your trouble. I must examine you.
D. Ich fürchte, dass irgend ein Hinderniss entweder in der Harnröhre oder Blase vorhanden ist, das diese Beschwerden veranlasst. Ich muss Sie untersuchen.

| | |
|---|---|
| *Please let me see your private parts.* | *D.* Bitte, lassen Sie mich Ihre Geschlechtstheile sehen. |
| *Don't be frightened, I will not hurt you, and it will be over in a moment.* | Fürchten Sie sich nicht; ich werde Ihnen nicht weh thun, und es wird bald geschehen sein. |
| *You have a stricture.* | Sie haben eine Strictur (eine Harnröhrenverengerung). |
| *The urine must be drawn off by means of a catheter.* | Der Urin muss mit dem Katheter abgelassen werden. |
| *You see, the urine flows out in a continuous stream; the bladder, I suppose, is now quite empty.* | Sie sehen, der Urin kommt in ununterbrochenem Strome heraus; die Blase muss fast leer sein. |
| *I will teach you how to pass the catheter and relieve yourself.* | Ich werde Sie lehren, den Katheter sich selbst einzubringen und sich selbst zu helfen. |
| *Is there any sediment, dregs, sand, or gravel in your urine?* | Macht der Urin einen Bodensatz, wie Hefen, Sand, Gries? |
| *There is some white sediment after it has stood a while.* | *P.* Es zeigt sich ein weisser Bodensatz darin, wenn er eine Zeit lang gestanden hat. |
| *Have you noticed any discharge when voiding your urine or at other times?* | *D.* Bemerkten Sie beim Uriniren oder sonst einen Ausfluss (Schleimabgang)? |
| *Just as I finish. At the beginning. At the same time; it is mixed with the urine.* | *P.* Gleich so wie ich fertig bin. Wenn ich anfange. Zu gleicher Zeit; beides ist vermischt. |
| *Is it troublesome? Yes, Sir, very much.* | *D.* Ist er lästig? *P.* Ja, sehr. |
| *I shall order you an injection.* | *D.* Ich muss Ihnen eine Einspritzung verordnen. |
| *I shall make it strong, and it may give you some pain.* | Ich werde dieselbe recht stark machen, und Sie werden beim Einspritzen etwas Schmerz haben. |
| *Well, how are you to-day?* | Nun, wie geht's heute? |
| *Did the injection pain you very much?* | Hatten Sie grossen Schmerz beim Einspritzen? |

O yes, Sir. Not very much; it was bearable.
Is the discharge still seen?
No, Sir, it is entirely stopped.
Still a little; it is less in quantity.

P. O ja. Nein, nicht sehr; es war erträglich.
D. Zeigt sich der Ausfluss noch?
P. Nein, er zeigt sich gar nicht mehr.
Noch etwas; es fliesst jetzt weniger aus.

### 45. *Hæmaturia.*

Is the bleeding very profuse?
Were large quantities of blood passed pure at first, or almost unmixed with urine?
Is there pain over the bladder, a frequent desire to pass water, or a stoppage in doing so?

Have you any pain in the lumbar region?
Did the bleeding follow active exertion, or a shock of the body from a fall?

Have you ever passed any gravel or stony substance with the urine?
Does the voided urine contain pus?

Are you subject to bleeding piles?

Have you any cardiac difficulty?

Take the prescribed tincture three or four times daily.

### 45. Blutharnen.

D. Ist die Blutung sehr stark?
Haben Sie viel Blut zuerst rein oder fast unvermischt mit Urin gelassen?
Existiert Schmerz über der Blase, häufiger Drang zum Urinieren, oder eine plötzliche Stockung während des Actes?

Haben Sie Schmerz im Kreuze?

Entstand die Blutung nach einer heftigen Anstrengung oder einer Erschütterung des Körpers durch einen Sturz?

Haben Sie je Nierengries oder eine steinige Masse mit dem Urin abgesondert?
Enthält die entleerte Flüssigkeit Eiter?

Leiden Sie gewöhnlich an blutenden Hämorrhoiden?

Haben Sie Herzbeschwerden?

Nehmen Sie von der verschriebenen Tinktur drei oder viermal täglich.

| | |
|---|---|
| Use injections of cold water (or water in which one of these powders has been dissolved), as I showed you yesterday, into the bladder or up the rectum. Also the cold hip-bath will do a good service. | D. Machen Sie sich Einspritzungen von kaltem Wasser (oder kaltem Wasser, worin eins dieser Pulver aufgelöst worden), in die Blase oder den After, wie ich Ihnen gestern gezeigt. Auch das kalte Hüftbad wird gute Dienste thun. |
| Remain in a quiet, recumbent posture. | Bleiben Sie in ruhiger Lage. |
| Your diet ought to be mild. | Ihre Diät muss eine milde sein. |

### 46. *Cystitis.* — 46. Blasenentzündung.

| | |
|---|---|
| Is the pain in the bladder severe? | D. Ist der Schmerz in der Blase heftig? |
| Yes, Sir, but not constant; it has intervals of cessation. | P. Ja, aber nicht andauernd; es giebt Zwischenräume (Intervalle), wo er aufhört (schweigt). |
| Can you urinate? | D. Können Sie Urin lassen? |
| Only drop by drop, and the constant desire to make water is not allayed thereby. | P. Nur tropfenweise, und das fortwährende Verlangen zu urinieren wird dadurch nicht gestillt. |
| Is the pain accompanied by any other sensation? | D. Ist der Schmerz noch von einem sonstigen Gefühle begleitet? |
| Yes, Sir, by a sense of constriction at or near the outlet of the bladder, and sometimes by tenesmus. | P. Ja, von einem Gefühl der Umstrickung an oder nahe dem Ausgang der Blase, und zuweilen von Stuhlzwang. |
| Have you drunk any young unfermented beer? | D. Haben Sie junges, ungegohrnes Bier getrunken? |
| Have you ever suffered from the presence of a stone? | Haben Sie je am Stein gelitten? |
| Please to bring me some of your urine (water) in a small vial to-morrow (or, Please keep some of . . . .). | Bringen Sie mir gefälligst morgen etwas Urin in einer kleinen Flasche mit (or, Heben Sie mir zu morgen etwas Urin auf). |

| | |
|---|---|
| *I recommend you hot linseed-meal poultices, to which you may add a little mustard or turpentine upon this region, here, with a hot hip-bath at night.* | D. Ich empfehle Ihnen heisse Leinsamenumschläge, denen Sie etwas Senf oder Terpentin beifügen können, auf diese Gegend hier, mit einem heissen Hüftbad des Abends. |
| *Introduce one of the suppositories into the rectum, every three or four hours, until the pain is relieved.* | Bringen Sie sich eins dieser Suppositorien (Stuhlzäpfchen) alle drei oder vier Stunden in den Mastdarm ein, bis der Schmerz sich gelegt hat. |
| *Drink freely of warm flax-seed tea.* | Trinken Sie reichlich heissen Flachssamenthee. |

## 47. Nephritis. — 47. Nierenentzündung.

| | |
|---|---|
| *Where do you complain of pain?* | D. Wo klagen Sie über Schmerz? |
| *Here, over my kidneys.* | P. In der Nierengegend. Hier, über den Nieren. |
| *How is the pain?* | D. Wie ist der Schmerz? |
| *Sometimes dull, at other times sharp and lancinating.* | P. Zuweilen dumpf, andere Male scharf und reissend. |
| *Is the pain confined to the kidney?* | D. Bleibt der Schmerz auf die Nierengegend beschränkt? |
| *No, it shoots from there down to my bladder?* | P. Nein, er schiesst von da abwärts nach der Blase zu. |
| *Is the pain influenced by movements, especially by stooping or any exertion?* | D. Wird der Schmerz durch Bewegung beeinflusst, besonders wenn Sie sich bücken oder irgend eine Anstrengung machen? |
| *The patient must remain at perfect rest.* | Der Patient muss in vollkommener Ruhe verbleiben. |
| *Apply hot fomentations or poultices.* | Wenden Sie heisse Bähungen oder Umschläge an. |

## 48. Nephralgia.

Did you ever before have similar attacks, during which with the urine a renal calculus was passed?

Do the attacks recur regularly?

When do the attacks generally take place?

Is the pain greatly increased by active exercise?

Does blood appear in the urine?

Is the pain also referable to other parts than to the region of the kidney and down the back?

*I feel it also here, and most on this side. The pain also extends to the testicle, which then becomes very sensitive and swells.*

Does the seat of the pain change?

Must you frequently make water?

*Yes, and it is often attended with pain at the end of my penis. Sometimes I discharge matter with the urine.*

## 48. Nierenschmerz.

D. Hatten Sie früher schon ähnliche Anfälle, bei denen mit dem Harn vielleicht ein Nierenstein entleert wurde?

Kehren die Anfälle zu regelmässigen Zeiten wieder?

Wann finden die Anfälle gewöhnlich statt?

Wird der Schmerz durch active, körperliche Anstrengung sehr vermehrt?

Erscheint Blut im Urin?

Empfinden Sie den Schmerz auch in andern Theilen als in der Nierengegend und den Verlauf des Rückens entlang?

P. Ich empfinde ihn auch hier, und meistens auf dieser Seite. Der Schmerz erstreckt sich auch auf den Hoden, der dann sehr empfindlich ist und aufschwillt.

D. Wechselt der Sitz des Schmerzes?

Müssen Sie häufig urinieren?

P. Ja, und es ist oft mit Schmerz an der Eichel verbunden. In einzelnen Fällen entleere ich auch Eiter mit dem Urin.

## 49. Acute Bright's Disease.

Have you been exposed to cold or especially to wet?

## 49. Acute Bright'sche Krankheit.

D. Haben Sie sich erkältet? Hat besonders feuchte Kälte auf Sie eingewirkt?

| | |
|---|---|
| Was the fever that set in, accompanied by nausea and by a dull pain in the region of both kidneys? | D. War das Fieber, das sich eingestellt, von Brechreiz und einem dumpfen Schmerz in der Gegend beider Nieren begleitet? |
| What is the color of the urine? | Was für Farbe hat der Urin? |
| It is generally natural, sometimes dingy. | P. Im Allgemeinen natürlich, manchmal rauchig. |
| It foams when passed into the chamber. | Er schäumt, wenn ich ins Nachtgeschirr lasse. |
| Is the quantity the same as when you were in good health? | D. Ist die Quantität dieselbe, wie in Ihren gesunden Tagen? |
| No, it is smaller. | P. Nein, sie ist geringer. |
| Did the dropsy commence with swelling of the face, and then become universal? | D. Hat die Wassersucht mit Anschwellung des Gesichts begonnen und sich dann allgemeiner verbreitet? |
| Has the sickness been preceded by a diminution in the urinary secretion? | Ist der Krankheit eine Abnahme der Harnabsonderung vorausgegangen? |
| Is your hearing impaired? | Ist Ihr Gehörsinn geschwächt? |
| Is there a loss of feeling in your lower limbs? | Empfinden Sie eine Gefühlsstumpfheit in den untern Gliedmassen? |
| Did the complaint come on suddenly? | Ist die Krankheit plötzlich aufgetreten? |
| No, it came on insidiously, and developed itself very slowly? | P. Nein, sie trat schleichend auf und entwickelte sich sehr langsam. |
| Do you live in a damp, ill-ventilated house? | D. Leben Sie in einem feuchten, schlecht ventilierten Hause? |
| Are you intemperate? | Sind Sie im Trinken unmässig? |
| Have you had an attack of dropsy prior to this? | Haben Sie vor einiger Zeit einen Anfall von Wassersucht gehabt? |
| Do you feel pain here? | Haben Sie Schmerz hier? |
| Is it augmented when lying on the affected side? | Ist der Schmerz gesteigert, wenn Sie auf der leidenden Seite liegen? |

| | |
|---|---|
| *You must carefully avoid every exposure to cold. Remain therefore at rest in bed, in a room of moderate uniform temperature.* | D. Sie müssen sorgfältig jede Erkältung vermeiden. Bleiben Sie daher in einem Zimmer mit gleichmässiger Temperatur ruhig im Bette. |
| *Your food should be scanty, consisting of gruel, arrow-root, milk or weak broth. When the tongue becomes clean, and the digestion improves, you may by degrees indulge in beef-tea and solid food. Pure water is the best drink; alcoholic fluids are not to be taken on any account.* | Ihre Kost muss schmal (kärglich) sein, bestehend aus Hafergrütze, Arrow-wurzel, Milch und schwacher Brühe. Wenn die Zunge rein und die Verdauung besser wird, können Sie nach und nach guten Beefthee und auch solidere Speisen geniessen. Reines Wasser ist das beste Getränk, alkoholische Getränke dürfen schlechterdings nicht genossen werden. |
| *Flannel must be worn next the skin.* | Tragen Sie Flanell auf dem blossen Leibe. |
| *When you have improved sufficiently, select a residence where the soil is sandy or chalky, where the air is mild and dry, so that as much open-air exercise may be taken as possible.* | Wenn Sie besser fühlen, so wählen Sie sich eine Wohnung, wo der Boden sandig oder kalkig und wo die Luft mild und trocken ist, so dass Sie sich möglichst viel Bewegung in freier Luft machen können. |
| *In case you should suffer from sleeplessness, take twenty drops of the mixture in a wine-glass full of water.* | Sollten Sie an Schlaflosigkeit leiden, so nehmen Sie vor dem Schlafengehen zwanzig Tropfen von der Mixtur in einem Weinglase voll Wasser. |
| *Take your principal meal in the middle of the day, but not later than three o'clock; and the last meal two or three hours before bedtime.* | Nehmen Sie Ihre Hauptmahlzeit Mittags, jedenfalls nicht später als um 3 Uhr; die letzte Mahlzeit 2 oder 3 Stunden vor Schlafengehen. |

After every loose stool, take 8 to 10 drops of this tincture in a little water.

D. Nach jedem flüssigen (diarrhöeartigen) Stuhlgang nehmen Sie 8 bis 10 Tropfen von dieser Tinktur in Wasser.

## 50. Diabetes.

Of what color is the urine?

It has very little color; it resembles water.

In what quantity is the urine passed?

The quantity is enormous.

Have you thirst?

I have thirst constantly, and a feeling of ravenous hunger and emptiness.

Have you lost in flesh?

O yes, Sir, steadily, for the last six months.

How is your tongue?

Continually dry.

Is your sight good?

Have you noticed any white spots upon your linen (body linen)?

## 50. Harnruhr.

D. Welche Farbe hat der Urin?

P. Er hat sehr wenig Farbe; er gleicht Wasser.

D. In welcher Quantität wird der Urin ausgeschieden?

P. Die Quantität ist enorm.

D. Haben Sie Durst?

P. Ich habe stets Durst und die Empfindung von Heisshunger und einer fortwährenden Leere.

D. Haben Sie an Gewicht verloren?

P. O ja, beständig seit den letzten sechs Monaten.

D. Wie ist Ihre Zunge?

P. Fortwährend trocken.

D. Ist Ihr Gesicht gut?

Bemerkten Sie weisse Flecke in Ihrer Leibwäsche?

## 51. Syphilis.

What is the matter with you?

I have the gonorrhœa, chancre, a bubo, a running sore on my privates, a swollen testicle.

## 51. Syphilis.

D. Was fehlt Ihnen?

P. Ich habe den Tripper, den Schanker, eine Leistenbeule (Bubo), ein offenes (eiterndes) Geschwür an meinen Geschlechtstheilen, einen geschwollenen Hoden.

| | |
|---|---|
| *Allow me to examine your penis.* | D. Erlauben Sie mir Ihr (männliches) Glied zu untersuchen (besichtigen). |
| *How long after sexual intercourse did you first perceive anything wrong?* | Wie lange nach dem geschlechtlichen (fleischlichen) Umgang bemerkten Sie, dass nicht Alles in Ordnung sei? |
| *Have you much pain in passing water?* | Haben Sie grossen Schmerz beim Wasser lassen? |
| *Are you troubled with painful erections at night?* | Werden Sie des Nachts sehr mit schmerzhaften Erectionen geplagt? |
| *I shall be obliged to open this bubo, which is suppurating, after which you must poultice it, until I see you again.* | Ich werde diesen Bubo (diese Leisten or Drüsenbeule) welcher (welche) eitert, öffnen müssen. Legen Sie dann Umschläge darauf, bis ich Sie wieder sehe. |
| *You need only apply this powder to the sores, covering them afterwards with finely picked oakum, to absorb the discharge.* | Sie brauchen nur dieses Pulver auf die wunden Stellen zu thun. Bedecken Sie dieselben sodann mit fein gezupftem Werg, um den Eiterausfluss aufzusaugen. |
| *You have syphilis, and will require a prolonged course of treatment to rid you of its effects.* | Sie haben die Syphilis, und es wird einer längern Behandlung bedürfen, um Sie von den bösen Folgen zu befreien. |
| *You have some trouble in the urethra; to locate it, I shall be obliged to pass this instrument.* | Sie haben ein Leiden (Übel) in der Harnröhre, und ich werde, um die Stelle ausfindig zu machen, dieses Instrument einführen müssen. |
| *I suspected as much! You have a stricture, and unless it is removed by an operation, you will soon be unable to pass your water at all.* | Ich habe es vermuthet! Sie haben eine Strictur (Verengerung), und wofern dieselbe nicht durch eine Operation entfernt wird, werden Sie bald kein Wasser lassen können. |

*You have an inflammation of the bladder, please bring me some of the water passed on rising: a small vial full will be sufficient.*

D. Sie haben eine Entzündung der Blase. Bringen Sie mir gefälligst etwas von Ihrem beim Aufstehen gelassenen Urin mit. Ein Fläschchen voll wird genügen.

## 52. *Anæmia.* — 52. Blutmangel.

*Have you had profuse and frequently recurring hemorrhages?*
*Have you had profuse chronic discharges?*
*Do bodily exercises immediately induce fatigue, dyspnœa and palpitation?*

D. Hatten Sie starke und häufig wiederkehrende Blutungen?
Hatten Sie profuse chronische Entleerungen?
Rufen körperliche Anstrengungen bei Ihnen sofort Ermüdung, Athembeschwerden und Herzklopfen hervor?

*Have you a persistent pain in the left side in the region of the spleen?*
*Nutritious substances must be supplied for diet, in the shape of easily digested meats and broths.*

Haben Sie einen permanenten Schmerz in der linken Seite, in der Gegend der Milz?
Es müssen nahrhafte Substanzen, in der Form von leicht verdaulichen Fleischspeisen und Brühen, dem Patienten gereicht werden.

*A change of air is absolutely necessary.*
*You ought to take iron largely diluted with water.*
*Do you also suffer from diarrhœa?*

Eine Luftveränderung ist absolut nothwendig.
Sie müssen Eisen verdünnt in Wasser nehmen.
Leiden Sie ebenfalls an Diarrhöe?

*Do you suffer also from menorrhagia?*
*Do you suffer also from leucorrhœa?*
*In any of these cases just mentioned, take the ordered medicine as directed.*

Leiden Sie zugleich an übermässiger Menstruation?
Leiden Sie auch am weissen Fluss?
In jedem der angegebenen Fälle, nehmen Sie das verordnete Medicament nach Anweisung.

## 53. Rheumatism.

Does the affection show any tendency to change its seat?

Did you previously have an attack of rheumatism?

Is the pain augmented by motion?

Is it more or less shifting?

Is it combined with stiffness, either of the muscles or of the joints?

Is it influenced by changes of temperature?

Does the pain shoot down to the groin, thigh, and testicle?

Do you perspire freely, and does the sweat have a sour smell?

Did either of your parents suffer from rheumatism?

How are the stools? Are they dark-colored and offensive?

Does the urine increase in quantity?

Are you subject to cardiac affections?

Are you rheumatic and suffering more or less persistently from rheumatism, though perhaps in a different part of the body from the one in which the acute affection has happened?

Have you had any strain or overwork?

## 53. Rheumatismus.

D. Zeigt das Leiden eine Neigung seinen Sitz zu wechseln?

Haben Sie schon früher einen Anfall von Rheumatismus gehabt?

Wird der Schmerz durch Bewegung vermehrt?

Wechselt der Schmerz mehr oder weniger seinen Sitz?

Ist der Schmerz mit Steifheit der Muskeln oder der Gelenke verbunden?

Hat der Temperaturwechsel einen Einfluss auf ihn?

Schiesst der Schmerz nach dem Schambug, dem Schenkel und dem Hodensacke hinab?

Schwitzen Sie viel und riecht der Schweiss sauer?

Hat Ihr Vater (Ihre Mutter) an Rheumatismus gelitten?

Wie sind die Stühle? Sind sie dunkel und übelriechend?

Nimmt die Harnabsonderung zu?

Sind Sie Herzbeschwerden unterworfen?

Sind Sie rheumatisch oder leiden Sie mehr oder minder beständig an Rheumatismus, obgleich vielleicht an einem andern Körpertheile als an dem, der von dem acuten Anfall ergriffen ist?

Haben Sie sich überarbeitet oder überangestrengt?

### 54 a. Gout.

*Do the early attacks recur with a certain amount of periodicity?*

*Do you live high or drink large quantities of wine?*

*Are you subject to indigestion, flatulency, pains and cramps, or palpitation of the heart?*

*Is the gout hereditary in your family?*

*Rest and low diet are absolutely necessary.*

*Warm anodyne lotions or fomentations may be used, and the part afterwards lightly covered or encased in flannel or fine wool, while the limb at the same time is kept elevated.*

### 54 a. Gicht.

D. Kehren die ersten Anfälle in einer gewissen periodischen Gleichmässigkeit wieder?

Leben Sie sehr gut, oder trinken Sie reichlich Wein?

Werden Sie häufig von Unverdaulichkeit, Flatulenz, Schmerzen und Krämpfe oder Herzklopfen geplagt?

Ist die Gicht in Ihrer Familie erblich?

Ruhe und schmale Kost sind unbedingt nothwendig.

Warme, schmerzstillende Abwaschungen oder Bähungen sind anzuwenden. Der Theil (das Glied) muss sodann leicht zugedeckt oder in Flanell oder feine Wolle eingeschlossen werden, das Glied muss auch in einer mässig erhöhten Lage verbleiben.

### 54 b. Chronic Gout.

*Vegetables with soup and meat must be allowed only once a day, not oftener.*

*Beer, wine and alcoholic fluids generally are injurious. You should take for drinks, water, pure water only, in quantities as large as possible.*

### 54 b. Chronische Gicht.

D. Gemüse, Suppe und Fleisch darf nur einmal des Tages genossen werden, nicht öfter.

Bier, Wein und alkoholartige Getränke sind im Allgemeinen schädlich. Sie sollten nur Wasser, reines Wasser trinken, in Quantitäten, wie Sie irgend können.

### 55. Fever.

*Have you ever before suffered from an intermittent fever?*

### 55. Fieber.

D. Haben Sie schon einmal am Wechselfieber gelitten?

| | |
|---|---|
| *Have you within a few months sojourned in a marshy, malarial neighborhood?* | D. Haben Sie sich innerhalb weniger Monate in einer sumpfigen, Miasmen erzeugenden Gegend aufgehalten? |
| *Have you over-exerted yourself?* | Haben Sie sich übermüdet? |
| *Have you committed any error in diet?* | Haben Sie einen Diätfehler begangen? |
| *Have you changed your mode of life?* | Haben Sie Ihre Lebensweise verändert? |
| *Has the heat of the sun or exposure to cold and moisture affected you?* | Hat die Sonnenhitze oder eine Erkältung auf Sie eingewirkt? |
| *Was the fever brought on by mental overwork or by grief?* | Sollte das Fieber durch geistige Überarbeitung oder durch Kummer bei Ihnen hervorgerufen sein? |
| *How long does each stage of the fever continue?* | Wie lange dauert jedes Fieberstadium? |
| *Of what does the patient complain during the interval of freedom from fever?* | Worüber klagt der Patient in der fieberfreien Zeit? |
| *Of loss of appetite, a spoiled stomach, lassitude.* | P. Über Appetitmangel, verdorbenen Magen, Mattigkeit. |
| *Does the fever with its various paroxysms recur daily?* | D. Kehrt das Fieber mit seinen verschiedenen Anfällen täglich wieder? |
| *During the cold stage apply external warmth and give warm drinks.* | Halten Sie während des Frosts den Patienten warm, und lassen Sie ihn Warmes trinken. |
| *During the hot stage, diluent drinks may be indulged in, if the patient desires them, with cooling diuretics.* | Während des heissen Zustandes können dem Patienten, wenn er danach verlangt, verdünnende Getränke mit kühlenden, harntreibenden Mitteln gereicht werden. |
| *Towards the end of the sweating stage, the first dose of quinine should be given him.* | Gegen das Ende des Schweissstadiums muss ihm die erste Dosis von Chinin gereicht werden. |

| | |
|---|---|
| *Are the patient's bowels open?* | D. Hat der Patient die Därme frei? *or* |
| *Has the patient had a large evacuation?* | Hat der Patient eine reichliche Ausleerung gehabt? |
| *Can the patient keep the medicine down?* | Kann der Patient die Medicin bei sich behalten? |
| *No, Sir, it is rejected.* | P. Nein, sie wird wieder ausgebrochen. |
| *Then give him the quinine by the rectum.* | D. Dann bringen Sie ihm das Chinin durch den After bei. |
| *After the sweating is over, make the patient change his linen cautiously.* | Wenn der Patient zu schwitzen aufgehört, lassen Sie ihn mit Vorsicht die Wäsche wechseln. |
| *The first and essential condition for a perfect recovery is the removal of the patient from the malarial region. He must not be permitted to remain under the influence of the malarial poison.* | Die erste und wesentliche Bedingung einer vollkommnen Wiederherstellung ist die Entfernung von der Malariagegend. Er darf nicht unter dem Einflusse des Malariagiftes gelassen werden. |

## *56. Typhoid Fever.*  56. Abdominaltyphus.

| | |
|---|---|
| *The patient is very weak and exhausted.* | D. Der Kranke ist sehr schwach und erschöpft. |
| *How are his stools?* | Wie sind seine Stuhlgänge? |
| *Always thin, of a yellow or dark brown color, and of offensive smell.* | P. Immer flüssig, von gelber oder dunkelbrauner Farbe und sehr übelriechend. |
| *How many evacuations occur during the 24 hours?* | D. Wie oft sind die Darmentleerungen (*or* Stühle) innerhalb 24 Stunden? |
| *Do the passages take place without the knowledge of the patient?* | Geschehen die Ausleerungen unbewusst? |
| *Have the stools contained blood, and in what quantity?* | Haben die Stuhlgänge Blut enthalten und in welcher Menge? |

| | |
|---|---|
| *Has a large intestinal hemorrhage taken place?* | D. Hat eine grossere Darmblutung stattgefunden? |
| *Does the patient complain of pain?* | Klagt der Patient sehr über Schmerz? |
| *Yes, Sir, particularly before the watery discharges.* | P. Ja, besonders unmittelbar vor den wässerigen Stuhlgängen. |
| *Is the patient always so restless?* | D. Ist der Patient stets so rastlos? |
| *Yes, he is delirious.* | P. Ja, er delirirt. |
| *When? Generally at night.* | D. Wann? P. Meistens des Nachts. |
| *Is it a wild delirium?* | D. Ist das Delirium wild? |
| *No, Sir, his mind is confused and his thoughts ramble.* | P. Nein, sein Verstand ist verwirrt, und die Gedanken wandern umher. |
| *Sometimes the delirium is attended with great restlessness. The sick man leaves his bed and wanders about the room.* | Zuweilen ist das Delirium von grosser Aufregung begleitet. Der Kranke verlässt das Bett und geht im Zimmer umher. |
| *Did the delirium occur early?* | D. Ist das Delirium früh aufgetreten? |

## 57. *Typhus.*     57. Flecktyphus.

| | |
|---|---|
| *How was he taken ill?* | D. Wie hat das Leiden begonnen? |
| *The sickness was preceded by lassitude and dejection. Whereupon a sudden chill, pain in the head and loins, followed.* | P. Es ging ihm Erschlaffung und niedergedrückte Stimmung voraus. Dann folgte ein plötzlicher Frostanfall, Schmerz im Kopfe und Kreuze. |
| *When did the lethargic state occur, early or late?* | D. Wann ist der lethargische Zustand eingetreten, früh oder spät? |
| *Does he always lie in this state?* | Liegt er stets in diesem Zustande da? |
| *Yes, he always lies in a state of half consciousness, very drowsy, weak and dull.* | P. Ja, er ist meist halb bewusstlos, sehr schläfrig, schwach und abgestumpft. |

| | |
|---|---|
| *How is the headache since delirium set in?* | D. Wie ist der Kopfschmerz, seitdem das Delirium eingetreten? |
| *It persists in the same degree.* | P. Er besteht in demselben Grade fort. |
| *Give him the medicine, a teaspoonful every six hours, and watch him carefully as to whether the headache abates.* | D. Geben Sie ihm die Medicin, alle sechs Stunden einen Theelöffel voll, und beobachten Sie ihn wohl, ob der Kopfschmerz nachlässt. |
| *If not, shave the hair off his head and cover the scalp with a rubber bag filled with crushed ice, or apply cold affusions.* | Wenn nicht, scheren Sie ihm das Haar ab und legen ihm einen mit zerstossenem Eise gefüllten Gummibeutel auf den Schädel, oder machen Sie ihm kalte Begiessungen. |
| *Give him a dose of this mixture every two hours. Commence about 9 P. M., and continue giving it until the patient sleeps.* | Geben Sie ihm alle zwei Stunden eine Dosis von dieser Mixtur. Beginnen Sie um neun Uhr Abends, und setzen Sie damit fort, so lange der Patient nicht eingeschlafen ist. |
| *How long has the patient been lying in this stupor?* | Wie lange liegt der Patient schon in dieser Betäubung? |
| *Almost uninterruptedly.* | P. Fast ununterbrochen. |
| *Apply mustard poultices to his loins, or* | D. Legen Sie ihm Senfumschläge um die Lenden, or |
| *Dip a thickly folded flannel into hot water, wring it out well, and bind it round the patient's loins, covered with a piece of waterproof cloth.* | Tauchen Sie dick übereinander gelegten Flanell in heisses Wasser, winden Sie ihn gut aus und legen ihn dem Patienten um die Lenden. Binden Sie darüber ein Stück wasserdichtes Zeug. |
| *Has the patient had a large evacuation from the bowels? No, Sir.* | Hatte der Patient eine reichliche Darmentleerung? P. Nein. |
| *Give him a turpentine enema.* | D. Geben Sie ihm ein Klystier von Terpentin, |

| | |
|---|---|
| *What has his state been since the cutaneous eruption made its appearance?* | D. Wie ist sein Zustand seit dem Hautausschlag? |
| *The confusion of mind and stupor have increased. The patient wanders and picks at his bedclothes.* | P. Die Gedankenverwirrung und der Stupor haben seitdem noch zugenommen. Der Kranke murmelt vor sich hin und zupft an seinem Bettzeuge. |
| *Does he drink much?* | D. Trinkt er viel? |
| *The amount of water drunk is large.* | P. Der Betrag des getrunkenen Wassers ist gross. |
| *Is the amount of the water voided as large in proportion?* | D. Ist der Betrag des entleerten Urins nach Verhältniss eben so gross? |
| *No, Sir, it is lessened.* | P. Nein, er ist geringer. |
| *The urine comes away drop by drop or is entirely retained.* | Der Urin wird tropfenweise gelassen oder ganz zurückgehalten. |
| *Withdraw all solids from the patient, but supply him freely with water, milk and water, coffee, weak broth, beef-tea, and especially with fresh air, and a cool (not cold) atmosphere.* | D. Entziehen Sie dem Patienten alle soliden (festen) Nahrungsmittel. Gewähren Sie ihm hingegen reichlich Wasser, Milch und Wasser, Kaffee, schwache Brühe, Beefthee, ganz besonders auch frische Luft und eine kühle (nicht kalte) Temperatur. |
| *You must give him nourishment often and at stated intervals, at least once every three or four hours. Even if the patient is asleep, or seems to be so, he must be roused at these intervals, to take his food or the stimulants ordered by me.* | Sie müssen ihm oft und zu festgesetzten Zwischenräumen Nahrung geben, wenigstens alle drei oder vier Stunden nach dem vierten Fiebertage. Sogar wenn der Kranke schläft oder zu schlafen scheint, muss er um diese Zeit ermuntert werden, um Nahrung oder die Stimulantien, die ich verordne, zu nehmen. |

## 58. Scarlet Fever.

*How do you feel?*
*I am very faint, have headache and nausea.*
*Have you vomited?* Yes.
*You are feverish.*
*Did the fever and vomiting come on suddenly?*
*When did the rash appear?*

*Early, on the second day of the disease.*
*It came out but late.*
*Did it come out simultaneously all over the body?*
*Is the skin itchy?*
*Have you a sore throat?*

*Yes, burning, difficulty in deglutition, and stiffness of the muscles and of the neck.*
*When did the sore throat show itself, before or after the eruption?*

*Has the child had any convulsions?*
*Yes, since the beginning of the illness.*
*Has it slept?*
*No, he (she) has been sleepless and delirious; he (she) grinds his teeth, and utters screams.*

*Admit as much fresh air to the room as possible.*

## 58. Scharlachfieber.

D. Wie fühlen Sie?
P. Ich bin sehr matt, habe Kopfschmerzen und Brechneigung.
D. Haben Sie gebrochen? P. Ja.
D. Sie sind auch fieberisch.
Ist das Fieber und Erbrechen plötzlich gekommen?
Wann ist der Ausschlag erschienen?

P. Früh, schon am zweiten Tage der Krankheit.
Er ist erst spät ausgebrochen.
D. Ist er gleichzeitig über den ganzen Körper ausgebrochen?
Juckt Ihnen die Haut?
Haben Sie Schmerzen im Halse?

P. Ja, Brennen, Schlingbeschwerde und Steifheit der Halsmuskeln.
D. Wann sind die Halsschmerzen gekommen, vor oder nach dem Ausschlage?

Hat das Kind Convulsionen gehabt?
P. Ja, gleich anfangs.

D. Hat es geschlafen?
P. Nein, es war schlaflos, und hat deliriert; es knirscht mit den Zähnen und stösst Schreie aus.

D. Lassen Sie so viel frische Luft ins Zimmer als möglich.

*Take care that the sick-room be kept at a uniform temperature, but it must not exceed 60° to 65° Fahrenheit.*

*Don't overload the patient with bedclothes.*

*Sponge over the skin several times daily with tepid water; but always expose one part of the body only, and dry it carefully with soft cloths.*

*Pure and cold water is the best drink for the patient.*

*His diet may consist of meat-soups, Liebig's extract of meat, stewed fruits, and milk. Give him the nourishment in small quantities, but frequently repeated.*

*Constipation is to be overcome by enemata of tepid water and salt.*

*Also give the patient wine, about four to six ounces in twenty-four hours. Give it but in small quantities mixed with two-thirds water.*

*Give him lumps of ice frequently, to be slowly dissolved in the mouth.*

D. Sorgen Sie für eine stets gleichmässige Temperatur des Zimmers; sie darf aber 60° bis 65° Fahrenheit nicht übersteigen.

Überladen Sie den Patienten nicht mit zu vielen Betten.

Fahren Sie mehrere Male täglich mit einem in laues Wasser getauchten Schwamme dem Patienten über die Haut hin; aber entblössen Sie immer nur einen Theil, und trocknen Sie denselben mit einem weichen Tuche sorgfältig ab.

Reines und kaltes Wasser ist das beste Getränk für den Patienten.

Seine Kost bestehe aus Fleischsuppen, Liebig's Fleischextract, gedämpftes Obst und Milch. Geben Sie ihm die Nahrungsmittel in kleinen Quantitäten, aber oft wiederholt.

Einer Stuhlverstopfung helfen Sie mit einem Klystiere von lauem Wasser und Salz ab.

Geben Sie dem Patienten auch Wein, etwa vier bis sechs Unzen in 24 Stunden. Geben Sie denselben nur in kleinen Quantitäten und mit zwei Drittel Wasser gemischt.

Geben Sie ihm oft Eisstückchen, die der Patient langsam im Munde zerschmelzen lassen soll.

*I have ordered a gargle. The patient may use it at pleasure; it will refresh him, and remove the offensive secretions.*

D. Ich habe ein Gurgelwasser verschrieben. Der Patient kann dasselbe nach Belieben brauchen; es wird ihn erfrischen und die unangenehmen Secretionen fortschaffen.

*The patient should also inhale the steam of hot water as long as his throat is sore.*

Der Patient soll auch, so lange sein Halsleiden dauert, heisse Wasserdämpfe einhauchen.

### 59. *Diseases of the Eye.* — 59. Augenkrankheiten.

*Do you have pain in your eye?*

D. Haben Sie Schmerzen im Auge?

*Does the pain extend from the eye into the head, or is it only in the ball of the eye?*

Breitet sich (*or* dehnt sich) der Schmerz vom Auge nach dem Kopfe aus, oder ist er nur im Augapfel?

*Does the light hurt your eyes?*

Thuts Ihnen weh, wenn Sie ins Licht sehen? *or*

Blendet Sie das Licht (*or* Schadet Ihnen das Licht)?

*Can you see well?*

Können Sie gut sehen?

*Does it hurt you to read?*

Haben Sie Schmerz beim Lesen (*or* Schadet Ihnen das Lesen)?

*My sight is blurred after reading half an hour.*

P. Mein Auge verdunkelt sich, wenn ich eine halbe Stunde lese.

*With which eye can you see best?*

D. Mit welchem Auge können Sie besser sehen?

*Do you see distinctly?*

Können Sie deutlich sehen?

*Have you had sore eyes lately?*

Haben Sie vor Kurzem schlimme Augen gehabt?

*Do you see better in a bright or in a dim light?*

Sehen Sie besser bei hellem oder bei mattem Licht?

*Open your eyes!*

Machen Sie die Augen auf!

*Shut your eyes!*

Machen Sie die Augen zu!

*Did the disease begin in this eye first?*

Fing die Krankheit zuerst in diesem Auge an?

| | |
|---|---|
| *How long has your vision been poor?* | D. Wie lange schon ist Ihr Gesicht schwach? |
| *Do you think it is getting worse?* | Glauben Sie, dass es schlimmer wird? |
| *Do you have pain, and is it worse at night?* | Haben Sie Schmerz, und ist er bei Nacht grösser? |
| *Is your vision worse in the twilight?* | Ist Ihr Gesicht schlimmer im Zwielicht? |
| *Can you see well at a distance?* | Können Sie gut in die Ferne sehen? |
| *Do you ever see the letters double?* | Sehen Sie je die Buchstaben doppelt? |
| *Do you ever see half an object only?* | Sehen Sie je einen Gegenstand nur halb? |
| *Do you see things surrounded by colored rings (with colored borders)?* | Sehen Sie die Gegenstände wie mit farbigen Ringen (Rändern) umgeben? |
| *I have a cloud, a haziness before my eyes. I am dim-sighted.* | P. Ich habe eine Wolke, einen Nebel vor den Augen. |
| *The bright light hurts (dazzles) my eye.* | Das grelle Licht blendet mich. |
| *The light hurts my eyes.* | Ich kann das Licht nicht vertragen. |
| *Look up, down!* | D. Sehen Sie nach oben, nach unten! |
| *Look at me!* | Sehen Sie mich an! |
| *Have you had styes on your eyes?* | Hatten Sie Gerstenkörner an den Augen? |
| *Do the lids stick together in the morning?* | Sind Ihnen die Augenlider des Morgens zusammen geklebt? |
| *Do you ever have crusts on the edges of the lids?* | Haben Sie je Krusten an den Augenlidrändern? |
| *Have the lids been swollen?* | Waren die Augenlider geschwollen? |
| *How long have they been swollen?* | Wie lange sind sie schon geschwollen? |

| | |
|---|---|
| *How long have you had that growth on the lid?* | D. Wie lange haben Sie schon dieses Gewächs an dem Augenlide? |
| *Can you not raise your lid?* | Können Sie das Augenlid nicht in die Höhe heben (erheben)? |
| *How long has your lid drooped?* | Seit wie lange ist das Augenlid gesunken? |
| *How long has your lid turned over (out)?* | Wie lange schon ist Ihr Augenlid umstülpt? |
| *How long have the hairs of the lid turned in?* | Wie lange waren die Augenlidhaare einwärts gekehrt? |
| *Do you feel better after the hairs are pulled out?* | Fühlen Sie besser, seitdem die Haare ausgezogen sind? |
| *Do your eyes feel as if there were sand in them?* | Fühlen Ihre Augen, als ob Sand in ihnen wäre? |
| *What is the character of the discharge, is it watery or yellow?* | Wie ist der Ausfluss, ist er wässerig oder gelb? |
| *Wash your eyes often with warm water. They must be kept perfectly clean.* | Waschen Sie die Augen oft mit warmem Wasser. Sie müssen vollkommen rein gehalten werden. |
| *Draw down the lower lid when you look up, and brush the inside of the lid with a camel's hair brush.* | Ziehen Sie das untere Augenlid, während Sie aufwärts blicken nieder, und fahren sich mit einem Kameelhaarpinsel leicht hin über das Innere des Augenlids. |
| *Fold up several pieces of linen cloth, about two inches square, and lay them on ice, and when they are cold, put them on the eye. They should be changed every five minutes for half an hour.* | Falten Sie mehrere Stücke Leinwand von ungefähr zwei Zoll ins Gevierte zusammen und thun sie aufs Eis; wenn dieselben kalt sind, legen Sie sie aufs Auge. Sie müssen alle fünf Minuten gewechselt werden, und das eine halbe Stunde lang. |
| *Soak linen cloths in very warm water, wring them and lay them on the eye.* | Tauchen Sie Leinwand in sehr warmes Wasser; winden Sie dieselbe aus und legen sie aufs Auge. |

| | |
|---|---|
| *Apply two leeches to the temple on the side of the affected eye.* | D. Lassen Sie sich zwei Blutegel an die Schläfe, auf der Seite des kranken Auges setzen. |
| *Smoke and dust are very injurious to your eyes.* | Rauch und Staub sind Ihren Augen äusserst schädlich. |
| *Do you use much tobacco or liquor?* | Rauchen oder trinken Sie stark? |
| *You must stop using tobacco, chewing as well as smoking.* | Sie müssen das Rauchen einstellen, das Rauchen sowohl als das Tabakkauen. |
| *Drop this liquid in the eye three times a day with a medicine dropper.* | Träufeln Sie ihm (ihr) dreimal des Tages von dieser Flüssigkeit mit einem Tropfglase ins Auge. |
| *You must wash off all the crust and afterward apply the ointment.* | Sie müssen alle Krusten abwaschen und dann die Salbe auflegen. |
| *Apply a piece of ointment the size of a pin-head, to the inside of the lid every evening.* | Gebrauchen Sie jeden Abend ein Stückchen Salbe von der Grösse eines Stecknadelkopfes innerhalb des Augenlides. |
| *You have an ulcer on your eye.* | Sie haben ein Geschwür an Ihrem Auge. |
| *There is a speck on your eye; you must have an operation performed, so as to make a new pupil.* | An Ihrem Auge ist ein Fleck; es ist eine Operation nöthig, um eine neue Pupille zu bilden. |
| *You have a foreign body on your eye and it must be taken off.* | Sie haben einen fremden Körper an Ihrem Auge; derselbe muss entfernt werden. |
| *You have a cataract, and when it is ripe it must be removed.* | Sie haben einen Staar, und wenn er reif ist, muss er entfernt werden. |
| *You have iritis; it is a severe inflammation within the globe, and may seriously impair your vision.* | Sie haben eine Regenbogenhautentzündung; es ist eine starke (schwere) Entzündung innerhalb des Augapfels, und sie kann ernstlich Ihr Gesicht beeinträchtigen. |

| | |
|---|---|
| *You have had a severe inflammation, that has entirely destroyed your vision.* | D. Sie haben eine schwere Entzündung gehabt, die Ihr Gesicht gänzlich zerstört hat. |
| *You have a sympathetic eye trouble, and the bad eye must be removed.* | Sie haben ein sympathetisches Augenleiden, und das kranke Auge muss entfernt werden. |
| *You have an inflammation of the optic nerve, and it will probably get entirely well.* | Sie haben eine Entzündung des Augennerven, der wahrscheinlich wieder ganz hergestellt werden wird. |
| *You must remain in bed in a dark room.* | Sie müssen in einem dunkeln Zimmer im Bette bleiben. |
| *You have a tumor of the iris, and the tumor must be taken out.* | Sie haben eine Geschwulst an der Regenbogenhaut, und die Geschwulst muss beseitigt werden. |
| *You have a cancer within the eye, and the eye must be taken out.* | Sie haben einen Krebs innerhalb des Auges, und das Auge muss herausgenommen werden. |
| *You have a cataract, it should be removed at once.* | Sie haben einen Staar, und der sollte sofort entfernt werden. |
| *Your cataract is not ripe enough to be removed.* | Ihr Staar ist nicht reif genug, um entfernt zu werden. |
| *Do you see floating bodies in the air? They are within the eyes, but will do no harm.* | Sehen Sie schwimmende Körper in der Luft? Sie sind innerhalb des Auges, werden jedoch nicht schaden. |
| *You have had a hemorrhage within the eye.* | Sie haben eine Blutung innerhalb des Auges gehabt. |
| *You have glaucoma; the eyeball is hard and the optic nerve is excavated.* | Sie haben den grünen Staar (das Glaukom). Der Augapfel ist hart, und der Augennerv ist ausgehöhlt. |
| *An operation will cure the glaucoma.* | Eine Operation wird das Glaukom heilen. |

| | |
|---|---|
| You are near-sighted, but with proper glasses you can see as well as any one, and probably the vision will not become worse. | D. Sie sind kurzsichtig, aber mit passenden Gläsern können Sie eben so gut als irgend einer sehen, und wahrscheinlich wird auch das Gesicht nicht schlimmer werden. |
| You should wear the glasses all the time. | Sie sollten die Gläser immer tragen. |
| All near-sighted people should have proper glasses. | Alle kurzsichtigen Leute sollten die geeigneten (*or* passende) Gläser haben. |
| You will not need glasses to read with when you become old. | Sie werden keine Gläser zum Lesen brauchen, wenn Sie alt sein werden. |
| There is a paralysis of one of the muscles. | Einer der Muskeln ist gelähmt. |
| The eyeball is pushed forward; there is a growth behind it. | Der Augapfel ist nach vorn gedrängt; es ist ein Gewächs dahinter. |
| There is a dead bone behind the eye. | Hinter dem Auge ist ein todter Knochen. |
| Do tears run down the cheeks? | Laufen Thränen die Wangen hinunter? |
| The tear-duct is closed up, and a probe must be passed through. | Der Thränengang ist verschlossen und eine Sonde (Senknadel) muss durchgeführt werden. |
| There is an abscess in the lachrymal sack. | Im Thränensack ist ein Geschwür. |
| There is a small polypus on the inside of the lid. | Auf der innern Seite des Augenlides ist ein Polyp. |
| You have a stye on the eyelid, and it will get well as soon as it opens. | Sie haben ein Gerstenkorn am Augenlide, und es wird wieder gut werden, so bald es sich öffnet. |
| You have granular lids, but they can be entirely cured by proper treatment. | Sie haben gekörnte (körnige) Augenlider, aber das kann gänzlich bei passender (geeigneter) Behandlung geheilt werden. |

| | |
|---|---|
| *An operation can be performed that will prevent the eyelashes from turning in.* | D. Es kann eine Operation unternommen werden, wodurch das Einwärtswenden der Augenwimpern verhindert werden wird. |
| *I wish to examine the interior of your eye. You have a serious disease within the eye that cannot be cured.* | Ich möchte das Innere Ihres Auges untersuchen. Sie haben eine bedenkliche (schwere) Krankheit innerhalb des Auges, die nicht geheilt werden kann. |
| *Your vision cannot be restored, but an operation will give entire relief from the pain.* | Ihr Gesicht kann nicht wieder hergestellt werden, aber eine Operation wird Sie von allem Schmerze befreien. |
| *All but two of the muscles of your eye are paralysed.* | Alle Muskeln Ihres Auges, mit Ausnahme von zweien, sind gelähmt. |
| *How long has the child been cross-eyed?* | Wie lange schon schielt das Kind? |
| *It became cross-eyed from the want of glasses.* | Es wurde schielend aus Mangel an Gläsern. |
| *An operation will make the child's eye straight, but it will always be obliged to wear glasses.* | Eine Operation wird des Kindes Auge gerade machen; aber es wird immer eine Brille tragen müssen. |
| *If the boy (girl) had had the proper glasses, he (she) would not have become cross-eyed.* | Hätte der Knabe (das Mädchen) die gehörigen Gläser (Brille) gehabt, so würde er (es) nicht schielend geworden sein. |
| *Does the child hide its eyes away from the light?* | Verbirgt das Kind das Auge vor dem Licht? |
| *You must dip its face into a dish (bowl) of cold water whenever it closes its eyes, and then it will open them.* | Sie müssen ihm das Gesicht in eine Schüssel mit kaltem Wasser tauchen, immer wenn es die Augen schliesst, und dann wird es sie öffnen. |

## 60. Ear Diseases.

## 60. Ohrenkrankheiten.

*I have a pain in my ear, am somewhat deaf, and there is a buzzing in the ear.*

*Is the pain severe?*

*Is the pain worse in the day-time or at night?*

*It is much worse at night.*

*The pain is increased by talking or eating*

*It also causes pain to swallow.*

*Have you a discharge from your ear, and of what color is it?*

*I have a yellow discharge, that sometimes smells very badly.*

*Did you have any pain before it discharged?*

*I had a very severe pain, but it subsided as soon as the discharge appeared.*

*Did it come during scarlet fever or measles?*

*No, it came during an attack of cold in the head, and the discharge is worse when I have a cold.*

*There is a perforation of the drumhead, and I will give you a powder to be blown in.*

P. Ich habe Schmerzen in meinem Ohre, bin etwas taub, und es summt mir im Ohr.

D. Ist der Schmerz gross?

Ist der Schmerz grösser bei Tage als bei (in der) Nacht?

P. Er ist weit schlimmer bei Nacht.

Der Schmerz wird durch Sprechen oder Essen erhöht (*or* verstärkt).

Auch das Schlucken (Schlingen) verursacht mir Schmerz.

D. Haben Sie einen Ausfluss aus dem Ohre und was für Farbe hat derselbe (*or* Läuft Ihnen das Ohr)?

P. Der Ausfluss sieht gelb aus, und riecht manchmal sehr schlecht.

D. Hatten Sie Schmerzen im Ohr, bevor (*or* ehe) es lief?

P. Ich hatte sehr grosse Schmerzen, aber sie legten sich (*or* liessen nach), sobald der Ausfluss erschien (*or* sich zeigte).

D. Kam derselbe während eines Scharlachfiebers oder der Masern?

P. Nein, er kam während eines Schnupfens, und er ist schlimmer, wenn ich erkältet bin.

D. Das vordere Trommelfell ist durchbohrt, und ich will Ihnen ein Pulver zum Einblasen geben.

| | |
|---|---|
| Syringe the ear every morning and blow the powder in afterward. | D. Machen Sie sich jeden Morgen eine Einspritzung (*or* Spritzen Sie sich jeden Morgen) ins Ohr, und blasen Sie dann das Pulver hinein. |
| There is a polypus in your ear that must be removed. | In Ihrem Ohre ist ein Polyp, der entfernt werden muss. |
| There is hardened wax in the ear that I will remove with a syringe. | Verhärtetes Ohrenschmalz ist in Ihrem Ohre, das ich mit einer Spritze entfernen will. |
| You have a boil in the ear that should be opened. | Sie haben eine Beule im Ohre, die geöffnet werden müsste. |
| You have eczema of the ear, but it will soon get well. | Sie haben einen Ohrenausschlag, der aber bald sich geben wird (*or* —besser werden wird). |
| You have a foreign body in the ear that can be easily removed. | Sie haben einen fremden Körper im Ohre, der leicht entfernt werden kann. |
| You have an acute inflammation of the middle ear with a collection of matter, and I must perforate the drum-head. | Sie haben eine heftige Entzündung im Mittelohre mit einer Ansammlung von Eiter, und ich muss das Trommelfell durchbohren. |
| You should apply three leeches behind the affected ear. | Sie sollten sich hinter dem kranken Ohre drei Egel setzen lassen. |
| I will inflate the ear every day. | Ich werde Ihnen jeden Tag das Ohr aufblähen. |
| Do you hear better after the ear is inflated? | Hören Sie besser, nachdem Ihnen das Ohr aufgeblasen wurde? |
| I hear better, and the pain is gone. | P. Ich höre besser, und der Schmerz ist weg. |
| You have a chronic catarrh of the middle ear, which causes the deafness and buzzings, and it cannot be cured. | D. Sie haben einen chronischen Katarrh im Mittelohr, der die Taubheit und das Summen verursacht (*or* zu Wege bringt), und er kann nicht geheilt werden. |

| | |
|---|---|
| *Have you much pain behind the ear?* | D. Haben Sie grossen Schmerz hinter dem Ohre? |
| *I have a great deal of pain, and there is a considerable swelling.* | P. Ich habe viel Schmerz, auch ist eine beträchtliche (bedeutende) Geschwulst daselbst. |
| *There is a collection of matter behind your ear, and I will be obliged to make an incision down to the bone.* | D. Es ist eine Ansammlung von Eiter hinter dem Ohre, und ich werde einen Einschnitt bis aufs Bein machen müssen. |
| *Sometimes there is dead bone in these cases.* | Manchmal ist in solchen Fällen ein todter (abgestorbener) Knochen vorhanden. |

## DISEASES OF WOMEN.  ## FRAUENKRANKHEITEN.

### 61. Menorrhagia.   ### 61. Menorrhagie.

| | |
|---|---|
| *How is your health in general?* | D. Wie ist Ihre Gesundheit im Allgemeinen? |
| *Have you an occasional hemorrhage?* | Haben Sie gelegentlich einen Blutfluss? |
| *Have you a constant discharge?* | Haben Sie einen beständigen Ausfluss? |
| *Has the discharge a peculiar color?* | Hat der Ausfluss eine eigenthümliche Farbe? |
| *Was the hemorrhage in consequence of a sudden alarm (or of an abortion)* | Kam die Ergiessung in Folge eines Schreckens (*or einer Fehlgeburt, Frühgeburt*)? |
| *Did it come on suddenly and quite independent of the menses?* | Kam dieselbe plötzlich und ohne allen Zusammenhang mit der Periode? |
| *Have you had an increased flow of the menstrual discharge?* | Hatten Sie einen stärkeren Erguss Ihrer monatlichen Reinigung? |
| *Was the actual quantity greater?* | War die Menge an sich grösser? |
| *Was the time of its duration longer?* | War Ihre Zeitdauer länger? |

| | |
|---|---|
| *Were the intervals shorter?* | D. Waren die Zwischenräume der Ruhe kürzer? |
| *I have bled very profusely.* | P. Ich habe ungeheuer viel Blut verloren. |
| *I have flooded.* | Ich schwimme in Blut. |
| *Have you had a fall lately?* | D. Haben Sie neulich einen Fall gethan? |
| *You must not sit upright at all, you ought to keep yourself in a recumbent posture for some time.* | Sie dürfen sich nicht aufsetzen. Sie müssen eine Zeit lang in liegender Stellung verbleiben. |

## 62. Amenorrhœa. — 62. Amenorrhöa.

| | |
|---|---|
| *How long has your monthly discharge (your menstruation) been retarded?* | D. Wie lange hat sich Ihre Periode (monatliche Reinigung *or* Ihr Monatliches) verzögert? |
| *How long have you missed your monthly sickness?* | Seit wann ist Ihre Regel ausgeblieben? |
| *Have the monthly discharges returned?* | Ist Ihr monatlicher Fluss zurückgekehrt (*or* Hat Ihr Monatliches sich wieder eingestellt)? |
| *Were they suppressed suddenly?* | Ist Ihre Regel plötzlich weggeblieben? |
| *Have they previously been regular or irregular?* | Ist Ihre monatliche Reinigung früher regelmässig oder unregelmässig gewesen? |
| *Is there a possible coexistence of pregnancy?* | Sollten Sie vielleicht zugleich in andern Umständen (*or* schwanger) sein? |
| *Have you had any attack of a severe illness?* | Hatten Sie einen Anfall von einer schweren Krankheit? |
| *Have you exposed yourself to cold, rain, or the damp air?* | Haben Sie sich der Kälte, dem Regen, der feuchten Witterung ausgesetzt? |
| *Have you ever had anæmia (or any severe illness) which was followed by scanty menstruation,* | Haben Sie jemals an Blutmangel (oder einer schweren Krankheit) gelitten, in dessen (deren) |

*terminating in complete suppression of the menses?*

*When did your monthly courses stop?*

*They stopped recently.*

*What is the character of the discharge as to color or quantity?*

*Is your discharge of the ordinary fluid, or are there coagula in it?*

*There are clots of blood in it.*

*Is the flux scanty or pale?*

*It is pale, white, glary, thick.*

*Is the fluid mixed with membranous shreds?*

*Are the discharges often attended with pain?*

*Is the pain accompanied by a diminution or by an excess of the catamenia?*

*Is the pain most intense at the commencement of the discharge?*

*Is the pain associated with other sensations of an analogous kind, headache, backache, etc.?*

*Have you ever had a disease of the liver?*

*Have you ever had the dropsy?*

Folge der monatliche Fluss spärlicher geworden und endlich ganz ausgeblieben ist?

Wann blieb Ihre Regel weg?

P. Sie blieb neulich (kürzlich) weg.

D. Wie ist Ihre Absonderung, was Farbe und Menge betrifft (*or* rücksichtlich der Farbe und Menge)?

Ist das Blut, das Sie absondern, wie gewöhnlich flüssig oder zum Theil geronnen?

P. Es gehen ordentliche Stücke Blutes mit ab.

D. Ist der Monatfluss spärlich oder blass?

P. Er ist blass, weiss, klebrig, dickflüssig.

D. Ist der Monatfluss mit häutigen Fetzen (Lappen) untermischt?

Ist der Monatfluss oft von Schmerz begleitet?

Ist der Schmerz von einer Verminderung oder einem Übermass der Blutabsonderung begleitet?

Ist der Schmerz am grössten, wenn die Absonderung beginnt?

Ist der Schmerz mit andern gleichartigen Empfindungen verbunden, wie Kopfweh, Rückenweh, etc.?

Hatten Sie je ein Leberleiden?

Hatten Sie je die Wassersucht?

| | |
|---|---|
| *Have you any disease of the kidney?* | D. Haben Sie eine Nierenkrankheit? |
| *Have you ever noticed milk or any other fluid in your breasts, by which they have been enlarged?* | Hat sich schon einmal Milch oder eine andere Flüssigkeit in Ihren Brüsten gezeigt, wodurch sie voller und strotzender geworden? |
| *Have you ever had any tumor in your abdomen? And where?* | Hatten Sie je eine Geschwulst an Ihrem Unterleibe? Und wo? |
| *Did this occur before you ceased to be regular or after it?* | Geschah dies, bevor Sie Ihr Monatliches nicht mehr bekamen oder nachher? |
| *Did your navel ever protrude from the abdomen?* | Stand Ihnen je der Nabel aus dem Unterleibe hervor? *or* War Ihnen je der Nabel blasig aufgetrieben? |
| *Did you ever menstruate while in the family-way?* | Hatten Sie je Ihre Regel zur Zeit, da Sie in andern Umständen waren? |
| *Do you have your regular monthlies while suckling your child?* | Haben Sie, während Sie stillen (*or* Ihr Kind nähren), regelmässig Ihr Monatliches? |
| *When did your last accouchment take place?* | Wann fand Ihre letzte Entbindung statt? |
| *Have you ever had a miscarriage?* | Hatten Sie einmal unrichtige Wochen (*or* Hatten Sie einmal eine Frühgeburt, Fehlgeburt)? |

## 63. Leucorrhœa.    63. Leucorrhöa.

| | |
|---|---|
| *What is the character of the secretion?* | D. Von welcher Beschaffenheit ist das, was Sie ausscheiden? |
| *Is it a thin discharge, or thick puriform matter?* | Ist es ein dünner Ausfluss oder ein dicker, eiterartiger Stoff? |
| *Was the discharge which is now thin and painless preceded by a copious secretion of thick puriform matter causing painful micturition?* | War dem Ausfluss, der jetzt dünn und schmerzlos ist, ein dicker und eiterartiger vorangegangen, der Ihnen Schmerz beim Harnen verursacht hat? |

| | |
|---|---|
| *Does the extensive discharge of mucus cause painful micturition?* | D. Verursacht Ihnen der übermässige Ausfluss von Schleim Schmerzen beim Harnen? |
| *What is the condition of your rectum?* | In welchem Zustande befindet sich Ihr Mastdarm? |

## 64. *Prolapsus Uteri.*    64. Der Vorfall der Gebärmutter.

| | |
|---|---|
| *Have you a sense of weight and bearing down?* | D. Haben Sie ein Gefühl der Schwere und des Drängens nach unten? |
| *Does this affection date from the time of your last pregnancy?* | Schreibt sich dieses Leiden von der Zeit Ihrer letzten Schwangerschaft her? |
| *Did you get up or walk about too soon after delivery?* | Sind Sie zu früh nach der Entbindung aufgestanden oder umhergegangen? |
| *Were you in an enfeebled state at the time when you were allowed to get up?* | Waren Sie zur Zeit, da Sie aufstehen durften, in einem geschwächten Zustande? |
| *I have never borne children.* | P. Ich habe nie Kinder geboren. |
| *Has there been any strain or violent effort that might have caused descent of the uterus?* | D. Hat eine übermässige Anstrengung den Vorfall (*or die Senkung*) der Gebärmutter veranlasst? |
| *Have you constantly carried heavy weights?* | Haben Sie beständig schwere Lasten getragen? |
| *Have you any annoyance in walking or making any exertion?* | Fühlen Sie irgend eine Beschwerde beim Gehen oder wenn Sie sich irgend anstrengen? |
| *Is there an external protrusion of the womb?* | Drängt sich die Gebärmutter äusserlich hervor? |
| *Have you any difficulty in evacuating the contents of the bladder and rectum?* | Haben Sie irgend eine Schwierigkeit beim Harnen oder Stuhlgang? |
| *Prolonged rest in the horizontal position is indispensable.* | Längere Ruhe in horizontaler Lage ist für Sie unerlässlich. |

You must receive, when in the recumbent position, injections of cold water (half a pint to a pint) into the canal leading to the uterus by means of a "fountain syringe" night and morning.

You will need to have an operation performed to restore the supports of your womb, torn away in your labor.

D. Sie müssen sich in liegender Stellung vermittelst einer elastischen Flasche langsame Einspritzungen von kaltem Wasser (eine halbe oder ganze Pinte) in den Canal, der zur Gebärmutter führt, machen lassen, und zwar des Morgens und Abends.

Es wird an Ihnen eine Operation vorgenommen werden müssen, um die Stütze für Ihre Gebärmutter wieder herzustellen, die durch die Wehen ist weggerissen worden.

## 65. Pregnancy, Childbed.

## 65. Schwangerschaft, Entbindung.

Are you pregnant?
Are you with child?

D. Sind Sie schwanger?
Sind Sie in anderen Umständen?

Are you in the family-way?

Sind Sie in gesegneten Umständen?

Are you near your time?

Sind Sie Ihrer Entbindung nahe?

How many months are you advanced in your pregnancy?
When do you expect to be delivered?
I am prepared for it every moment.
I am within one month of my time.
Have you ever had a child?

Wie viele Monate sind Sie schwanger?
Wann denken Sie niederzukommen?
P. Ich bin jeden Augenblick darauf gefasst.
Ich habe noch einen Monat bis zu meiner Entbindung.
D. Haben Sie schon ein Kind gehabt?

I have two alive, the third died; besides I have had a premature birth, and one was still-born.

P. Ich habe zwei am Leben, das dritte starb; ausserdem hatte ich eine Frühgeburt, und eins wurde todt geboren.

| | |
|---|---|
| *Have you a nurse (wet nurse)?* | D. Haben Sie schon eine Wärterin (Amme)? |
| *I have arranged everything?* | P. Ich habe alles angeordnet. |
| *Doctor, please come very soon, my wife is in labor.* | Bitte, Herr Doctor! kommen Sie recht bald, meine Frau ist in Kindesnöthen. |
| *When did her pains come on?* | D. Wie lange hat sie schon Wehen? |
| *Are they light (weak)?* | Sind dieselben leicht (schwach)? |
| *They are frequent.* | P. Sie folgen schnell auf einander. |
| *Have the waters come away?* | D. Ist das Wasser (Fruchtwasser) schon abgegangen? |
| *Is she doing well?* | Wie geht's mit ihr? |
| *She seems doing badly.* | P. Es scheint nicht besonders mit ihr zu gehen. |
| *She was delivered prematurely.* | D. Sie ist vorzeitig niedergekommen. |
| *She was brought to bed of a son.* | Sie ist mit einem Sohne niedergekommen. |
| *A fine boy.* | Ein hübscher Junge. |
| *Give me some thread, some linen.* | Geben Sie mir einen Faden und ein leinenes Läppchen. |
| *What shall I give the child?* | P. Was soll das Kind haben? |
| *A little water and sugar.* | D. Ein wenig Zuckerwasser. |
| *Has the afterbirth come away?* | Ist die Nachgeburt schon abgegangen? |
| *Will the child nurse his (her) mother?* | Wird das Kind an der Mutterbrust saugen? *or* Wird die Mutter das Kind selbst nähren (*or* stillen)? |
| *Let the child be put to its mother's breast immediately.* | Legen Sie das Kind sogleich an (*supply* die Mutterbrust). |
| *She has no milk.* | Sie kann nicht nähren, *or* Sie hat keine Nahrung. |
| *She cannot nurse the baby.* | Sie kann das Kind nicht säugen. |

| | |
|---|---|
| *We must get a wet nurse for the child.* | D. Wir müssen eine Amme für das Kind haben. |
| *We must feed him (her) with a spoon.* | Wir müssen es pappeln. |
| *We must feed him (her) with a bottle.* | Wir müssen ihm die Flasche geben. |
| *We shall bring up the child by hand.* | Wir werden das Kind mit dem Saugglass aufziehen. |
| *Can you give the breast to the child?* | Können Sie dem Kinde die Brust geben? |
| *Let her change her linen.* | Lassen Sie sie die Wäsche wechseln. |
| | Lassen Sie ihr andere Wäsche anziehen. |
| *Has her milk come?* | Hat Sie schon Milch? |
| *Yes, the baby nurses well.* | P. Ja, das Kind nimmt die Brust gut. |
| *No, her breasts are very hard, tender, painful.* | Nein, die Brüste sind ganz hart, empfindlich, schmerzhaft. |
| *The child will not take hold (of the nipple).* | Das Kind will nicht anfassen. |
| *Her nipple is sore.* | Die Brustwarze ist wund. |
| *You must have artificial nipples as a protection.* | D. Sie müssen Warzenschützer haben. |
| *How is the mother? how is the child?* | Wie geht's mit Mutter und Kind? |
| *Both are doing well.* | P. Es geht mit Beiden gut. |

## 66. Diseases of Children. 66. Kinderkrankheiten.

| | |
|---|---|
| *What is the matter with the child?* | D. Was fehlt dem Kinde? |
| *What ails the child?* | |
| *The child is sickly (poorly).* | P. Das Kind ist kränklich. |
| *There is always something the matter with him (her).* | Es fehlt ihm immer etwas. |
| *It coughs, vomits, purges.* | Es hustet, bricht, laxiert. |
| *How old is this child?* | D. Wie alt ist das Kind? |

| | |
|---|---|
| *Has he (she) any teeth yet?* | D. Hat es noch keine Zahne? |
| *It is cutting an upper (a lower) tooth.* | P. Es bekommt oben (unten) einen Zahn. |
| *Has the child had any sickness before?* | D. Ist das Kind schon einmal krank gewesen? |
| *Have you ever before lost any children?* | Haben Sie schon Kinder verloren? |
| *Of what did they die?* | Woran sind dieselben gestorben? |
| *Has the child when sleeping in a cold room uncovered himself (herself)?* | Hat sich das Kind im Schlafe bei kalter Luft aufgedeckt? |
| *Has he (she) drunk anything cold?* | Hat es Kaltes getrunken? |
| *Keep the child very warm.* | Halten Sie das Kind recht warm. |
| *Apply very warm poultices upon his (her) abdomen.* | Machen Sie ihm recht warme Umschläge auf den Leib. |
| *Give him (her) warm, mucilaginous drinks.* | Geben Sie ihm warme, schleimige Getränke. |
| *Does the child get enough nutriment?* | Bekommt das Kind genug Nahrung? |
| *Is it not suffering for want of proper food (lit. does it not perish with hunger)?* | Verhungert es nicht? |
| *Is the nourishment suitable?* | Ist die Nahrung zweckmässig? |
| *Is the child very thirsty?* | Trinkt das Kind sehr viel? |
| *Yes, but it vomits up the milk immediately.* | P. Ja, aber es bricht die Milch gleich wieder aus. |
| *Please, let me see how the child nurses.* | D. Bitte, lassen Sie mich doch einmal sehen, wie das Kind die Brust nimmt. |
| *The child is already weaned from the breast.* | P. Das Kind ist schon entwöhnt. |
| *How old was it when you weaned it?* | D. Wie alt war es, als Sie es entwöhnten? |

| | |
|---|---|
| *It was but eight months old; but I had not sufficient milk for it.* | P. Es war erst acht Monate alt, aber ich hatte nicht Nahrung genug. |
| *What food do you give it?* | D. Womit nähren Sie es? |
| *I give milk and water, half and half.* | P. Ich gebe halb Milch, halb Wasser. |
| *Does the child have a tendency to convulsions?* | D. Zeigt das Kind eine Neigung zu Krämpfen? |
| *Has it any fits (or spasms)?* | Hat es oft krampfhafte Zufälle? |
| *Do these convulsions return periodically without the presence of any known cause?* | Kommen die Krämpfe periodisch immer wieder, ohne dass eine erregende Ursache bekannt wäre? |
| *Does the child suffer much during dentition?* | Leidet das Kind viel beim Zahnen? |
| *It is always ill when teething.* | P. Es ist immer krank an den Zähnen. |
| *How old was it when it cut its first tooth?* | D. Wie alt war es, als es den ersten Zahn bekam? |
| *Does it sleep soundly?* | Hat es einen gesunden Schlaf? |
| *It is very fretful; it often cries out in its sleep.* | P. Es ist sehr unruhig; es schreit oft auf im Schlafe. |
| *You must guard the child from exposure during dentition, and regulate his (her) bowels, if necessary, by injections.* | D. Sie müssen das Kind während des Zahnens vor Erkältungen schützen und seinen Stuhlgang, wenn nothwendig, durch Klystiere regeln. |
| *Are its bowels right?* | Hat es gehörige Leibesöffnung? |
| *No, Sir, they are costive.* | P. Nein, es ist verstopft. |
| *No, Sir, they are too loose.* | Nein, es hat den Durchfall. |
| *What has he (she) eaten?* | D. Wann hat es zu essen bekommen? |
| *It has refused to taste anything.* | P. Es will durchaus nichts zu sich nehmen. |
| *Is it suffering from thrush?* | D. Hat es Schwämmchen im Munde? |
| *Has it a sore mouth?* | |

| | |
|---|---|
| *Does the child make water regularly* (lit. *Does it wet his (her) swaddling clothes fairly*)? | D. Macht das Kind die Windeln gehörig nass? |
| *Has the child been subject to coughs and colds?* | Ist das Kind Husten und Erkältungen unterworfen? |
| *Does the cough end in copious expectoration?* | Endet der Husten mit reichlichem Auswurf? |
| *Has the child had whooping-cough?* | Hatte das Kind Keuchhusten? |

### 67. *Hydrocephalus.*    67. Wasserkopf.

| | |
|---|---|
| *Is he (she) fretful?* | D. Ist er (sie) reizbar (ärgerlich)? |
| *Has he (she) any aversion to light?* | Ist ihm (ihr) das Licht widerwärtig? *or* Scheut er (sie) das Licht? |
| *Does he (she) turn away peevishly when you attempt to amuse him (her) or occupy his (her) attention?* | Wendet er (sie) sich verdriesslich ab, wenn Sie versuchen, ihn (sie) zu belustigen (unterhalten) oder seine (ihre) Aufmerksamkeit zu fesseln? |
| *Has he (she) any dislike to the erect posture?* | Ist es ihm (ihr) widerwärtig, aufrecht zu stehen (*or* Mag er (sie) nicht stehen)? |
| *Is thirst urgent?* | Hat er (sie) viel Durst? |
| *Does he (she) vomit habitually after taking food?* | Bricht es gewöhnlich, wenn es was gegessen? |
| *Is there a tendency to diarrhœa?* | Hat er (sie) eine Neigung zum Durchfall (*or* Leidet er (sie) viel an Durchfall)? |

### 68. *Thrush.*    68. Schwämmchen.

| | |
|---|---|
| *Has the child diarrhœa, colic, much thirst?* | D. Hat das Kind Diarrhöe, Kolik, viel Durst? |
| *Let me look into his (her) mouth.* | Lassen Sie mich ihm (ihr) in den Mund sehen. |

| | |
|---|---|
| *The child has the thrush.* | D. Das Kind hat den Soor (Mundschwamm *or* die Schwämmchen). |
| *That will soon pass away, if you keep his (her) mouth perfectly clean, and wash it out in all parts with lukewarm water after drinking.* | Das wird sich bald geben, wenn Sie ihm (ihr) den Mund streng reinlich halten und denselben jedesmal nach dem Trinken allenthalben (*or* in allen Theilen) mit reinem, lauwarmen Wasser auswaschen. |
| *Also be careful to give him (her) fluid, mild food (milk, broth), pure air and clean clothing.* | Sorgen Sie auch für flüssige, milde Nahrung (Milch, Fleischbrühe), reine Luft und reine Wäsche.. |

## 69. False Croup. — 69. Falscher Croup.

| | |
|---|---|
| *Was the child unwell last night?* | D. War das Kind gestern Abend unwohl? |
| *No, Sir, it went to bed well.* | P. Nein, es war beim Schlafengehen gesund. |
| *It had a slight cold.* | Es hat an einem leichten Katarrh gelitten. |
| *It was somewhat indisposed from teething.* | Es war etwas unpässlich vom Zahnen. |
| *When did the attack occur?* | D. Wann bekam es den Anfall? |
| *At night.* | P. In der Nacht. |
| *How long did the paroxysm continue?* | D. Wie lange dauerte der Anfall? |
| *For about an hour.* | P. Ungefähr eine Stunde. |
| *Did the child then fall asleep quietly? Yes.* | D. Schlief das Kind dann ruhig ein? P. Ja. |
| *When was the attack renewed?* | D. Wann kam der Anfall wieder? |
| *Towards morning.* | P. Gegen Morgen. |
| *Did the child in the time between the paroxysms have the same symptoms, difficulty in breathing, the ringing cough, and the husky voice?* | D. Hatte das Kind in der Zeit zwischen den Anfällen dieselben Symptome, die Athemnoth, den bellenden Husten und die heisere Stimme? |

No, voice and respiration were natural.
Was it feverish in these intervals?
No, or very little.
The child has caught a severe cold.

P. Nein, Stimme wie Athmen waren natürlich.
D. War es fieberisch in dieser Zwischenzeit?
P. Nein, oder sehr wenig.
D. Das Kind hat eine starke Erkältung erfahren.

### 70. True Croup.

Was the attack a sudden one?
No, Sir, a slight cold with fever, and some hoarseness, preceded for a few days.
Did the present symptoms, the high fever, difficulty in breathing, the peculiar cough and altered voice disappear in the intervals between the attacks?
No, they did not cease, and were more or less severe; always present.
Has the child much thirst? Yes.

Has it any appetite? Very little.

Has it any difficulty in swallowing? No.
When did the child's voice appear changed to you?
Almost from the beginning of the sickness.
Does not the child expectorate?
Yes, it sometimes vomits or coughs up solid masses of membrane.

### 70. Die häutige Bräune.

D. Kam der Anfall plötzlich?
P. Nein, leichtes Fieber, Katarrh und etwas Heiserkeit gingen einige Tage lang voraus.
D. Waren die gegenwärtigen Symptome, die Athemnoth, der eigenthümliche Husten und die veränderte Stimme in der Zeit zwischen den Anfällen geschwunden?
P. Nein, sie hatten nicht aufgehört, und waren bald stärker, bald weniger stark ununterbrochen vorhanden.
D. Hat das Kind viel Durst? P. Ja.
D. Hat es Appetit? P. Nur sehr geringen.
D. Hat es Beschwerden beim Schlucken? P. Nein.
D. Wann kam Ihnen die Stimme des Kindes verändert vor?
P. Vom ersten Beginne der Krankheit.
D. Wirft das Kind nichts aus?
P. Zuweilen werden von ihm feste Massen von Haut ausgeworfen oder ausgebrochen.

*Is the child always so restless?* — D. Ist das Kind immer so unruhig?

*Yes, Sir, it tosses about in bed and cannot find any rest; it also throws its head back.* — P. Ja, es wirft sich hin und her im Bette und kann keine Ruhe finden, auch wirft es den Kopf nach hinten zurück.

*Give the child the powder every two hours.* — D. Geben Sie dem Kinde alle zwei Stunden das Pulver.

*Maintain an even and moist temperature in the room, and therefore continually admit the steam from a kettle of boiling water into it.* — Erhalten Sie eine feuchte Temperatur im Zimmer, und lassen Sie daher den Dampf eines fortwährend kochenden Wasserkessels in dasselbe zu.

## 71. Measles. — 71. Die Masern.

*How long has the child been sick?* — D. Seit wann ist das Kind krank?

*It has not been quite well for the last eight or ten days; but about four days ago a fever and a bad catarrh made their appearance.* — P. Es war schon seit acht bis zehn Tagen nicht recht gesund, aber vor vier Tagen zurück stellte sich Fieber ein und ein starker Katarrh.

*When did the rash come on?* — D. Wann ist der Ausschlag erschienen?

*This morning.* — P. Heute früh.

*The child has the measles. It must remain in bed, and you must watch it night and day to prevent it from lying uncovered and catching cold.* — D. Das Kind hat die Masern. Es muss im Bette bleiben, und Sie müssen es Nacht und Tag bewachen, dass es sich nicht aufdecke und erkälte.

*The temperature ought to be regulated by the thermometer and kept at 60° to 65° Fahrenheit.* — Die Temperatur muss nach dem Thermometer geregelt und auf 60° bis 65° Fahrenheit erhalten werden.

*Keep a current of fresh air circulating in the room without exposing the child to it.* — Bringen Sie täglich eine Strömung frischer Luft ins Zimmer, ohne das Kind einer Erkältung auszusetzen.

The diet must be easily digestible and fluid, given often, but in small quantities.

The child ought to remain in bed, as long as fever is present and desquamation is going on, and it must not leave the room so long as catarrh is present.

Keep the room darkened, to protect the inflamed eyes of the child.

D. Die Kost sei leicht verdaulich und flüssig, öfters, aber in kleinen Quantitäten verabreicht.

Das Kind muss durchaus im Bette bleiben, so lange das Fieber dauert und die Abschuppung stattfindet, und es darf nicht aus dem Zimmer, so lange noch Katarrh vorhanden.

Verdunkeln Sie auch wegen der entzündeten Augen des Kindes das Zimmer.

## 72. Worms.

Does the child pick his nose?

Does it scratch its anus?
Has it a ravenous appetite?
Has he (she) passed any worms?

Have worms or portions of them come away with the fæces?

What sort of worms came out?

Give him (her) this powder tonight and a dessertspoonful of the medicine in the morning before breakfast.

## 72. Würmer.

D. Stochert sich das Kind in der Nase?

Kratzt es sich am After?
Hat es einen Heisshunger?
Sind ihm Würmer abgegangen?

Gingen ihm Würmer oder Theile derselben mit dem Stuhle ab?

Was für Würmer sind abgegangen?

Geben Sie ihm (ihr) heute Abend dieses Pulver und des Morgens vor dem Frühstück einen Desertlöffel voll von der Medicin.

## 73. Growth above the Palate.

Does the child usually sleep with the mouth open?

## 73. Gewächs über dem Gaumen.

D. Schläft das Kind gewöhnlich mit offenem Munde?

*He (she) has slept with his mouth open for two years, and usually has a cough.*  P. Es hat zwei Jahre lang mit offenem Munde geschlafen und hat gewöhnlich Husten.

*There is quite a growth above the palate that should be removed.*  D. Es hat ein Gewächs (eine Geschwulst) über dem Gaumen, das entfernt werden muss.

*Give him the medicine to-night.*  Geben Sie ihm heute Abend die Medicin.

*The medicine is all used.*  P. Die Medicin ist verbraucht.

*Get some more of the same kind.*  D. Lassen Sie dieselbe noch einmal machen.

*Is it possible for him (her) to recover?*  P. Ist noch Besserung möglich?

*I hope to find him (her) better to-morrow.*  D. Ich hoffe ihn morgen besser zu finden.

*Note.* For SCARLET FEVER *see* p. 234, and for DIPHTHERIA, *see* p. 180.

## 74. Toothache.    74. Zahnweh.

*Good morning, Sir (Madam), what can I do for you?*  D. Guten Morgen, mein Herr (Madame), womit kann ich Ihnen dienen?

*I have been suffering from toothache for some time. Yesterday the pain became so severe that I spent a sleepless night.*  P. Ich leide seit einiger Zeit an Zahnschmerzen. Seit gestern jedoch sind dieselben so heftig geworden, dass ich die Nacht schlaflos zugebracht.

*Do you know which tooth it is?*  D. Wissen Sie, welcher Zahn es ist?

*I cannot tell you exactly, for all ache, and in fact I believe that it is rheumatism.*  P. Genau weiss ich es nicht, denn alle schmerzen, und glaube ich, dass es Rheumatismus ist.

*Do both rows of teeth ache? Yes, Sir.*  D. Sind beide Zahnreihen vom Schmerze befallen? P. Ja.

*Does the pain change its seat? Yes, Sir.*  D. Wechselt der Schmerz seinen Sitz? P. Ja.

| | |
|---|---|
| *You must use a sweating cure, footbaths. You must have poultices made.* | D. Sie müssen eine Schwitzcur, Fussbäder brauchen. Sie müssen sich Umschläge machen lassen. |
| *No, Sir, I always feel the pain in one place, particularly in this part.* | P. Nein, ich fühle den Schmerz immer nur an einer Stelle, besonders in dieser Gegend. |
| *Please to let me examine it.* | D. Bitte, lassen Sie mich dieselbe untersuchen. |
| *Be kind enough to seat yourself on this chair.* | Setzen Sie sich gefälligst auf diesen Stuhl hier. |
| *Is this the right one?* | Ist dies der rechte? |
| *The next one I believe.* | P. Ich glaube der daneben. |
| *That cannot be, for the next one is sound.* | D. Das kann nicht sein, der daneben ist gut. |
| *But it aches also.* | P. Aber der thut mir auch weh. |
| *That may be. But the pain is sympathetic, really emanating from this one, and extending to the sound teeth.* | D. Das mag sein. Aber der Schmerz ist sympathisch; er rührt wirklich nur von diesem einen her, und erstreckt sich auch auf die gesunden (or wird auf die gesunden übertragen). |
| *Do you feel more pain when I beat upon this or that one?* | Empfinden Sie mehr Schmerz, wenn ich auf diesen oder jenen klopfe. |
| *It is this one without doubt, and I also see that it is hollow.* | Es ist ohne Zweifel dieser, und ich sehe auch, er ist hohl. |
| *In my opinion, it is not necessary to extract the tooth, which might be of service to you many years longer. I can fill it for you with gold, silver or platina.* | Meiner Meinung nach wäre es nicht nöthig, den Zahn auszuziehen. Er könnte Ihnen noch viele Jahre gute Dienste leisten. Ich kann ihn mit Gold- Silber- oder Platinablättchen plombiren (ausfüllen). |
| *No, I would rather have it extracted, for I fear I might get toothache again.* | P. Nein, ich möchte ihn lieber ausziehen lassen, denn ich befürchte, dass ich wieder Zahnweh bekommen könnte. |

| | |
|---|---|
| *You need have no fears on that score. The nerve can be made insensible, or be protected against irritation.* | D. Das haben Sie keineswegs zu befürchten. Der Nerv kann unempfindlich gemacht, oder er kann vor Reizung geschützt werden. |
| *What would you advise me to do?* | P. Wozu rathen Sie mir? |
| *I would advise you to have it filled, and to preserve it by this means.* | D. Ich rathe Ihnen, denselben plombieren zu lassen und ihn dadurch sich zu erhalten. |
| *The bad tooth must be removed: it is too badly decayed, and cannot be filled.* | Der schadhafte Zahn muss entfernt werden; er ist zu sehr angefressen und kann nicht mehr plombiert werden. |
| *You have an ulcer under this tooth.* | Sie haben unter diesem Zahn ein Geschwür. |
| *You have a fistula, which has opened outwardly on the gums (or on the cheek). It comes from this tooth (or this root). As soon as the tooth (or the root) is removed, and the abscess opened, the fistula will close, and the pain will cease.* | Sie haben eine Zahnfistel, die sich nach aussen am Zahnfleisch (or auf der Backe) geöffnet. Sie rührt von diesem Zahne (or von dieser Zahnwurzel) her. Sobald der Zahn (or die Wurzel) entfernt und der Abscess geöffnet ist, wird die Fistel sich schliessen und auch der Schmerz aufhören. |
| *Please to keep still; it will soon be done!* | Bitte, verhalten Sie sich ruhig; es wird bald geschehen sein. |
| *Would you like me to use chloroform or æther?* | Wünschen Sie, dass ich Chloroform oder Aether anwende? |
| *Yes, whichever you think best.* | P. Ja, was Sie für's Beste halten. |
| *No, Sir, that causes my head to ache for days, and as I hear, is not altogether without danger.* | Nein, es verursacht mir das tagelang Kopfweh, auch ist es, wie ich gehört, nicht ganz ohne Gefahr. |
| *No, I prefer the pain to an interruption of the mental faculties.* | Nein, ich ziehe Schmerz der Unterbrechung der Geisteskräfte vor. |
| *Oh! that was a frightful pain!* | O das war ein schrecklicher Schmerz! |

| | |
|---|---|
| *I believe you! But just look at the tooth. A tooth with three such roots!* | D. Das glaube ich! Aber sehen Sie sich den Zahn an. Ein Zahn mit solchen drei Wurzeln! |
| *Rinse out your mouth, please.* | Bitte, spülen Sie sich den Mund aus. |
| *Do you wish cold or warm water?* | Wollen Sie kaltes oder warmes Wasser? |
| *It is well that it bleeds so profusely.* | Es ist gut, dass es so stark blutet. |
| *Press the gum together, please.* | Bitte, drücken Sie das Zahnfleisch zusammen. |
| *If you will permit me, I would advise you to take better care of your teeth. They are black and more or less hollow. This is caused by teeth animalculæ and fungi, which destroy the teeth and cause the aching.* | Wenn Sie mir erlauben, würde ich Ihnen rathen, etwas mehr Sorgfalt auf Ihre Zähne zu verwenden. Ihre Zähne sind schwarz und zum Theil mehr oder weniger hohl. Es kommt das von mikroskopischen Zahnthierchen und Zahnpilzchen her, welche die Zähne zerstören und die Zahnschmerzen verursachen. |
| *You ought to have the hollow teeth filled.* | Sie sollten sich die hohlen Zähne füllen lassen. |
| *You must also rinse your mouth properly after every meal.* | Auch müssen Sie sich nach jeder Mahlzeit den Mund ordentlich ausspülen. |
| *Furthermore, you must brush your teeth thoroughly in the morning, and at first also in the evening with strong spirits (to which is to be added some acetic æther) or with cologne.* | Ferner des Morgens und anfangs auch des Abends die Zähne von allen Seiten mit starkem Spiritus (dem etwas Essigäther zuzusetzen ist) oder mit Kölnischem Wasser tüchtig abbürsten. |
| *The deposit of tartar may be prevented by occasional scouring with finely powdered pumice (stone) or charcoal.* | Dem Ansetzen von Zahnstein können Sie durch zeitweiliges Abscheuern mit feingepulvertem Bimstein oder Kohlenpulver entgegenwirken. |

| | |
|---|---|
| Here is a very good dentifrice, that I can warmly recommend to you. | D. Hier ist ein sehr gutes Zahnpulver, das ich Ihnen sehr empfehlen kann. |
| Here is also an excellent tincture for toothache arising from a cold. | Hier ist auch eine Tinktur, die gegen Zahnweh, das in Folge einer Erkältung entstanden, vorzüglich ist. |
| What do I owe you, please? | P. Bitte, was habe ich zu entrichten (or Was bin ich Ihnen shuldig) ? |
| For extracting . . . . and for the rest . . . . | D. Für das Zahnausziehen . . und für das Übrige . . . . |
| Pardon me, madam, but you have a tooth here which disfigures you somewhat. You ought to have it extracted. | Sie haben hier, Madame, einen Zahn, der — Sie verzeihen — Sie sehr entstellt. Sie sollten ihn herausziehen lassen. |
| And what should I do without it? | P. Und was soll ich ohne denselben thun ? |
| I can put a beautiful artificial one in its place. | D. Ich kann ihn durch einen sehr schönen, kunstvollen ersetzen. |
| How much would it cost? | P. Wie viel würde es kosten ? |
| Only . . . . | D. Nur . . . . |
| What do you charge for a full set? | P. Wie viel kostet ein ganzes Gebiss ? |
| The price varies, from 20 to 50 dollars. | D. Das ist verschieden, von 20 bis 50 Thaler. |

## 75. At a Druggist.   75. Beim Apotheker.

| | |
|---|---|
| Good morning, Sir. Here is a prescription. Can you make it up for me quickly? | P. Guten Morgen! Ich bringe hier ein Recept; können Sie es mir schnell besorgen ? |
| I am sorry, Madam, you will have to wait a little. It will take a good half-hour to prepare this prescription. | D. Es thut mir leid, Madame, Sie werden etwas warten müssen. Die Anfertigung wird wohl eine gute halbe Stunde dauern. |

| | |
|---|---|
| *I cannot wait so long, and would prefer to call again, as I have some other errands to do.* | P. Ich kann nicht so lange warten, und ich werde lieber wiederkommen, da ich noch einige andere Aufträge zu besorgen habe. |
| *I think I can wait, but you would oblige me greatly by preparing it as soon as possible.* | Ich kann wohl warten; doch würden Sie mich verbinden, wenn Sie das Medicament so schnell wie möglich bereiten wollten. |
| *Pray be seated.* *Please to take a seat.* | D. Nehmen Sie gefälligst Platz, *or* Bitte, setzen Sie sich. |
| *The doctor explicitly advised me to try to get the medicine fresh and unadulterated, and therefore I have come to you, although I do not live in this neighborhood any longer.* | P. Der Doctor hat mir ausdrücklich empfohlen, die Arzenei frisch und unverfälscht zu bekommen, und daher kam ich hierher, obwohl ich jetzt nicht mehr in Ihrer Nähe (*or* Nachbarschaft) wohne. |
| *How must I take this medicine (or How must this remedy be applied or used)?* | Wie muss die Arzenei genommen (*or* Wie muss das Mittel gebraucht) werden? |
| *Ten drops to be taken three times a day in a tumbler of sugar-water.* | D. Dreimal des Tages zehn Tropfen in einem Glase Zuckerwasser. |
| *I do not like a medicine that has to be taken by drops, for fear that I should not count the drops accurately, and give (take) either more or less.* | P. Ich liebe nicht ein Medicament, das in Tropfen verabreicht wird, da ich immer besorge, die Tropfen nicht genau zu zählen und bald mehr, bald weniger zu geben. |
| *You should procure a dropping-tube, which costs but a trifle.* | D. Sie sollten sich einen Tropfenzähler anschaffen, was nur wenig kostet. |
| *Put a tablespoonful of this tea in a pint of boiling water and drink it at pleasure.* | Sie sollen einen Esslöffel voll von diesem Thee in einem Quart Wasser aufkochen lassen und nach Belieben davon trinken. |
| *Commence by taking one pill every morning and evening, augment-* | Sie sollen damit anfangen, eine Pille morgens und abends zu |

| | |
|---|---|
| ing the dose by one pill every ten days. | nehmen und alle zehn Tage um eine Pille höher gehen. |
| *Have this powder blown into your ear, three or four times a day.* | Lassen Sie sich von diesem Pulver täglich drei bis viermal ins Ohr einblasen. |
| *Snuff some of this powder five or six times a day.* | Schnupfen Sie fünf bis sechsmal täglich von diesem Pulver. |
| *Put the plaster on your breast, and let it remain there for 24 hours.* | Legen Sie das Pflaster auf die Brust, und lassen Sie es 24 Stunden darauf liegen. |
| *Take about half a teaspoonful of this ointment and rub the aching part well.* | Nehmen Sie etwa einen halben Theelöffel von dieser Salbe, und reiben Sie sich damit die schmerzhafte Stelle gut ein. |
| *Wash the shingles three times a day with this solution, and let the moistened place dry without wiping.* | Waschen Sie den Ausschlag dreimal des Tages mit diesem Wasser, und lassen Sie die so befeuchtete Stelle von selbst trocknen. |
| *Let two drops fall into the affected eye every morning. Make use of the medicine-dropper for this purpose.* | Lassen Sie jeden Morgen zwei Tropfen in das kranke Auge fallen. Bedienen Sie sich dabei des Tropfglases. |
| *Will this medicine keep a long time, without deteriorating ?* | P. Lässt sich die Arzenei für längere Zeit aufbewahren ohne zu verderben? |
| *Yes, but the vial must be hermetically closed and kept cold.* | D. Ja, Sie müssen sie aber luftdicht verschliessen und kalt stehen lassen. |
| *Do you think this glass stopper (cork) closes hermetically ? O yes.* | P. Glauben Sie, dass dieser gläserne Stöpsel luftdicht verschliesst? D. O ja! |
| *The doctor also recommended wine, and explicitly said that it must be good and neither diluted nor adulterated.* | P. Der Doctor empfiehlt auch Wein und bemerkt ausdrücklich, dass derselbe gut und nicht verdünnt oder verfälscht sein darf. |

| | |
|---|---|
| *Would you be kind enough to write the direction for use down for me?* | P. Wollten Sie so gut sein, und mir die Gebrauchsanweisung aufschreiben? |
| *I have done so, you will find the direction upon the box (or the bottle).* | D. Das ist geschehen; Sie werden die Aufschrift auf der Schachtel (oder der Flasche) finden. |
| *How must I use the solution?* | P. Wie habe ich das Wasser zu gebrauchen? |
| *Rinse out your mouth with it, and retain it in the mouth for several minutes before ejecting it.* | D. Spülen Sie sich damit den Mund aus, und behalten Sie es einige Minuten im Munde, ehe Sie es ausspeien. |
| *This old woman's remedy was also recommended to me, it is said to be approved (or to hold good). What do you think of it?* | P. Es wurde mir auch dieses Hausmittel anempfohlen, das probat sein (or das sich bewähren) soll. |
| *I cannot say; try it to see if it will do you good.* | D. Ich kann nicht sagen; versuchen Sie, ob Ihnen das gut thun wird. |
| *I would not advise you to doctor yourself.* | Ich möchte Ihnen nicht rathen, an sich selbst zu doctern. |
| *Your physician is a skillful and experienced man, and the bare confidence that such a man inspires often contributes much toward recovery.* | Ihr Arzt ist ein geschickter und erfahrner Mann, und schon das Vertrauen, das ein solcher Mann einflösst, trägt oft viel zur Wiederherstellung bei. |
| *You are perfectly right (or I share your opinion). I will leave quackery to others. I, for my part, can have confidence in a conscientious, scientific physician only.* | P. Sie haben vollkommen Recht (or ich theile ganz Ihre Meinung). Ich will das Quacksalbern Anderen überlassen. Ich, für meine Person vertraue nur einem gewissenhaften und wissenschaftlich gebildeten Arzte. |

# ALPHABETICAL INDEX

OF

# GERMAN WORDS AND TERMS.

## A

Aal, 4003
abbürsten, 5569
Abdominaltyphus, 3119
Abdominalvergrösserung, 3009
Abendbrod, 3927
Abendessen, 3928
abendlich, 5179
abführen, 4191
Abführmittel, 4194
abgeflacht, 2472
abgehen, 5487
abgemagert, 2438
abgestumpft, 5391
abgezehrt, 2439
abhelfen, 5412
ablagern, 930
Ablagerung, 931
ablassen, 5313
ablegen, 4750
abmagern, 2433
Abmagerung, 2435
Abmagerungscur, 2447
Abnahme, 5338
abnehmen, 2424
Abnehmen, 2425
Abneigung, 2163
abnorm, 2292
Absatz, 5310
(sich) abschälen, 1935
abscheren, 5393
abscheuern, 5575
Abschuppung, 1934
absondern, 2275
Abspannung, 2637
absteigen, 676
Absteigen der Aorta, 678
absterben, 5474
abstossen, 2144

abstumpfen, 5390
abtreiben, 3296
Abtritt, 4597
abtrocknen, 5409
abwärts, 5334
Abwaschung, 5372
abwechseln, 2114
Abweichen, 2243
(sich) abwenden, 5525
abzehren, 2434
Abzehrung, 2436
abziehen, 561
Abzieher, 565
Accoucheur, 3829
Achillesflechse, 621
Achillessehne, 620
Achsel, 400
Achselbein, 401
Achselfaser, 821
Achselgrube, 409
Achselhaar, 971
Achselhöhle, 408
Acht, 4840
(in) Acht nehmen, 4841
achtlos, 2514
acut, 1628
Adamsapfel, 1463
Addison'sche Krankheit, 3068
Ader, 662
(zur) Ader lassen, 4251
Aderbinde, 3593
Adergeflecht, 682
Aderhaut, 1344
Aderlasszeug, 3761
ähnlich, 68
ändern, 4820
Ärmel, 4689
Äther, 4290
ätzen, 3314

Ätzmittel, 4223
Ätzstifthalter, 3681
aficieren, 4999
After, 1133
Afterdarm, 1135
Aftergegend, 1134
Afterschmerz, 1848
gichtischer ——, 1849
Agrypnie, 2591
Alaun, 4307
Albinismus, 3214
Alkohol, 4305
allenthalben, 5531
allgemein, 1974
allmählig, 1653
Allopath, 3836
allopathisch, 4116
Aloë, 4306
Alp, 2607
Alpdrücken, 2606
alt, 4901
Altheesyrup, 4329
Altheewurzel, 4328
Alter, 1242
alternierend, 1666
alterschwach, 1587
Alterschwäche, 1588
Ambos, 1409
Ambulanz, 3587
Ameise, 1767
Ameisenkriechen, 1769
Ameisenlaufen, 1770
Amenorrhöa, 3265
Amme, 1298
Amt, 3811
Anämie, 3066
Anästhesie, 2622
Anästheticum, 4289
Ananas, 4055
andauernd, 1727

ander, 4917
Anfall, 2212
anfangen, 4794
anfangs, 5097
anfassen, 5510
anfertigen, 5589
anfressen, 2897
angeben, 4963
angefressen, 2898
(sich) ängstigen, 5129
ängstlich, 2506
anhalten, 4941
anhaltend, 4980
Anisöl, 4400
Anlage, 32
anlegen, 31
(ein Kind) anlegen, 5509
anliegen, 5208
anordnen, 5505
anreizen, 4215
Ansammlung, 5469
anschaffen, 5603
anschwellen, 2399
Anschwellung, 1761
ansehen, 5582
ansetzen, 5571
Anstalt, 3867
anstecken, 3502
ansteckend, 1638
(nicht) ———, 1640
Ansteckung, 3504
Ansteckungsgift, 4532
Ansteckungsstoff, 4533
sich anstrengen, 4950
Anstrengung, 4985
Antagonist, 583
anweisen, 5364
Anweisung, 5365
anwenden, 4961
Anwendung, 5055
anwesend, 2323
Anzeichen, 5024
anziehen, 562; 4746
Anzieher, 566
Anzug, 4680
anzünden, 4614
Aorta, 673
Apfel, 4044
Apfelwein, 4096
Apoplexie, 2763
Apotheke, 3859
Apotheker, 3849
Apothekergehilfe, 3850
Apothekerlehrling, 3851
Apothekerladen, 3860
Apothekerwaaren (pl.), 3854
Appetit, 2148
Appetit bekommen, 2149
(den) Appetit reizen, 2150
Appetitlosigkeit, 2153
Appetitmangel, 2154

Apricose, 4048
arg, 4834
ärgerlich, 5522
Arm, 410
Armgeflecht, 890
Arsenik, 4520
Art, 70
Arterie, 669
arteriell, 714
Arterienerweiterung, 2846
arterienartig, 71
articulieren, 2575
Arzenei, 3857
Arzeneikräuter, 3858
Arzeneimittel, 4185
Arzeneipflanzen (pl.), 3858
Arzeneitrank, 4845
Arzt, 3805
ärztlich, 3806
Aschenkasten, 4625
Askaride, 3552
Ast, 794
Asthma, 2817
Athem, 985
(kurzer) Athem, 2014
Athem holen, 988
Athem schöpfen, 989
(ausser) Athem, 994
(den) Athem anhalten, 995
Athembeschwerde, 5035
athemlos, 2017
Athemnoth, 2023
Athemzug, 993
(schwere) Athemzüge (pl.), 2018
athmen, 984
(oberflächliches) Athmen, 2026
(schwer) athmen, 991
(tief) athmen, 990
Athmung, 4151
Athmungsbeschwerde, 2032
Athmungsheilung, 4152
Athmungsmuskeln, 1004
Athmungsprocess, 993
Atlas, 361
aufathmen, 2015
aufbewahren, 5609
aufbinden, 4754
aufblähen, 5170
aufblasen, 2405
aufbrechen, 3223
aufdecken, 5512
aufdunsen, 2406
Aufenthalt, 5168
aufessen, 3881,
auffahren im Schlafe, 2609
aufgeblasen, 1871

Aufgeblasenheit, 2408
aufgedunsen, 1872
Aufgedunsenheit, 2407
aufgelegt, 4819
aufgeregt, 2509
aufgetrieben, 2403
(sich) aufhalten, 5378
Aufhebebinde, 3603
aufheben, 1312; 5181
Aufheber, 1313
aufhören, 5238
aufknöpfen, 4758
aufkochen, 5604
auflegen, 5438
(sich) aufliegen, 3336
Aufliegen, 3333
auflösen, 4221
Auflösungsmittel, 4222
aufmachen, 4756
aufmerksam, 4935
Aufmerksamkeit, 5530
aufrecht, 2340
aufregen, 2640
Aufregung, 2641
aufreissen, 3381
aufschrecken im Schlafe, 2608
aufschnüren, 4764
aufschreiben, 5613
aufschreien, 5519
Aufschrift, 5614
aufschwellen, 5336
(sich) aufsetzen, 4776
aufspringen, 3339
aufstehen, 4872
aufsteigen, 677
aufsteigende Aorta, 679
aufstossen, 2201
Aufstossen, 2202
Auftrag, 5590
auftreiben, 2402
auftreten, 2033
aufwachen, 4804
Aufwachen, 4805
aufweichen, 5274
aufweisen, 5257
aufziehen, 1303
——mit der Flasche, 1304
Augapfel, 1336
Auge, 296
Augenarzt, 3824
Augenblick, 2693
(lichte) ——e, 2694
Augenbraue, 1318
Augenbraune, 1318
(die) Augenbrauen in die Höhe ziehen, 1482
(die) —— senken, 1483
Augenbrauenrunzler, 1320
Augenbutter, 1335
Augenentzündung, 3245

Augenfluss, 3435
Augenhaut, 1337.
(harte) ——, 1338
(weisse) ——, 1339
(schwarze) ——, 1345
Augenhöhle, 1311
Augenkammer, 1361
(hintere) ——, 1363
(vordere) ——, 1362
Augenkrebs, 3441
Augenlid, 1321
Augenlidband, 1324
Augenlidknorpel, 1323
Augenlidkrampf, 3444
Augenmittel, 4271
Augenmuskel, 589
Augenmuskelnerv (äusserer), —— 869
(gemeinschaftlicher) ——, 867
Augennerv, 5444
Augenpulver, 4273
Augensalbe, 4272
Augenschmerz, 1787
Augenthränenfistel, 3440
Augentriefen, 3431
Augenwasser, 4270
Augenweh, 1788
Augenzahn, 297
ausarten, 4842
ausathmen, 997
Ausathmung, 2013
ausbleiben, 4976
ausbrechen, 5406
(sich) ausbreiten, 5410
ausdehnen, 743
Ausdehnung, 744
Ausdruck, 2638
ausdrücklich, 5593
ausdruckslos, 2521
ausdünsten, 1965
Ausdünstung, 1966
auseinander, 2729
ausfinden, } 5359
ausfindig machen, }
Ausfluss, 2063
ausfüllen, 5555
Ausgang, 5330
ausgebreitet, 1731
aushöhlen, 5448
aushusten, 5126
auskleiden, 4753
Ausleerung, 5198
auslöschen ein Licht, 4616
Ausnahme, 5461
(sich) ausruhen, 2377
Aussatz, 3217
ausscheiden, 974
Ausscheidung, 2333
Ausscheidungsdrüsen, 975
Ausschlag, 1900
ausschliesslich, 5254

ausschweifen, 5162
ausschwitzen, 1967
Ausschwitzen, } 1968
Ausschwitzung, }
aussehen, 2322
Aussehen, 1589
(krankhaftes) —— } 1590
(ungesundes) ——, }
aussetzen, 4952
aussetzend, 2003
äusser, 922
äusserlich, 1721
äussern, 5182
(sich) ——, 5183
ausserordentlich, 2174
äusserst, 5396
aussickern, 3383
ausspeien, 2083
Ausspeien, 2086
ausspreiten, 4781
ausspreizen, 4782
ausspritzen, 1198
Ausspritzen, 4247
Ausspritzung, 1199
Ausspritzungsgang, 1200
ausspucken, 2084
ausspülen, 5563
ausstossen, 2484
ausstrahlend nach allen Seiten, 1732
ausstrecken, 2337
ausströmen, 1983
Auster, 4011
auswärts, 572
Auswärtsdreher, 576
——wender, 577
——roller, 578
auswaschen, 5532
auswerfen, 2078
auswinden, 5193
auswischen, 5308
Auswuchs, 3389
Auswurf, 2079
Auszehrung, 2437
ausziehen, 3883; 4743

**B**

Backe, 267
backen, 3939
Backenbein, 269
Backenmuskel, 531
Backenzahn, 295
Backofen, 4652
Backwerk, 3950
Bad, 3764
Badecur, 4119
Bademutter, 3831
Bader, 3843
bähen, 3317
Bahn, 724
Bahre, 3570

Bähung, 4233
Baldrianöl, 4411
Baldriansäure, 4204
Balggeschwulst, 3225
Balsam, 4282
(peruvianischer) ——, 4420
Band, 105
Bandagist, 3840
Bandfuge, 106
Bandwurm, 3541
Bank, 4635
Barbier, 3845
Barsch, 3999
Bart, 969
Bartdoctor, 3844
Bartfinne, 3192
Barthaar, 970
Bau, 11
Bauch, 378
Bauchbinde, 3597
Bauchbruchband, 3609
Bauchfellentzündung, 2968
Bauchgrimmen, 1815
Bauchmuskel, 606
Bauchmuskelwand, 607
Bauchpresse, 609
Bauchschmerz, 1813
Bauchspeichel, 1082
Bauchspeicheldrüse, 1083
Bauchspeicheldrüsengeschwulst, 3014
Bauchspeicheldrüsenkrebs, 3015
Bauchwassersucht, 3010
Bauchweh, 1814
Bauchwirbel, 380
Bauchwirbelnerv, 897
bauen, 10
Baum, 4488
Baumöl, 4386
Baumwolle, 4489
Becher, 4665
Becken, 449
Beckenknochen, 450
Beckenmuskeln, 610
bedecken, 2128
bedenklich, 1612
bedeuten, 4868
(nichts zu) bedeuten, 4869
bedeutend, 1611
Bedeutung, 5060
(sich) bedienen, 5172
Bedrückung, 5144
bedürfen, 5357
beeinflusst, 1740
beeinträchtigen, 2031
Beeinträchtigung, 2677
beengen, 5133
befallen, 5545
befeuchten, 5056
Befinden, 1503

(sich) befinden, 1502
(sich wohl) befinden, 1506
befragen, 1985
befreien, 5358
befriedigen, 3889
befürchten, 5557
begatten, 1180
Begattung, 1181
Begattungsorgan, 118
begehen, 4994
Begiessung, 5396
beginnen, 4927
begleiten, 4992
begreifen, 2123
begrenzen, 5004
begrenzt, 1731; 5004
Begriff, 2663
begriffen sein in, 2124
behaglich, 1600
behalten, 5200
behandeln, 3807
Behandlung, 3808
beifügen, 5311
Bein, 91
beinähnlich, 94
beinartig, 95
(auf den) Beinen sein, 4889
beinern, 93
Beinerv, 875
beinicht, 93
Beinkleider, 4685
Beinweh, 1773
beissen, 291
Beistand, 4786
beitragen, 5624
bekleiden, 4742
beklemmen, 5185
Beklemmung, 2639
Beklommenheit, 5139
bekommen, 322; 4862; 4876
bekunden, 2347
Beleg, 2131
belegen, 3948
belegt, 2126
belieben, 5415
(nach) Belieben, 5416
bellen, 2076
belustigen, 5527
bemerken, 4940
benöthigt sein, 4789
beobachten, 4811
beölen, 5305
berauben, 2378
beraubt sein, 2379
beräuchern, 4234
Beräucherung, 4235
(sich) berauschen, 3911
berauscht, 2706
bereiten, 5592
Bergamotöl, 440

Bernstein, 4394
Bernsteinöl, 4395
bersten, 5083
Beruhigungspulver, 4367
Berührung, 1744
(sich) besaufen, 3910
beschädigen, 3305
Beschädigung, 3307
beschaffen, 14
Beschaffenheit, 15
beschauen, 2304
beschleunigen, 2028
beschränken, 5037
(sich) beschränken, 4882
beschränkt sein, 5038
beschreiben, 4936
Beschwerde, 1547
(sich) beschweren, 1546
beschwerlich, 2370
beschwert, 2030
besehen, 2302
beseitigen, 5446
besichtigen, 2303
Besen, 4629
besinnungslos, 2675
Besinnungslosigkeit 2676
besonder, 5258
besonders, 4799
(nicht) besonders, 5507
besorgen, 3385; 5587
bespritzen, 4244
Bespritzen, 4245
(sich) bessern, 1560
beständig, 2341
Bestandtheil, 2282
bestehen, 1917
bestimmt, 4984
bestreuen, 4256
besuchen, 4851
betäuben, 4105
Betäubung, 2453
Betäubungsmittel, 4206
beträchtlich, 5473
Betrag, 5400
betreffen, } 5484
(was) betrifft, }
(sich) betrinken, 3909
betrunken, 2707
betrunken sein, 3912
Bett, 3572
(das) Bett hüten, 1575
Bettdecke, 4564
Bettgardine, 4580
bettlägerig, 1574
Bettlaken, 4562
Bettlinnen, 4559
Bettpfanne, 4582
Bettstatt, } 4555
Bettstätte, }
Bettstelle, 4554
Bettuch, 4563

Bettüberzug, 4560
Bettvorhänge (pl.), 4579
Bettwärmflasche, 4584
Bettwärmer, 4583
Bettwäsche, 4558
Bettzeug, 4557
beugen, 559
Beuger, 563
Beule, 2415
beunruhigen, 4844
Beutel, 544
bewachen, 5538
(sich) bewähren, 5618
bewegen, 139
beweglich, 140
Bewegung, 523
Bewegungskraft, 2369
Bewegungsnervenfaser, 885
bewusst, 2594
bewusstlos, 2595
Bewusstlosigkeit, 2597
Beziehung, 5178
biegen, 419
Bier, 3905
bieten, 5501
bilden, 27
Bildung, 28
biliös, 2958
Bilsenkraut, 4335
Bimstein, 5577
Binde, 3588
(grosse) Binde, 3589
Bindegewebe, 540
Bindehaut, 3433
binden, 98
Birne, 4045
Biscuit, 3954
bisher, 4960
(ein) Bissen Brod, 3938
bitte, 4803
bitter, 2180
Bittersalz, 4451
blähen, 2193
blähend, 2194
Blähung, 2196
Bläschen, 1909
Bläschenflechte, 3172
Bläschenrothlauf, 3162
Blase, 915
Blasen ziehen, 1910
——bekommen, 1911
(nässende) Blase, 3173
blasen, 2404
Blasenausschlag, 3182
Blasenentzündung, 3034
Blasenhals, 1159
Blasenkrampf, 3037
Blasenkrebs, 3044
Blasenpflaster, 4426
Blasenrose, 3163
Blasenschmerz, 1842

Blasenschwanzwurm, 3558
Blasenwurm, 3556
blass, 1591
Blässe, 1593
Blatt, 866
Blattern, 3136
Blatternarbe, 3142
blatternarbig, 3144
Blatterngift, 3499
blau, 1889
bläulichroth, 1891
Blausäure, 4522
Blei, 4422
bleiben, 4812
bleich, 1881
Bleichsucht, 3067
Bleikolik, 2960
Bleipflaster, 4424
Bleisalbe, 4423
Blendehaut, 1354
blenden, 1353
blind, 1116
Blinddarm, 1117
Blinddarmbruch, 2970
Blindheit, 3467
Blindsack, 1080
blitzgleich, 1709
Blödsinn, 2779
bloss, 5273; 5342
Blumenkohl, 4034
Blut, 683
(dunkles) Blut, 715
(venöses) Blut, 716
(arterielles) Blut, 717
(hellrothes) Blut, 718
Blutader, 692
Blutarmuth, 3065
Blutbahn, 727
(grosse) Blutbahn, 730
(kleine) Blutbahn, 731
Blutbrechen, 5138
Blutdrüse, 978
Blutegel, 5435
bluten, 684
Bluterbrechen, 5138
Bluterguss, 2091
blutfarbig, 690
Blutfleckenkrankheit, 3075
Blutfluss, 5476
Blutgang, 3261
Blutgefäss, 694
Blutgefässknoten, 979
Blutgerinnsel, 696
Blutgeschwür, 2418
blutgestreift, 2265
Blutharnen, 3029
Bluthusten, 2835
blutig, 685
Blutigel, 5435
Blutklumpen, 699

Blutkörperchen, 702
Blutkrämpfe, 2367
Blutkreislauf, 725
Blutkuchen, 700
Blutkügelchen, 2324
blutleer, 687
blutlos, 686
Blutmangel, 5051
Blutmasse, 695
Blutpfropf, 3359
blutroth, 691
Blutschwär, 2419
Blutschwitzen, 1975
Blutstreifen, 2101
Blutsturz, 5175
Bluttheilchen, 703
Blutüberfüllung, 2993
Blutumlauf, 726
Blutung, 3357
Blutvergiftung, 4538
Blutwasser, 703
Boden, 5344
Bodensatz, 5319
Bogen, 420
bogenförmig, 680
Bogengänge des Labyrinths, 1412
Bohnen, 4031
bohrend, 1696
Bonbons, 4060
Bordeaux, 4089
Borke, 3175
bösartig, 1647
Brand, 3395
(feuchter) Brand, 3399
(heisser) Brand, 3397
(kalter) Brand, 3398
(trockner) Brand, 3400
Brandblase, 3403
brandig, 3241
brandig werden, 3406
Brandjauche, 3402
Brandschorf, 3365
Brandwunde, 3382
Branntwein, 4099
braten, 3966
Braten, 3967
brauchen, 4780
braun, 1892
bräunlichroth, 1893
braunroth, 2492
braunschäumend, 2311
Brausegeräusche, 2564
brausen, 2562
Brausepulver, 4362
(abführendes) Brausepulver, 4365
(englisches) Brausepulver, 4363
Brechdurchfall, 2090
brechen, 321; 2181
Brechfieber, 3106

Brechgefühl, 5140
Brechlust, 2187
Brechlust haben, 2947
Brechmittel, 4197
Brechreiz, 2189
Brechreiz verursachen, 2946
Brei, 1085
breiig, 5276
Breiumschlag, 4283
breit, 3243
Brennen, 1821
brennend, 1691
Brennholz, 4626
Brett, 3578
Brille, 5464
Brod, 3932
(altbacknes) Brod, 3943
(geröstetes) Brod, 3944
(hausbacknes) Brod, 3941
(neubacknes) Brod, 3942
Brodkrumme, 3936
Brodrinde, 3937
Bronchialkatarrh, 2803
Bronchienerweiterung, 2805
Bronzekrankheit, 3209
Bröse, 1035
Bruch, 3407
Bruchband, 3607
Bruchbandmacher, 3841
Bruchbinde, 3608
Brücke, 845
Brühe, 3971
Brunnencur, 4120
Brust, 1231; 365
(weibliche) Brust, 377
(die) Brust geben, 1296
Brustbein, 367
Brustbeinschmerz, 1808
Brustbeklemmung, 2811
Brustbeschwerde, 2810
Brustbinde, 3602
Brustb'att, 368
Brustdrüse, 983
Brustentzündung, 2827
Brustfellentzündung, 2842
Brustgeschwür, 2812
Brustsaft, 4327
Brusthöhle, 372
Brustkasten, 374
Brustknochen, 369
Brustknorpel, 370
Brustkrampf, 2815
Brustkrebs, 2813
Brustmittel, 4226
Brustmuskeln, 600
Brustnerven, 892
Brustpulver, 4325
Brustschmerz, 1806

Brustseite, 2470
Brustsyrup, 4326
Brustthee, 4344
Brusttropfen, 4324
Brustwand, 2466
Brustwarze, 1232
Brustwassersucht, 2814
Brustweh, 1807
Brustwirbel, 371
Buchstabe, 5427
buchstabieren, 4898
(sich) bücken, 2356
Bündel, 206
Burgunder, 4090
Burgundernase, 3190
Bürste, 4591
Busen, 1234
Butter, 1334
Butterbrod, 3947
(belegtes) Butterbrod, 3949

## C

Cachexie, 2461
Capillaren, 772
capilläre Bronchitis, 2804
capriciös, 2159
Carbolsäure, 4297
Carcinom, 2796
Catalepsie, 2767
central, 1659
Champagner, 4091
Charpie, 3694
(geschabte) Charpie, 3700
(gezupfte) Charpie, 3698
Charpiebäuschchen, 3370
Charpiewatte, 3696
Chinapulver, 4361
Chinarinde, 4358
Chinin, 4357
Chirurg, 3821
Chlorwasser, 4311
Chocolade, 4080
Chorioidea, 1346
chronisch, 1630
Chymification, 1091
Chymus, 1087
Circulation, 2844
Circulationsstörung, 2845
Citronensäure, 4299
Closet, 4598
Coma, 2586
comatös, 5023
Commode, 4640
complicirt, 3344
Confect, 4064
Confusion, 2673
constant, 1725
Contagionsgift, 4534
contagiös, 1639

contrahiert, 2540
Cornea, 1341
Crotonöl, 4409
Croup (falscher), 2790
Croup (wahrer), 2791
Cur, 4112
(eine) Cur gebrauchen, 4114
(in der) Cur sein, 4115
curieren, 4111

## D

Dach, 276
daher, 5596
Dampf, 4531
Dampfbad, 3774
dämpfen, 5411
daneben, 5548
danieder, 2446
daniederliegen, 2447
Daniederliegen der Kräfte, 2448
Darm, 454
Darmbein, 455
Darmblutung, 2250
Darmeinklemmung, 2979
Darmeinschiebung, 2980
Darmeinstülpung, 2981
Darmentleerung, 2249
Darmentzündung, 2965
Darmerweichung, 2973
Darmfellentzündung, 2967
Darmgeschwür, 2971
Darmgicht, 2986
Darmgrimmen, 2953
Darmkanal, 1092
Darmkatarrh, 2966
Darmkolik, 2952
Darmkrebs, 2972
Darmruhr, 2989
Darmsaft, 1093
Darmschwindsucht, 2974
Darmstrenge, 1829
Darmthätigkeit, 2228
Darmverengerung, 2983
Darmverhärtung, 2984
Darmverschliessung 2985
Darmverschlingung 2982
Darmweh, 1828
Dauer, 49
dauern, 48
dauernd, 1726
Daumen, 442
Deckel, 1477
decken, 1476
deliriren, 2615
Delirium, 2616
Deltamuskel, 613
denken, 2646
Denkfähigkeit, 2647

deprimiert, 2505
destilliren, 4308
deutlich, 5421
Diaphragma, 604
Diarrhöe, 2239
Diastole, 745
Diät, 4177
Diätfehler, 4993
dick, 484
Dickbein, 485
Dickdarm, 1112
Dickleibigkeit, 2388
dilatiert, 2539
dienen, 4784
diesmal, 4926
Docht, 4613
doctern, 5619
Doctor, 3812
Doppelbier, 4098
doppelt, 5428
doppelschlägig, 1995
Doppelsehen, 2560
Dotter, 4020
Douche, 3777
Douchebad, 3778
drahtförmig, 2008
Drang, 2245
drängen, 2246
Droguen, 3855
Droguenhandlung, 3856
Droguist, 3853
Druck, 1745
Druckempfindung, 1749
drücken, 2274
drückend, 1690
Drüse, 934
Drüsenbeule, 5353
dumpf, 1688
dunkel, 710
dunkelbraun, 2135
dunkelgefärbt, 2491
dunkelroth, 1884
dünn, 1867
Dünndarm, 1094
Dünndarmgekröse, 1035
Dunstbad, 3775
durchaus, 5541
durchbrechen, 324
durchbringen, 5116
durchbohren, 5467
Durchfall, 2240
(den) Durchfall haben, 2242
durchführen, 5454
durchliegen, 3335
Durchmässung, 5017
durchschlafen, 4857
Durst, 2171
(den) Durst stillen, 3903
(über den) Durst trinken, 3907
Durstgefühl, 2173

durstig, 2172
Durstmangel, 2179
Durstsucht, 2178

# E

Eczem, 3170
Efflorescenz, 1902
Egel (see Blutegel)
Egelwurm, 3542
ehe, 4989
Ei, 1211
Eibischsyrup, 4330
Eibläschen, 1218
Eichel, 1188
Eier (*pl. of* Ei), 1211
 (harte) Eier, 4014
 (weiche) Eier, 4015
Eier auf Butter, 4019
Eierkuchen, 4016
Eierstock, 1211
Eierstocksgeschwulst, 3024
eigenthümlich, 5089
(sich) eignen *or* geeignet sein, 1545
Eileiter, 1223
Eimer, 4655
einander, 3392
(auf) einander, 5506
einathmen, 996
Einathmung, 2012
einblasen, 4239
Einblasung, 4240
einbringen, 5317
eindringen, 5210
einfach, 1652
einfallen, 2467
einfältig, 2522
einflössen, 5623
Einfluss, 5226
einführen, 5076
Eingemachtes (in Zucker), 4025
Eingemachtes (in Salz, Essig), 4026
Eingenommenheit, 1781
eingesunken, 2497
Eingeweide, 919
Eingeweidewurm, 5279
eingewurzelt, 1635
einhauchen, 5159
einhergehen, 2351
einkeilen, 145
Einkeilung, 146
einklemmen, 2975
einmal, 4920 (*after imperative* 4971)
Einnahme, 5229
einnehmen (den Kopf), 1780; 4807; 5060

einreiben, 4848
Einrenkung, 3418
einrichten, 3416
Einrichtung, 3417
einschieben, 2976
einschlafen, 2478
einschläfern, 4199
Einschläferungsmittel, 4200
einschliessen, 5190; 5374
einschneiden, 1023
Einschnitt, 1024
einschnüren, 4762
einseitig, 1729
(sich) einschränken, 4881
einsinken, 2465
Einspeichelung, 1084
einspritzen, 3730
Einspritzen, 4246
Einspritzer, 3732
Einspritzung, 4236
einstellen, 5436
(sich) einstellen, 4849
einstülpen, 2977
einträpfeln, 4237
Einträpfelung, 4238
einwachsen, 3562
Einwachsen, 3563
einwärts, 571
einwärts kehren, 5432
Einwärtsdreher, 573
Einwärtsroller, 575
einwärts wenden, 5459
Einwärtswender, 574
einweichen, 5205
einwickeln, 4180
Einwickelung, 4182
einwirken, 5337
einziehen, 2338
Eis, 4062
Eisen, 4126
Eisencur, 4127
eisig kalt, 1879
Eisstücke, 5237
Eiter, 1922
Eiterabfluss, 5355
eiterartig, 5496
Eiterbeule, 2416
Eiterbläschen, 1912
Eiterblase, 1912
Eiterblatter, 1913
Eiterharnen, 3030
Eiterherd, 3311
eiterig, 2096
eitern, 2409
Eiterung, 2410
Eiterungsprocess, 5030
Eitervergiftung, 3072
Ekel, 2162
Elbogen, 421
Ellenbogen, 421
Elbogengelenk, 424

Elbogenhöcker, 423
Electricität, 3072
electrisch, 1737
Elephantenaussatz, 3219
Elle, 417
Eltern, 4914
empfehlen, 5247
empfehlenswerth, 5167
empfinden, 828
empfindlich, 829
Empfindlichkeit, 830
Empfindung, 1752
Empfindungsfaser, 832
Empfindungslosigkeit, 2620
Empfindungsvermögen, 2644
Emphysem, 2808
emporheben, 4771
Ende, 185
endemisch, 1642
enden, 184
eng, 998
Engathmigkeit, 2024
entarten, 5295
entbinden, 1258
Entbindung, 1262
(sich) entblössen, 4751
Ente, 3984
Entenbraten, 3994
entfernen, 3366
Entfernung, 3367
entgegenarbeiten, 5095
entgegenwirken, 5573
enthalten, 5240
(sich) enthalten, 4886
entkleiden, 4744
entkräften, 1556
Entkräftung, 1557
(sich) entledigen, 4752
entleeren, 5155
Entleerung, 2244
entrichten, 5578
(sich) entschlagen, 5175
entsetzlich, 2496
entstehen, 4997
entstellen, 5583
entwickeln, 5340
(sich) entwickeln, 5340
entwöhnen, 1305
Entwöhnung, 1306
entziehen, 5402
entzündet, 2137
entzündlich, 1636
Entzündung, 2714
Entzündungsfieber, 3111
epidemisch, 1641
Epiglottis, 1479
Epilepsie, 2771
epileptisch, 2773
Epistaxis, 2800
erblich, 1648

erbrechen, 2182
(sich) erbrechen wollen, 2184
Erbrechen, (Neigung zum), 2188
Erbsen, 4029
erdfahl, 1888
Erdgrindpilz, 3523
erfahren, 5077; 5620
(sich) erfreuen, 4922
erfrieren, 3506
erfrischen, 4224
Erfrischungsmittel, 4225
ergeben, 5235
ergiessen, 2090
ergreifen, 5369
ergriffen sein, 5370
Erguss, 5492
erhaben, 1916
erhalten, 5535
erheben, 2373
(sich) erheben, 2374
erhitzen, 5225
Erhitzung, 5131
erhöhen, 1756
(sich) erinnern, 4933
(sich) erkälten, 4957
erkältet sein, 5466
Erkältung, 4813
erkennen, 2422
erlauben, 5349
erleichtern, 4847
Erleichterung, 5195
erleiden, 5028
erlöschen, 2577
ermangeln, 4852
ermatten, 2441
Ermattung, 2442
ermüden, 2444
Ermüdung, 2445
ermuntern, 5405
ernähren, 748
Ernährung, 749
Ernährungsflüssigkeit, 750
erneuen, 5194
erneuern, 5194
ernstlich, 5442
erregen, 2231
erregt, 2508
erreichen, 2119
erscheinen, 1920
erschlaffen, 2380
Erschlaffung, 5389
erschöpfen, 1585
erschöpfend, 1714
Erschöpfung, 1586
erschreckt, 2507
erschüttern, 2724
Erschütterung, 5327
erschwert, 2034
ersetzen, 5585

erst, 5101; 5099
Erstarrung, 2457
Ersticken, 1766
Erstickung, 1766
(sich) erstrecken, 5335
ertragen, 5202
erträglich, 5196
erwachen, 5152
erwärmend, 4353
erweichen, 4219
Erweichung, 2717
Erweichungsmittel, 4220
erweitern, 1001
Erweiterer, 1365
Erweiterung, 1003
erzeugen, 1173
erzielen, 5199
essen, 3877
Essen, 3883
Essig, 4066
essigsauer, 4523
Essigsäure, 4296
Essigäther, 5568
Esslöffel, 4671
Esslust, 3885
eustachische Trompete, 1406
exanthematischer Typhus, 3122
Excess, 5013
Excrement, 2258
existieren, 5029
Exsudat, 1969
Extremitäten, 398

# F

Fäcalmasse, 2221
Faden, 3545
fadenförmig, 2007
Fadenwurm, 3546
fähig, 44
fahl, 1887
Fall, 4944; 5156
(einen) Fall thun, 4945
(im) Falle, 5156
fallen, 1583
Fallen, 2121
Fallsucht, 2772
falsch, 156
Falte, 1349
falten, 5207
Faltenkranz, 1351
Familie, 4924
Farbe, 243
färben, 5221
Farbestoff, 941
farbig, 689
farblos, 251
Färbung, 2513

Fasan (sáhn), 3987
Faser, 201
(centrale) Faser, 820
Faserband, 205
Faserbündel, 207
faserig, 204
Faserknorpel, 548
Faserstoff, 203
Fass, 4660
fassen, 4825
(sich) fassen, 4827
fast, 2282
faul, 2199
faulicht, 2205
faulig, 2205
Federbett, 4565
Federmatratze, 4567
fehlen, 1552
Fehlgeburt, 1270
Feige, 4056
Feigwarze, 3242
Feile, 3718
feingepulvert, 5576
Feldapotheke, 3863
Feldlazareth, 3876
Fell, 601
Felsen, 235
Felsenbein, 236
Fenchelholz, 4453
Fenchelthee, 4347
Fenster, 1401
Fensterladen, 4632
Fensterscheibe, 4631
Ferne, 5425
(in der) Ferne, 5426
ferner, 5275
Ferse, 506
Fersenbein, 507
fertig, 5323
fertig sein, 5324
fesseln, 5529
fest, 2354
festzetzen, 5404
Fett, 927
fett, 926
Fettablagerung, 932
Fettgewebe, 929
Fetthaut, 928
Fettherz, 2876
fettig, 2268
Fettleber, 3001
fettleibig, 2394
Fettleibigkeit, 2395
Fettniere, 3032
Fettsalbe, 4267
Fetzen, 1936
feucht, 1372
Feuchtigkeit, 1373
(wässerige) Feuchtigkeit, 1374
Feuerherd, 4627
Feuermal, 3213

Feuermaser, 3134
Fichtenöl, 4390
Fieber, 3087
(täglich wiederkehrendes) Fieber, 3089
(einmal täglich――――)Fieber, 3090
(zweimal täglich――――) Fieber, 3091
(dreitägiges) Fieber, 3092
(viertägiges) Fieber, 3093
Fieber haben, 3094
(vor) Fieber zittern, 3097
Fieberanfall, 3088
fieberfrei, 5381
Fiebergefühl, 3095
Fiebergeruch, 3096
fieberhaft, 1633
fieberisch, 4802
Fiebermittel, 4356
fiebern, 3094
fiebernd, 2011
Fieberrinde, 4359
Fieberschauder, 3098
Fieberschauer, 3098
Filzlaus, 3529
Finger, 437
(der kleine) Finger, 447
(böser) Finger, 3229
Fingerband, 438
Fingerbein, 441
Fingergelenk, 439
Fingerglied, 440
Fingerhut, 4518
Fingerwurm, 3228
Finne, 3184
Finnenausschlag, 3187
Finnenwurm, 3557
finnig, 3186
Fisch, 3997
Fischschuppenkrankheit, 3206
Fistel, 3230
Fistelgeschwür, 3232
flach, 178
Fläche, 179
(glatte) Fläche, 181
(überknorpelte) Fläche, 182
flachgedrückt, 2471
Flachssamenthee, 4351
Flanell, 5203
Flasche, 1700
(kleine) Flasche, 4662
Flatulenz, 2197
flau, 2949
Flechse, 541
Flechte, 1945
(nasse) Flechte, 1946
(nässende) Flechte, 1947
(trockne) Flechte, 1948
flechten, 681

Flechtengrind, 3524
Fleck, 1369; 2557; 5440
(gelber) Fleck, 1370
(weisser) Fleck, 3443
Fleckensehen, 2558
Flecktyphus, 3121
Fleisch, 319; 3955
(gekochtes) Fleisch, 3964
(wildes) Fleisch, 3235
Fleischbrühe, 3965
Fleischfaser, 539
fleischig, 556
fleischlich, 5351
fleischlos, 537
Fleischmasse, 538
Fleischwärzchen, 3373
fleischwasserähnlich, 2266
Fliederblume, 4331
Fliederthee, 4348
Fliege, 2548
(die spanische) Fliege, 4319
fliegend, 1708
Fliegennetz, 4586
Fliegensehen, 2552
fliessen, 210
Flimmerbewegung, 2556
Flocke, 2550
Flockensehen, 2554
Floh, 3532
Flohstich, 3533
flüchtig, 2683
Flügel, 1027
Fluss, 2369
(weisser) Fluss, 2370
flüssig, 211
Flüssigkeit, 5074
Folge, 4871
Folgeleistung, 5272
folgen, 4998
Forderung, 5266
Forelle, 4005
Form, 2272
fort, 2359
fortbestehen, 5392
Fortsatz, 1121
fortschaffen, 5417
fortsetzen, 1120
fortwährend, 2165
Frauenhemd, 4734
Frauenkleider, 4717
frei, 2343
fremd, 3368
fremdartig, 2580
frequent, 1990
fressen, 2896
fressendes Geschwür, 3878
(sich) freuen, 5082
Friesel, 3130
frisch und gesund, 1509

Frost, 2112
(vor) Frost zittern, 3099
Frostanfall, 2114
Frostballen, 3508
Frostbeule, 3507
frösteln, 2116
Frösteln, 2117
Frucht, 1219; 4042
Fruchthalter, 1220
Fruchtwasser, 1291
früh, 4816
früher, 4919
frühgebären, 3294
Frühgeburt, 3294
Frühlingsfieber, 3107
Frühstück, 3922
frühstücken, 3884
Fuge, 103
fügen, 102
Fugengelenk, 173
fühlbar, 1997
fühlen, 1984
fühllos, 2452
führen, 5265
Füllsel, 3996
Function, 2652
functionell, 1660
Funke, 59
Funkensehen, 2555
Furche, 848
fürchten, 4785
(sich) fürchten, 4785
furibund, 1655
Furunkel, 2417
Fuss, 451
Fussbad, 3771
Fussbank, 4636
Fussbett, 3583
Fussbinde, 3606
Fussdecke, 4576
Fussgicht, 3083
Fussknochen, 452
Fussmachine, 3584
Fussmuskel, 620
Fussrücken, 514
Fussschmerz, 1857
Fusssohle, 513
Fusswurzel, 502
Futter, 4692

# G

Gabel, 1021
Gabelfrühstück, 3923
Gabelung der Luftröhre, 1022
gähren, 2214
Gährung, 2215
Gallapfel, 4447
Galle, 1099
gallenbitter, 1100

Gallenblase, 1104
Gallenbrechen, 2216
Gallenfieber, 3109
Gallengang, 1102
Gallenkanal, 1103
Gallenkolik, 2957
Gallenstein, 5255
gallig, 2136
gallopieren, 2823
Gallsäure, 1101
Gang, 2361
Ganglien, 815
Ganglienkette, 907
Ganglienkugeln, 824
Gangliensystem, 816
Gans, 3983
Gänsebraten, 3992
Gänsebrust, 3993
Gänsehaut, 1950
Gänseklein, 3995
gänzlich, 2578
gar, 4964
gar nicht, 4791
Gardine, 4578
Gaslampe, 4612
Gastmal, 3921
Gaumen, 264
Gaumenabscess, 3493
Gaumenbein, 265
Gaumenblutung, 2885
Gaumenbogen, 1453
Gaumengeschwulst, 2905
Gaumensegel, 1452
Gebäck, 3951
gebären, 1263
geboren werden, 1271
Gebärmutter, 1212
Gebärmutterhals, 1213
Gebärmutterhöhle, 1214
Gebärmutterschmerz, 3281
Gebäude, 73
Gebein, 92
(sich) geben, 5107
gebildet, 5633
Gebiss, 292
falsches Gebiss, 293
Gebrauch, 5187
gebrauchen, 4113
Gebrauchsanweisung, 5612
Gebrechen, 1579
gebrechlich, 1577
Gebrechlichkeit, 1578
gebückt, 2357
Geburt, 1264
(unzeitige)Geburt, 1269
(schwere) Geburt, 1265
Geburtsarzt, 3827
Geburtshelfer, 3828
Geburtsschmerzen, 1846
Geburtswehen, 1846

Gedanke, 2664
Geduld, 5063
geeignet, 5452
Gefahr, 4832
gefährlich, 1614
Gefährlich krank sein, 1530
gefälligst, 5106
gefärbt, 2252
Gefäss, 622
Gefässdrüse, 973
Gefässgewebe, 1623
Gefässhaut, 1343
Gefässmuskel, 611
gefasst auf, 5504
Geflügel, 3982
Gefrornes, 4061
Gefühl, 1500
(ein) G. erkennen, 1501
Gefühllosigkeit, 2621
Gefühlsnerven, 835
Gefühlsstumpfheit, 5339
Gefühlsvermögen, 2658
gefurcht, 2139
Gegend, 388
Gegengift, 4285
Gegenstand, 4991
gegenwärtig, 4840
gegenwirken, 570
gegenwirkende Muskeln 580
Gegner, 583
Geheimfach, 4642
Geheimrath, 3839
gehen, 2350
(sich) gehen lassen, 4828
Gehirn, 808
(das grosse) Gehirn, 842
(das kleine) Gehirn, 843
Gehirnhaut, 2752
Gehirnkrankheit, 2711
Gehirnlappen, 855
Gehirnnerv, 810
Gehirnnervensystem, 811
Gehirnrheumatismus, 3077
Gehirnsymptome, 2613
Gehirnwassersucht, 2722
Gehör, 1383
(ein gutes) Gehör haben 1491
(ein schlechtes) Gehör haben, 1496
Gehörgang, 1389
(der äussere) Gehörgang 1389
gehörig, 5163
Gehörknöchelchen, 1407
Gehörnerv, 1416
Gehörorgan, 1381
Gehörsinn, 5002

Gehörsteinchen, 1413
Gehörtäuschungen, 2561
Gehörwerkzeug, 1384
Geist, 57
Geisteskraft, 5561
geisteskrank, 2757
geistesschwach, 2755
geistesstumpf, 2756
Geistesstumpfheit, 2674
Geistesträgheit, 2660
Geistesstörung, 2661
Geistesverwirrung, 2670
Geisteszerrüttung, 2672
geistig, 2629
gekörnt, 5457
Gekrösdrüsen, 1113
Gekröse, 1108
Gekrösendärme, 1109
Gelatine, 4448
gelb, 1894
gelbbraun, 2260
gelbgrün, 1895
gelbgestreift, 2092
Gelbsucht, 2992
Gelée, 4022
geléeartig, 2267
gelegentlich, 5475
Gelenk, 167
(das straffe) Gelenk, 220
Gelenkband, 169
Gelenkentzündung, 3078
Gelenkfläche, 183
Gelenkfügung, 171
Gelenkhöhle, 172
Gelenkkapsel, 218
Gelenkknorpel, 168
Gelenkknorren, 414
Gelenkkopf, 487
Gelenkpfanne, 463
Gelenkring, 170
Gelenkrolle, 415
Gelenkschmerz, 1854
Gelenkschmiere, 215
Gelenkschmierkapsel, 217
gelind, } 1757
gelinde, }
gelüsten, 1254
Gelüst, 1255
gemächlich, 5289
gemeinschaftlich, 866
Gemüse, 2170
Gemüth, 2627
Gemüthsaufregung, 5132
Gemüthsstimmung, 2628
genau, 4965
geneigt zu, 4975
genesen, 1517
Genesung, 4106
Genick, 359
Genickschmerz, 1797

geniessen, 4879
Genossen, 582
gentigen, 5361
gentigend, 5218
Genuss, 5047
gerade, 5463
gerathen, 5052
Geräusch, 655
geräuschvoll, 2035
gering, 1687
gerinnen, 660
Gerippe, 88
gern, 4664
gern haben, 4885
Gerste, 4484
Gerstenkorn, 3442
Gerstenmehl, 5285
Geruch, 1421
(keinen) Geruch haben, 1499
(ein übler) Geruch, 2042
geruchlos, 5233
Geruchsnerv, 864
Geruchsorgan, 1419
Geruchsphantasmen, 2570
Geruchssinn, 1422
Geruchswerkzeug, 1423
Gerüst, 76
Gesäss, 458
Geschäft, 4907
geschehen, 5312
geschickt, 3835
Geschlecht, 2
(männliches) Geschlecht, 1162
(weibliches) Geschlecht, 1163
geschlectlich, 5012
geschlechtliche Unreife, 1170
Geschlechtsorgan, 1161
Geschlechtsreife, 1169
Geschlechtstheile, 1164
Geschlechtstrieb, 1166
Geschmack, 1434
Geschmacksorgan, 1432
Geschmacksphantasmen, 2571
Geschmackssinn, 1435
Geschmackswärzchen, 1442
Geschwister, 4916
geschwollen, 1870
geschworen, 2141
Geschwulst, 2400
Geschwür, 2414
geschwürig, 2411
geschwürig werden, 2412
Gesicht, 241
(ein gutes) Gesicht haben, 1492
Gesichtsausdruck, 2502

Gesichtsfarbe, 244
Gesichtsgrind, 3174
Gesichtshälfte, 248
Gesichtsknochen, 242
Gesichtskrampf, 2775
Gesichtsmuskel, 588
Gesichtsnerv, 871
Gesichtsrose, 3161
Gesichtsschmerz, 1784
Gesichtszüge, 248
gespalten, 2134
gespannt, 1863
Gestalt, 25
gestalten, 24
Gestank, 2093
gesund, 1507
gesund und munter, 1510
Gesundheit, 1511
(bei guter) Gesundheit, 1512
Gesundheitsamt, 3812
Gesundheitspolizei, 3815
Gesundheitszustand, 1514
Getränk, 2178
(geistige) Getränke, 4097
Gevierte (ins Gevierte), 5434
Gewächs, 5070
Gewähren, 5217
Gewand, 4679
Gewebe, 85
Gewicht, 2321
gewissenhaft, 5631
gewöhnlich, 1650
Gicht, 3082
Gichtfieber, 3086
Gichtschmerzen, 55
gierig, 2160
Giessbad, 3779
giessen, 5157
Giesskanne, 1467
Gift, 3494
Gifthahnenfuss, 4511
giftig, 4492
Giftkräuter, 4508
Giftkresse, 4513
Giftkunde, 4496
Giftlattich, 4512
Giftlehre, 4497
giftlos, 4493
Giftmehl, 4521
Giftmittel, 4495
Giftpflanzen, 4507
Giftpilze, 4509
Giftschlangen, 4499
Giftschwämme, 4510
Giftstoff, 4494
Giftthiere, 4498
glänzend, 2529
Glas, 3915
(zu tief ins) Glas gucken, 3916

Gläsern, 2536
Glasfeuchtigkeit, 1378
Glasflüssigkeit, 1379
Glashaut, 1380
Glaskörper, 1377
Glasur, 316
glatt, 180
glauben, 4831
Glaubersalz, 4418
gleich, 2384; 5098
gleichartig, 5492
gleichen, 2257
gleichförmig, 5171
gleichgültig (gegen), 2636
gleichmässig, 5006
gleichzeitig, 5008
Glied, 392; 5350
(das männliche) Glied, 1185
Gliedbad, 3769
Gliederschmerz, 1856
Gliedmassen, 393
(die oberen) Gliedmassen, 396
(die untern) Gliedmassen, 397
Glüheisen, 3364
Glycerin, 4449
Glycosurie, 3050
Goldblatt, 5556
graben, 337
Grad, 2103
greifen, 2122
grell, 5431
Grenze, 905
Grenzstrang, 906
Gries, 2299
Grimm, 1123
Grimmdarm, 1125
grimmen, 1124
Grimmen, 2950
Grind, 1938
(nässender) Grind, 1949
grindicht, 1940
grindig, 1940
Grippe, 3105
grob, 5206
gross, 1987
grossblasig, 2049
Grösse, 2387
Grube, 338
Grund, 3391
grünlichgelb, 2309
Grünspan, 4526
gucken, 1987
Gummibeutel, 5394
Gummipflaster, 4433
Gummiröhrchen, 3679
Gurgelwasser, 4268
Gurken, 4023
Gürtel, 3178; 4738
Gürtelausschlag, 3179

Gürtelflechte, 3180
Gürtelrose, 3181
gutartig, 1646
(sich) gütlich thun, 5280
gymnastisch, 4174

# H

Haar, 770
Haarbürste, 4593
Haargefäss, 771
haarig, 957
Haarkeim, 956
Haarkopf, 3554
Harrschaft, 962
Haarseil, 4248
Haarseillegen, 4249
Haarwurzel, 960
Haarzwiebel, 959
habituell, 2987
Hacke, 505
Hafer, 4480
Hafermehl, 4482
Haferschleim, 4071
haften, 5176
halb, 246
Halbkugel, 852
Halblähmung, 2738
Halbsehen, 2559
halbseitig, 2737
Halbstiefel, 4702
halbstündig, 5252
halbverdaut, 5141
Hälfte, 247
(die rechte) Hälfte, 627
(die linke) Hälfte, 628
Hallucination, 2545
Hals, 310
Halsbein, 355
Halsbinde, 4696
Halstuch, 4697
Halsbräune, 2921
Halsentzündung, 2022
Halsgeflecht, 889
Halsgelenk, 356
Halshöhle, 357
Halsleiden, 2781
Halsmuskel, 596
Halsnerv, 888
Halsschmerzen, 4974
Halsweh, 1793
Halswirbel, 354
Halswirbelbein, 355
halten, 817
(sich) halten, 4887
halten auf, 5148
halten für, 5559
Hammelbraten, 3970
Hammelcotelett, 3973
Hammelfleisch, 3958
Hammelkeule, 3972

Hammelrippchen, 3974
Hammer, 1408
hämmernd, 1693
Hämorrhoidalknoten, 5287
Hämorrhoiden, 2991
Hand, 425
(die rechte) Hand, 428
(die linke) Hand, 428
Handbad, 3770
Handfläche, 431
Handgelenk, 432
Handgicht, 3084
Handmuskel, 616
Handrücken, 429
Handtuch, 4596
Handwurzel, 433
Hanfsamen, 4443
Hangband, 3617
Hangbandage, 3618
Häring, 4008
Harn, 1138
(den) Harn lassen, 1138
Harn absondern, 2276
Harnabfluss, 3061
Harnabgang, 2332
Harnabsonderung, 1143
Harnapparat, 1141
harnartig, 2291
Harnbeschauung, 2306
Harnbesichtigung, 2305
Harnbestandtheil, 2283
Harnblase, 1153
Harnblasenerweichung, 3043
Harnblasengeschwür, 3042
Harnblasenschwindsucht, 3045
Harnbrennen, 3059
Harndrang, 2330
harnen, 1139
Harngang, 1154
Harngangentzündung, 3053
Harngefäss, 1142
Harngries, 2300
Harnkanälchen, 1155
Harnleiterentzündung, 3054
Harnlosigkeit, 3056
Harnphosphor, 2288
Harnröhre, 1160
Harnröhrengeschwür, 3060
Harnröhrenschmerz, 1841
Harnröhrenentzündung, 3052
Harnröhrenverengerung, 3055
Harnruhr, 3046

Harnsalze, 2285
Harnsand, 2301
Harnsatz, 2291
Harnsäure, 1145
harnsaure Salze, 2286
Harnsperre, 3057
Harnsteine, 3039
Harnstoff, 1144
Harnstrenge, 3058
harntreibend, 5382
Harnvergiftung, 3031
Harnverhaltung, 2326
Harnverstopfung, 2327
Harnweg, 1153
Harnzwang, 2328
hart, 1858
(einen) harten Leib haben, 2237
harter Stuhlgang, 2236
Härte, 1762
Hartgummi, 3734
barthörig, 3480
Harthörigkeit, 3481
hartleibig, 2238
hartnäckig, 1634
Haselnuss, 4054
Haube, 4735
Hauch, 55
hauchen, 54
hauen, 3318
häufig, 2329
Häufigkeit, 2027
Hauptbinde, 3622
Hauptlappen, 1026
Hauptsache, 5108
hauptsächlich, 5259
Hausapotheke, 3865
Hauskleid, 4719
Hausmannskost, 3931
Hausmittel, 5616
Haut, 916
(äussere) Haut, 923
(seröse) Haut, 917
Hautabschürfung, 1927
Hautausschlag, 1901
Hautbeissen, 1953
Hautblüthe, 1903
Hautbrand, 1961
Hautbrennen, 1960
Hautfarbe, 245
Hautfinne, 3185
häutig, 2792
Hautjucken, 1952
Hautkrebs, 3218
Hautreiz, 1955
Hautröthe, 2475
Hautschwiele, 1962
Hebamme, 3830
heben, 1312
heben ein Übel, 5054
heben eine Krankheit, 5054

Heber, 1113
Hecht, 4000
Hefen, 5320
heftig, 1626
Heftigkeit, 5147
Heftnadel, 3662
Heftpflaster, 3703
Heil, 4166
Heilanstalt, 3868
Heilbad, 3765
heilbar, 1623
Heilbein, 468
heilen, 4104
heilig, 465
(das) heilige Bein, 467
Heilkunde, 3803
Heilkunst, 3803
Heilmittel, 4167
Heilpflaster, 3704
Heilung, 4105
Heilungsmittel, 4168
Heirath, 1240
heirathen, 1239
heirathsfähig, 1241
heirathsfähiges Alter, 1243
Heirathsfähigkeit, 1244
heiss, 1875
heiser, 2034
Heiserkeit, 2085
heissen, 4897
Heisshunger, 2161
heiter, 4822
hektischroth, 1885
helfen, 5318
hell, 2307
hellgelb, 2308
hellroth, 1713
Hemd, 4708
Hemiplegie, 2739
hemmen, 5297
herabgehen, 2107
herauslaufen, 5073
herausspringen, 5142
heraustreten, 3472
Heraustreten, 3284
herbeiführen, 5224
Herbstfieber, 3108
Herd, 2065; 4651
Herr, 4813
herrühren, 5549
(sich) herschreiben, 5497
herstellen, 1561
Herstellung, 1562
herunter lassen, 4773
herunter machen, 4772
(sich) hervordrängen, 5303
hervorragen, 2145
hervorrufen, 5041
hervorstehend, 2473
Hervorwölbung, 2430

Herz, 624
Herzabscess, 2874
Herzatrophie, 2873
Herzbein, 637
Herzbeklemmung, 2854
Herzbeschwerung, 2855
Herzbeutel, 638
Herzbeutelentzündung, 2860
Herzbeutelwassersucht, 2882
Herzbewegung, 648
Herzblausucht, 2881
Herzbräune, 2858
Herzbrennen, 2856
Herzbündel, 641
Herzdrücken, 1810
Herzdrüse, 643
Herzentzündung, 2859
Herzerweichung, 2869
Herzerweiterung, 2866
Herzfehler, 2849
Herzfell, 640
Herzfleischentzündung, 2862
Herzgegend, 645
Herzgekröse, 647
Herzgeräusch, 657
Herzgerinnsel, 661
Herzgeschwulst, 2877
Herzgeschwür, 2878
Herzgrube, 644
Herzhaut, 642
Herzhöhle, 646
Herzhöhlenerweiterung, 2867
Herzhypertrophie, 2865
Herzkammer, 630
Herzklappe, 634
Herzklopfen, 2851
Herzkrampf, 2857
Herzkrankheit, 5050
Herzlähmung, 2853
Herzleiden, 2850
Herzmuskelentzündung, 2861
Herzohr, 650
Herzöhrchen, 651
Herzpochen, 742
Herzpolyp, 2879
Herzpuls, 738
Herzröhre, 670
Herzsack, 639
Herzschall, 658
Herzschlag, 739
Herzschrumpfung, 2872
Herzschwäche, 2848
Herzspitze, 649
Herzstoss, 640
Herzverengerung, 2868
Herzverhärtung, 2870
Herzverknöcherung, 2871

Herzwassersucht, 3062
Herzweh, 1809
Herzzittern, 2852
Heufieber, 3101
Hexenschuss, 3080
Hiebwunde, 3319
Himbeeressig, 4295
Himbeersaft, 4074
Himbeersyrup, 4075
Himbeerzunge, 2147
hinabschlucken, 4783
hinaufschiessen, 5079
Hinderniss, 5113
hindurch, 5119
hinfahren über, 5408
hinfällig, 1584
Hinfälligkeit, 1584
hingegen, 5403
hinken, 2363
hinlänglich, 5189
(sich) hinlegen, 4775
hinreichend, 5499
hinsiechen, 1569
hinter, 222
Hinterhaupt, 228
Hinterhauptsbein, 229
Hinterkopf, 224
Hinterzahn, 306
hinunterschlucken, 4783
hinzufügen, 5191
Hirn, 807
Hirnanämie, 2726
Hirndruck, 2716
Hirnentzündung, 2715
Hirnerschütterung, 2725
Hirnerweichung, 2718
Hirnfläche, 861
Hirnflüssigkeit, 862
Hirnhaut, 2752
(harte) Hirnhaut, 857
(weiche) Hirnhaut, 861
Hirnhöhle, 841
Hirnkrankheit, 2711
Hirnnerv (der dreigetheilte, 870
Hirnnervenfasern, 863
Hirntumor, 2778
Hirnverhärtung, 2720
Hirnverletzung, 2712
Hirnwassersucht, 2722
Hirschhorn, 4413
Hirschhornsalz, 4414
Hirschhorngeist, 4415
Hirse, 1918
hirsenförmig, 1919
Hitzbläschen, 1923
Hitzblatter, 1924
Hitze, 2206
hitzig, 1627
Hobelspänbinde, 3590
hochgestellt, 2314
hochschwanger, 1253

Höcker, 422
Hode, 1191
Hoden, 1191
Hodenanschwellung, 3246
Hodensack, 1192
hoffentlich, 4818
Hoffmannsche Tropfen, 4419
Höhe, 2110
(in die) Höhe gehen, 2111
hohl, 118
Hohlader, 774
Höhle, 121
höhlen, 119
Hohlgeschwür, 3231
Hohlhand, 430
Hohlmuskel, 558
holen, 986
Höllenstein, 4316
Höllensteinhalter, 3680
Hollunderblüthe, 4332
Holzessig, 4294
Holzkohle, 4321
Holzthee, 4352
Homöopath, 3837
homöopathisch, 4117
Honig, 4458
Honigharnruhr, 3048
Honorar, 5580
honorieren, 5581
hören, 1382
horizontal, 1754
Horn, 942
hornartig, 946
hörnern, 943
Hörnerv, 872
Hornhaut, 1342
hornig, 944
hornicht, 945
Hornplatte, 948
Hosen, 4648
Hosenträger, 4707
Hospital, 3404
Hospitalbrand, 3405
hübsch, 5508
Hüftbad, 5333
Hüftbein, 383
Hüftblatt, 453
Hüfte, 382
Hüftgegend, 3020
Hüftgelenkschmerz, 1838
Hüftknochen, 384
Hüftnerv, 899
Hüftschmerz, 1837
Hügel, 488
Huhn, 3985
Hühnerauge, 3560
Hühneraugenoperateur, 3842
Hülle, 529
hüllen, 528
Hülsenfrüchte, 4028

Hülsenwurm, 3559
Hummer, 4010
Hungercur, 4145
Hungerpest, 3124
Hungerschwitzcur, 4146
Hungertyphus, 3123
hungrig, 2151
hüpfend, 1999
hüsteln, 2068
husten, 2067
Husten, 2069
Hustenanfall, 2070
Hustenfieber, 3102
Hut, 4693
hüten, 1575
(sich) hüten, 4814
(sich) hüten vor, 4815
hydropatisch, 4118
Hyperämie, 2998
hypogastrisch, 3021
Hysterie, 2776
hysterisch, 1665

I

Icterus, 2992
impfen, 3149
Impfen, 3152
Impfinstrument, 3677
Impfnadel, 3676
Impfstoff, 3153
Impfung, 3151
Incisionslancette, 3642
Incisionsmesser, 3641
Incisionsschere, 3644
Ingwer, 4068
Injection, 3777
Injectionsspritze, 3733
injicieren, 2528
Innere, 5460
innerhalb, 4978
innerlich, 1720
Insomnie, 2590
intellectuell, 2654
Intelligenz, 2648
Intensität, 5005
intensiv, 1718
intermittierend, 1664
irgend ein, 4938
Iris, 1357
Irre, 2758
Irritans, 4292
Irrreden, 2682
Irrsinn, 2759
irrsinnig, 2524
Isländisches Moos, 4355

J

ja, 4867; 5109

Jacke, 4731
Jahreszeit, 5170
je, 5265
jedenfalls, 4859
jetzig, 4972
Joch, 266
Jochbein, 268
jucken, 1951
juckend, 1876
Jungfer, 1225
Jungfernhäutchen, 1226
Jungfrau, 1224

K

Kaffee, 4078
Kaffeelöffel, 4673
Kaffeesatz, 2218
Kaffeesatzartig, 2219
Kahlgrind, 3523
Kaisergeburt, 3291
Kaiserschnitt, 3292
Kalbfleisch, 3957
Kalbsbraten, 3969
Kalbscotelett, 3975
Kalk, 4309
kalkig, 5345
Kalkwasser, 4310
Kalmus, 4402
Kalmusöl, 4403
kalt, 1878
Kaltwassercur, 4121
Kamillenthee, 4346
Kamm, 4594
kämmen, 4595
Kammer, 629
Kampfer, 4404
Kampfergeist, 4435
Kampferöl, 4405
Kampferspiritus, 4434
Kappe, 4695
Kapsel, 116
Kapselstaar, 3454
Karbunkel, 2490
Karpfen, 3998
Kartoffeln, 4040
Käse, 2129
käseähnlich, 2130
Kasten, 373
Katamenien, 3260
Katarrh, 2798
katarrhalisch, 2828
(die schnelle) Katharine, 2241
Katheter, 3674
Katzenjammer, 2933
kauen, 593
kaum, 1907
Kaumuskel, 594
Kegel, 1444
kegelförmig, 1445

Kehldeckel, 1478
Kehldeckelentzündung, 2793
Kehle, 1009
Kehlentzündung, 2783
Kehlgeschwulst, 2789
Kehlkopf, 1462
Kehlkopfbräune, 2785
Kehlkopfentzündung, 2784
Kehlkopfhöhle, 1472
Kehlkopfödem, 2788
Kehlkopfschwindsucht, 2795
Kehlkopfsknorpel, 1464
Kehlkopfspiegel, 3755
Kehlkopfstasche, 1481
Kehlschwindsucht, 2794
keichen, 2071
Keil, 144
Keilbein, 232
Keim, 955
keimen, 954
Keimlager, 1217
keineswegs, 4878
Kelch, 1149
kelchförmig, 1450
Keller, 4654
kennen, 2421
kerben, 2489
Kern, 123
kerngesund, 1508
Kerze, 4610
Kessel, 4658
Kettensäge, 3656
keuchen, 2071
Keuchhusten, 2072
Kiefer, 253
Kiefergelenk, 254
Kieferknochen, 255
Kieferwinkel, 5078
Kind, 4923
Kindbetterin, 1280
Kindbettfieber, 2969
Kinderkrankheit, 1522
Kinderpulver, 4336
(in) Kindesnöthen sein, 1290
Kinn, 260
Kinnbacken, 261
Kinnbackenbein, 263
Kinnbackenknochen, 263
Kinnbinde, 3595
Kinnlade, 262
Kinntuch, 3596
Kirsche, 4047
Kitzel, 1954
kitzeln, 1228
Kitzeln, 5103
Kitzler, 1229
klaffen, 3350
klagen, 4931

Klappenkrankheit, 2883
Klapperschlange, 4500
klar, 2087
Klaue, 2487
Klauenfett, 4374
klauenförmig, 2488
kleben, 3363
kleberig, 1897
Kleid, 4718
kleiden, 4741
Kleiderlaus, 3528
Kleiderschrank, 4638
Kleidung, 4677
Kleidungsstück, 5134
Kleie, 3196
Kleienflechte, 3197
Kleiengrind, 3198
Kleienschwinde, 3199
Kleiensucht, 3200
klein, 1988
Klima, 4148
Klimacur, 4149
klingen, 2579
Klinik, 3869
klonisch, 2745
klopfen, 2728
klopfend, 1692
Klumpen, 697
Klystier, 3739
klystieren, 3738
Klystierröhre, 3740
Klystierschlauch, 3741
Klystierspritze, 3742
knapp, 4178
knappe Diät, 4179
knarren, 5044
knastern, 2057
knattern, 2057
Knattergeräusch, 2058
kneifen, 5278
kneifend, 1717
kneipend, 1718
Knie, 491
Kniebinde, 3605
Kniegicht, 3085
Kniekehle, 494
Kniescheibe, 493
Knieschmerz, 1853
knirschen, 2480
knistern, 2052
Knöchel, 63; 508
Knöchelchen, 62
Knöchlein, 62
Knochen, 61
knochenähnlich, 69
knochenartig, 72
Knochenbildung, 90
Knochenbrand, 3396
knochenbrandig, 3401
Knochenbruch, 3408
(der lange) Knochenbruch, 3409

(der quere) Knochenbruch, 3410
(der schiefe) Knochenbruch, 3411
Knochenfeile, 3719
Knochenfuge, 104
Knochengebäude, 74
Knochengerippe, 89
Knochengerüst, 77
Knochengewebe, 86
Knochenhaut, 111
Knochenhöhle, 122
Knochenkapsel, 117
Knochenkern, 124
Knochenknorpel, 154
Knochenkopf, 126
Knochenleiden, 4979
Knochenmark, 115
Knochenmasse, 83
Knochennaht, 109
Knochenpfanne, 113
Knochenrand, 132
Knochenrand mit Zacken, 136
Knochensäge, 3651
Knochenschaber, 3721
Knochenschale, 130
Knochenschere, 3652
Knochenschmerz, 1771
Knochenschneidewerkzeug, 3659
Knochensubstanz, 81
Knochensystem, 79
Knochenverbindung, 101
(die bewegliche) Knochenverbindung, 142
(die unbewegliche) Knochenverbindung, 143
Knochenwand, 128
Knochenweh, 1772
Knochenzange, 3715
knöchern, 65
knochicht, 67
knochig, 67
knollig, 2270
Knopf, 4690
Knopfloch, 4691
Knorpel, 147
(die wahren) Knorpel, 157
(die falschen) Knorpel, 158
(die rundlichen) Knorpel, 1471
knorpelähnlich, 151
knorpelartig, 150
Knorpelband, 159
Knorpelfuge, 166
Knorpelhaut, 160
knorpelicht, 149
knorpelig, 148
Knorpelmasse, 153
Knorpelpolster, 164

Knorpelring, 165
Knorpelüberzug, 177
Knorren, 413
Knötchen, 1904
Knoten, 1907
Knotenbinde, 3632
kochen, 3963
Kohl, 4053
Kohlen, 4628
Kohlensäure, 4208
Kolik, 2951
(einfache) Kolik, 2954
kolikartig, 1816
kollern, 2229
Kollern, 2230
Kompressionsverband, 3333
Kopf, 125
(etwas im) Kopfe haben, 3913
Kopfbinde, 3591
Kopfgrind, 1939
Kopfhaar, 968
Kopfküssen, 4569
Kopfnicker, 598
Kopfrose, 3160
Kopfsäge, 3653
Kopfschmerz, 1778
Kopfweh, 1779
Korb, 4659
Koriandersamen, 4444
Korn, 3512
körnig, 5456
Körper, 8
Körperbau, 12
Körperbeschaffenheit, 16
Körperbildung, 29
Körperblutbahn, 735
Körperkraft, 18
körperlich, 19
Körperpulsader, 674
Körpertheil, 5227
Körperumfang, 2427
Körperwärme, 2105
Korsett, 4725
Kost, 3930
kosten, 5586
Koth, 2220
Kothbrechen, 2222
Kothstück, 5282
Kraft, 17
kräftig, 1554
kräftigen, 1555
Kräftigung, 4162
Kräftigungsmittel, 4164
krähen, 2059
Kragen, 3250
(der spanische) Kragen, 3252
Krampf, 1765
krampfähnlich, 1712
Krampfanfall, 2748

krampfartig, 1711
krampfhaft, 1710
Krampfkolik, 2955
Krampfkrankheit, 2743
Krampflachen, 2751
krank, 1519
krank aussehen, 1520
krank sein, 1527
(sich) krank fühlen, 1528
krank werden, 1529
Krankenbett, 1572
(vom) Krankenbett aufstehen, 1576
Krankenhaus, 3873
Krankenhospital, 3872
Krankenpfleger, 3847
Krankenpflegerin, 3848
Krankenstube, 4547
Krankenstuhlwagen, 3586
Krankenthee, 4264
Krankenträger, 3577
Krankenwärter, 3846
Krankenwärterin, 3847
Krankenzimmer, 4545
krankhaft, 3437
Krankheit, 1521
Krankheitserscheinung, 5102
Krankheitssitz, 1564
kränklich, 1525
Kränklichkeit, 1526
Kranz, 1350
Krätze, 1944
kratzen, 1942
krätzig, 1943
Krätzmilbe, 3525
Krätzwunde, 1956
Krauseminze, 4337
Krauseminzöl, 4383
Kraut, 4035
Kräuterarzenei, 4184
Kräuterbad, 3796
Kräutercur, 4150
Kräuterküssen, 4183
Kräuterthee, 4345
Krebs, 4009; 3655
Kreide, 4490
Kreis, 720
kreisen, 719
Kreislauf, 720, 721
(grosser) Kreislauf, 728
(kleiner) Kreislauf, 729
Kreissäge, 3655
Kreuz, 464
Kreuzbein, 466
Kreuznerv, 898
kreuzen (die Arme), 4773
Kreuzschmerzen, 1835
kreuzweis, 3631
Kreuzwirbel, 469
kribbelnd, 1706

kriechen, 1768
kritisch, 2248
Krone, 308
Kröte, 4506
Krückenzange 3713
Krug, 4656
krumm, 1107
Krummdarm, 1107
Kruste, 1937
Küche, 4648
Kuchen, 698; 3952
Küchenkräuter, 4027
Küchenschrank, 4649
Kuckuck, 474
Kuckucksbein, 476
Kugel, 187
Kugelbohrer, 3725
Kugelzange 3714
kühl, 1877
kühlen, 4217
Kühlmittel, 4218
Kühltrank, 4263
Kuhpocken, 3146
Kummer, 4952
(sich) kümmern, 2635
Kunde, 3802
Kunst, 3801
kunstvoll, 5584
Kupfer, 4501
Kupferkolik, 2961
Kupferoxyd, 5424
Kupferrose, 3189
Kupferschlange, 4502
Kur, see Cur
kurz, 1991
(vor) Kurzem, 5422
Kurzathmigkeit, 2025
kurzsichtig, 1498
Kurzsichtigkeit, 3446
Küssen, 4568

L

laben, 4158
Labemittel, 4160
Laboratorium, 3862
Labung, 4159
Labungsmittel, 4161
lachen, 2749
Lachen, 2750
Lachs, 4004
lädieren, 3806
Lage, 1735
Lageveränderung, 2884
lahm, 2382
lahm gehen, 2383
lähmen, 5453
Lähmung, 2734
Lähmungskrankheiten, 2742

Laken, 4561
Lakritzensaft, 4471
Lampe, 4604
lancinierend, 1697
Land, 4891
lange, 4904
Längenbruch, 3412
Längenspalt, 851
Längenwunde, 3346
langsam, 1632
lankwierig, 1631
Lanzette, 3638
Läppchen, 1025
Lappen, 854
Larve, 1644
Läsion, 3309
Lassband, 3634
Lassbinde, 3635
lästig, 5068
Laterne, 4618
Laubad, 3767
laufen, 721
Laune, 2631
launenhaft, 2632
Launenhaftigkeit, 2633
launisch, 2158
Laus, 3526
laut, 2037
lauwarm, 5246
Lavendel, 4377
Lavendelblüthe, 4442
Lavendelöl, 4378
Laxieren, 4192
Laxiermittel, 4195
Lazareth, 3874
(fliegendes) Lazareth, 3875
leben, 4913
Leben, 4918
lebendig, 46
Lebensdauer, 50
lebensfähig, 45
Lebensfunken, 60
lebensgefährlich, 1615
Lebensgeist, 58
Lebenshauch, 56
Lebenskraft, 47
Lebensmittel, 3919
Lebenswärme, 53
Lebensweise, 4949
Leber, 1097
Leberabscess, 3000
Leberegel, 3545
Lebererentzündung, 2990
Lebererweichung, 3008
Leberflecken, 3210
Leberhyperämie, 2994
Leberkrebs, 3004
Leberlappen, 1098
Leberleiden, 5018
Leberschmerz, 1831
Leberschwindsucht, 3007

Leberthran, 4373
Leberüberzug, 2997
Lebervergrösserung, 2999
Leberverstopfung, 2995
Leberwassersucht, 3064
Leder, 933
Lederhaut, 934
leer, 688
Leere, 2166
Leerdarm, 1106
Lefze, 1210
legen, 30
(sich) legen, 5298
Lehnsessel, 4549
Lehnstuhl, 4548
lehren, 5316
Leib, 20
Leibchen, 4726
Leibarzt, 3817
Leibbinde, 3598
Leibesbeschaffenheit, 22
Leibesbeschwerde, 1582
Leibesfehler, 1581
Leibesfrucht, 1282
Leibesgebrechen, 1580
Leibesgestalt, 26
Leibeskraft, 23
Leibesöffnung, 5137
Leibesschaden, 1599
Leibesverstopfung, 2234
leiblich, 21
Leibschmerz, 1811
Leibstuhl, 3761
Leibwäsche, 5347
Leibweh, 1812
Leiche, 2518
leichenblass, 1592
Leichenblässe, 1594
Leichengift, 4536
leichenhaft, 2493
leicht, 1609
leid thun, 5588
leiden, 1548
Leiden, 1549
langes Leiden, 1550
leidend, 1551
Leinöl, 4380
Leinsamen, 4379
Leinwand, 5057
leise, 2046
leisten, 5271
Leistenbeule, 5348
Leiter, 1224
Leitsonde, 3669
Leitungssonde, 3670
Lende, 379
Lendengeflecht, 896
Lendengegend, 3019
Lendennerv, 894
Lendenschmerz, 1836
Lendenwirbel, 381
lesen, 5419

Lethargie, 2460
Leuchter, 4617
Leukämie, 3070
Leukorrhöa, 3271
Licht, 4609
Lichtschirm, 4619
lieben, 5599
lieber, 5277
liegen, 2192
Ligatur, 3630
Limonade, 4073
Lindenblüthe, 4333)
lindern, 1673
lindernd, 4213
lindernde Mittel, 4214
Linderung, 4864
Linderungsmittel, 4212
link, 427
Linse, 1375
Linsen, 4082
Linsenkapsel, 1376
Linsenknöchelchen, 1410
Linsenstaar, 3455
Lippe (lip), 1208
Liqueur, 4100
livid, 1890
local, 1661
localisiert, 1728
Loch, 278
löchern, 1485
Lochien, 3280
locker, 1861
locker machen, 4768
Löffel, 4670
losbinden, 4765
löschen, 3904
lose machen, 4767
lösen, 4770
losmachen, 4766
Lücke, 313
Luft, 1005
(die freie) Luft, 5346
Luftbad, 3792
Luftbett, 3585
Luftbläschen, 1043
Luftbrust, 2843
luftdicht, 5610
lüften, 5124
Luftgeschwulst, 2807
Luftheilung, 4153
Luftkanal, 1011
Luftkissen, 4570
Luftröhre, 1010
Luftröhrenast, 1019
Luftröhrenkopf, 1012
Luftröhrenstamm, 1020
Luftvergiftung, 4539
Luftveränderung, 4895
Luftweg, 1006
Luftzellen, 1045
Luftzug, 5043
Lunge, 732

Lungenabscess, 2832
Lungenapoplexie, 2833.
Lungenatrophie, 2839
Lungenbläschen, 1042
Lungenblatt, 1013
Lungenblutader, 773
Lungenblutbahn, 733
Lungenblutsturz, 2834
Lungenblutung, 2836
Lungenbrand, 2825
Lungendrüse, 1039
Lungenemphysem, 2809
Lungenentzündung, 2826
(katarrhale) Lungenentzündung, 2828
(croupeuse) Lungenentzündung, 2829
Lungenerweichung, 2837
Lungenfell, 1030
Lungenflügel, 1029
Lungengefäss, 1038
Lungengeflecht, 1037
Lungengeschwür, 2841
Lungengewebe, 1036
Lungenkammer, 1040
Lungenkrampf, 2816
lungenkrank, 5177
Lungenkreislauf, 733
Lungenlähmung, 2831
Lungenlappen, 1028
Lungenmagennerv, 874
Lungenödem, 2830
Lungenschall, 1046
Lungenschmerz, 1830
Lungenschrumpfung, 2838
Lungenschwindsucht, 2819
Lungenspitze, 1041
Lungensucht, 2818
lungensüchtig, 2824
Lungenwassersucht, 2840
Lungenzellen, 1044
Lust, 4801
Lustseuche, 3236
Lympfbahn, 760
Lympfcoagulum, 763
Lympfdrüse, 756
Lymphe, 752
Lymphgefäss, 753
Lympfknoten, 758
Lympfkörperchen, 761
Lympfplasma, 762
Lympfweg, 759

**M**

Macaroni, 4058
machen, 1672
(es) macht nichts, 4846
Made, 3549

Madenwurm, 3550
Magen, 1067
Magenbrei, 1088
Magenbrennen, 2207
Magenbruch, 2944
Magendrücken, 2930
Magendrüse, 1074
Magenentzündung, 2938
Magenerweichung, 2940
Magenerweiterung, 2939
Magengefäss, 1071
Magengeflecht, 1072
Magengegend, 1068
Magengekröse, 1076
Magengeschwulst, 2934
Magengeschwür, 2936
Magengrube, 1822
Magengrund, 1070
Magenhaut, 1075
Magenkatarrh, 2932
Magenkrampf, 1820
Magenkrebs, 2935
Magenlähmung, 2942
Magenmund, 1078
Magennerv, 1073
Magenpförtner, 1079
Magenpulver, 4323
Magenpumpe, 3745
Magensaft, 1089
Magensäure, 2209
Magenschmerz, 1819
Magenschwäche, 1826
Magenschwindsucht, 2943
Magenspritze, 3744
Magentropfen, 4322
Magenverhärtung, 2941
Magenwand, 1069
Magenweh, 2211
mager, 2431
Magerkeit, 2432
magnetisch, 4137
Mahl, 5216
Mahlzeit, 3920
Majoranöl, 4381
Makrele, 4006
Makrone, 4059
Mal, 3211
Malz, 4457
Mandeln, 1454; 1455
Mandelbräune, 2786
Mandelemulsion, 4370
Mandelmilch, 4371
Mandelöl, 4399
Mangel, 4951
(aus) Mangel an, 5462
(wegen) Mangels an, 5462
mangelhaft, 2157
Mannestollheit, 3286
Mannheit, 1171
männlich, 375
Manntollheit, 3286
Mantel, 4086

Marasmus, 2440
Mark, 114
markhaltig, 818
markiert, 1715
marklos, 819
Markmasse, 966
Markscheide, 822
Marksubstanz, 1157
Marter, 1608
martern, 1607
marternd, 1716
martervoll, 5248
Masern, 3132
Masse, 82
mässig, 1994
Mast, 1126
mast, 1127
Mastdarm, 1129
Mastdarmgefäss, 1131
Mastdarmgeflecht, 1130
Mastdarmknoten, 1132
Mastdarmwurm, 3551
mästen, 1128
Matratze, 3579
matt, 1565; 5423
Mattigkeit, 1566
Medicament, 4277
Medicin, 3803
medicinal, 3806
Medicinalbad, 3797
Medicinalcollegium, 3813
Mediciner, 3816
Meerrettig, 4486
Mehl, 4479
Mehlspeise, 4884
meiden, 5215
Meinung, 4838
Meiranöl, 4382
Meissel, 3323
Meisselwunde, 3324
meistens, 5387
Melissenblätter, 4336
Meliturie, 3049
Menge, 2256
(acute) Meningitis, 2753
(cerebro-spinale) Meningitis, 2754
Mensch, 1
Menschengeschlecht, 3
Menschenrace, 5
Menschheit, 6
menschlich, 7
Menses, 3259
Menstruation, 3254
(vicariirende) Menstruation, 3266
Menstrualkolik, 3268
Menstruationsstörung, 3264
menstruieren, 3253
merklich, 4962
Merkur, 4131

Merkurialcur, 4133
mesenterial, 3018
Messer, 3637
metallisch, 2077
Metallkolik, 2959
Miasma, 5376
miasmatisch, 5377
Mieder, 4724
Milch, 300
Milchborke, 3177
Milchbrod, 3945
Milchcur, 4122
Milchdrüse, 1230
Milchfieber, 3298
Milchgang, 1236
Milchgefäss, 1238
Milchkanal, 1237
Milchkanalentzündung, 3300
Milchleiter, 1235
Milchpumpe, 3746
Milchruhr, 3299
Milchsäure, 4300
Milchschorf, 3176
Milchverhaltung, 3301
Milchzahn, 301
Milchzucker, 4462
mildern, 1753
Militärarzt, 3818
Militäroberarzt, 3819
Milz, 1105
Milzweh, 1834
mindern, 4986
Mineralien, 4519
Mineralwasser, 4103
Minute, 1986
mischen, 5125
Miserere, 2223
Missbehagen, 5231
Missgeburt, 3289
mit (adv.), 5488
Mitbewegung, 838
Mitempfindung, 836
Mitesser, 3188
Mittagsbrod, 3926
Mittagsessen, 3925
Mittagsmahl, 3924
Mittagsschläfchen, 4204
mittel, 434
Mittel, 3918
Mittelfell, 1071
Mittelfellhöhle, 1032
Mittelfinger, 445
Mittelfuss, 509
Mittelgehirn, 844
Mittelhand, 435
Mittelknochen, 436
Mittelohr, 5468
Möbel, 4633
Mobilien, 4634
möglich, 4883
Mohn, 4387

Mohnkopf, 4463
Mohnöl, 4388
Mohnsamen, 4464
Möhren, 4038
Molke, 4123
Molkencur, 4124
momentan, 1658
Monat, 4892
(monatliche) Reinigung, 3256
(das) Monatliche, 3258
Monatsfluss, 3257
mondsüchtig, 2690
Monomanie, 2700
Moos, 4355
Morgen, 4987
morgen, 4874
Morgenhaube, 4737
Morgens, 4988
Morphium, 4530
Moschus, 4459
Moselwein, 4087
Moskitonetz, 4585
Mücke, 2547
Mückensehen, 2551
müde, 2443
Mühe, 2375
mühevoll, 2029
mühsam, 2039
Mund, 335
Mundathmen, 2019
Mundentzündung, 2888
Mundfäule, 2886
Mundfäulniss, 2886
Mundhöhle, 336
Mundkrebs, 2890
Mundmuskel, 592
Mundrachenhöhle, 1007
Mundschleimhautentzündung, 2889
Mundschmerz, 1785
Mundschwamm, 2901
Mundspiegel, 3753
Mündung, 633
Münze, 2097
münzenartig, 2098
murmeln, 5399
mürrisch, 2503
Mus, 4260
Muscatnuss, 4069
Muskel, 515
Muskelband, 521
Muskelbewegung, 524
Muskelbinde, 522
Muskelbündel in den Herzkammern, 652
Muskelfaser, 519
Muskelfaserstoff, 550
Muskelgewebe, 518
Muskelhaut, 520
Muskelhülle, 530
muskelig, 516

Muskelkraft, 531
Muskelrheumatismus, 3079
Muskelschicht, 527
Muskelschmerz, 1774
muskelstark, 533
Muskelstärke, 534
Muskelstarre, 2770
Muskelübung, 4175
Muskelzucken, 2474
Muskulatur, 535
musculös, 517
müssen, 5315
Muth, 4823
(zu) Muthe sein, 5302
(guten) Muthes sein, 4824
Mutterband, 1216
Mutterbeschwerde, 2375
Mutterblutfluss, 3279
Mutterfieber, 3297
Mutterkorn, 4474
Mutterkrampf, 2376
Mutterkrankheit, 2374
Mutterkrebs, 2377
Muttermal, 3212
Mutterscheide, 1207
Mutterscheidenentzündung, 2378
Mutterschmerz, 1845
Mutterspiegel, 3760
Mutterspritze, 3750
Muttertrompete, 1222
Muttervorfall, 3283
Mutterwuth, 3285
Mütze, 4694
Myrrhe, 4460

N

Nabelbinde, 3599
Nabelbruchband, 3600
Nabelgegend, 3016
Nabelstrang, 1202
nach und nach, 2108
Nachbarschaft, 5597
Nachcur, 4136
nachgebend, 1993
Nachgeburt, 1293
Nachkrankheit, 1523
nachlassen, 4858
Nachlüssigkeit, 5269
Nacht (bei), 4982
Nachtbecken, 3763
Nachtgeschirr, 4600
Nachthaube, 4736
Nachthemd, 4710
Nachtjacke, 4711
Nachtkappe, 4713
Nachtlampe, 4605
Nachtmahl, 3929
Nachtmütze, 4712

Nachtruhe, 5018
Nachtschatten, 4515
Nachtschweiss, 1977
Nachtstuhl, 4601
Nachttopf, 4599
Nachtwandeln, 2612
Nachtzeug, 4709
Nachverdauung, 1115
Nachweh, 3288
Nacken, 358
Nackenmuskel, 599
Nackenweh, 1796
Nadel, 3661
Nadelsonde, 3668
Nagel, 951
Nagelgeschwür, 3227
Nagelkörper, 952
Nagelpilz, 3522
Nagelwurzel, 953
nagend, 1702
Nähe, 5598
nähen, 107
nähern, 3393
nähren, 1056
nahrhaft, 1058
Nährstoff, 1060
Nahrung, 1057
Nahrungsmittel, 1059
Naht, 108
Name, 4899
Narbe, 3139
Narkose, 2587
Narkosis, 2587
Nase, 272
(durch die) Nase sprechen, 1487
näseln, 1489
Nasenathmen, 2020
Nasenbein, 273
Nasenbinde, 3594
Nasenbluten, 2800
Nasendach, 277
Nasenflügel, 1426
Nasenhöhle, 1008
Nasenknochen, 1423
Nasenknorpel, 290
Nasenloch, 278
Nasenmuschel, 281
Nasenmuschelbein, 282
Nasenmuskel, 590
Nasenpolyp, 2801
Nasenrücken, 1425
Nasenscheidewand, 284
Nasenschleim, 1427
Nasenschleimhaut, 1428
Nasenspiegel, 3757
Nasenspitze, 275
Nasenspritze, 3747
Nasentriefen, 3484
Nasenwurzel, 1424
nüssen, 3171; 3387
Natter, 4503

Natur, 2464
Nebel, 5430
Nebenhoden, 1193
nehmen, 2390
Neid, 3564
Neidnagel, 3565
Neigung, 2188
Nelke, 4406
Nelkenöl, 4407
Nerv, 779
nervenartig, 806
Nervenast, 794
Nervenbündel, 787
Nervenendigung, 797
Nervenfaser, 786
Nervenfieber, 3120
Nervengeflecht, 783
Nervengewebe, 782
Nervenhaut, 791
Nervenhülle, 792
Nervenkern, 801
Nervenknoten, 814
Nervenkörperchen, 802
nervenlos, 805
Nervenmark, 789
Nervenmasse, 780
Nervennetz, 785
Nervenreiz, 826
Nervensaft, 790
Nervenschicht, 788
Nervenschmerz, 1775
Nervenstamm, 793
Nervensubstanz, 781
Nervenverbreitung, 796
Nervenwarze, 799
Nervenwärzchen, 800
Nervenwurzel, 798
Nervenzelle, 823
nervicht, 803
nervig, 803
nervös, 804
Nessel, 3165
Nesselausschlag, 3166
Nesselfieber, 3168
Nesselfriesel, 3169
Nesselsucht, 3167
Netz, 784
Netzhaut, 1368
Neubildung, 3215
neugeboren, 1283
neulich, 5481
Neuralgie, 1776
Neuralgie des Gesichts, 2780
neuralgisch, 1656
nichts-desto weniger, 4855
nichtssagend, 2525
Nicken, 597
niederdrücken, 5021
niedergeschlagen, 2504
niederknien, 4777
niederkommen, 1260; 1281

Niederkunft, 1261
(sich) niederlegen, 4981
niederschlagen, 4368
niederschlagendes Pulver, 4369
niederziehen, 1314
Niederzieher, 1315
Niednagel, 3566
Niere, 1146
Nierenabscess, 3035
Nierenbecken, 1151
Nierenbeckenentzündung, 3036
Nierenentzündung, 3026
Nierenkelch, 1149
Nierenkolik, 3041
Nierenkrebs, 3027
Nierenleiden, 5049
Nierenleiter, 1152
Nierenpyramide, 1148
Nierenschmerz, 1840
Nierenstein, 3028
Nierenwassersucht, 3063
Nierenweh, 1839
Nierenwurzel, 1147
Niesemittel, 4228
niesen, 2061
Niesen, 2062
Nikotin, 4528
Nisse, 3531
Norm, 2109
Noth, 1289
nöthig, 4787; 4788
nothwendig, 5188
nüchtern, 5228
nun, 5121
Nuss, 4052

O

ober, 394
Oberarm, 411
Oberarmbinde, 3601
Oberarmkopf, 412
Oberarmmuskel, 614
Oberdecke, 4575
Oberfläche, 1763
oberflächlich, 1722
Oberhäutchen, 925
Oberkiefer, 256
Oberkieferknochen, 257
Oberkörper, 5173
Oberleib, 5174
Oberschenkel, 483
Oberschenkelbein, 486
Oberschenkelmuskel, 617
Obst, 4043
Ödem der Glottis, 2787
ödematös, 1873
Oel, 967

Oelseife, 4467
Ofen, 4621
Ofenloch, 4624
Ofenschirm, 4622
Ofenthür, 4623
offen, 2530
offener Leib, 4973
Officin, 3861
öffnen, 5455
oft, öfters, 4880
Ohnmacht, 2598
(in) Ohnmacht fallen, 2590
ohnmächtig, 2600
ohnmächtig werden, 2601
Ohr, 1385
(das mittlere) Ohr, 1397
(die) Ohren klingen, 1484
Ohrblatt, 1387
Ohrenarzt, 3825
Ohrenbrausen, 2565
Ohrenfluss, 3478
Ohrenklingen, 2568
Ohrensausen, 2566
Ohrenschmalz, 1391
Ohrenschmerz, 1789
Ohrensingen, 2567
Ohrenspiegel, 3758
Ohrenspritze, 3748
Ohrentönen, 2569
Ohrentrommel, Trommel, 1393
Ohrenwachs, 1392
Ohrenweh, 1790
Ohrfinger, 448
Ohrgeschwür, 3477
Ohrhöhle, 1386
Ohrkrystall, 1414
Ohrläppchen, 1388
Ohrloch, 1386
Ohrlöcher stechen, 1486
Ohrmuschel, 1387
Ohrsand, 1415
Ohrschmalz, 1391
Ohrspiegel, 3759
Ohrspritze, 3749
Ohrtrompete, 1405
Ohrwachs, 1392
Olivenöl, 3485
Opiat, 4288
Opium, 4529
Orange, 4050
ordentlich, 5489
Ordnung, 2373
Organ des Gesichts, 1310
Ort, 4967
örtlich, 5292
oscilieren, 2537
Ovarialtumoren, 3025
oxalsauer, 2298

# P

paar, 4873
Panacee, 2484
Pantoffel, 4716
Papel, 1905
Papille, 2146
Paralyse, 2735
Paraplegie, 2741
Parasiten, 3515
Parese, 2736
passen, 5149
passend, 5450
Pastete, 4021
Pastille, 4278
Patent-Charpie, 3695
Pauke, 1394
Paukenfell, 1396
Paukenhöhle, 1399
Pechpflaster, 4425
Pein, 1606
peinigen, 1605
Peitschenwurm, 3553
Pelz, 4688
pelzig, 2132
Pelzkragen, 4740
penetrierend, 3348
Penis, 1187
peptisch, 2987
Perforation, 2964
Periode, 3263
peripher, 1662
perlenartig, 2535
perniciös, 3069
peruvianischer Balsam, 4420
Pest, 3126
Pestgift, 4535
Petersiliensamen, 4445
Pfanne, 112
Pfannkuchen, 4017
Pfeffer, 4065
Pfeffergurken, 4024
Pfefferminze, 4338
Pfefferminzthee, 4349
pfeifen, 2040
Pfirsich, 4049
pflanzlich, 3519
Pflaster, 3692
(englisches) Pflaster, 3705
Pflaume, 4046
pflegen, 5034
Pflugschar, 288
Pflugscharbein, 289
Pfortader, 776
Pfortaderblut, 777
Pfortaderblutumlauf, 778
Pforte, 775
Pförtner, 1077
Pförtnerklappe, 1081
Pfropf, 3358
Pfropfen, 4663

Pfropfenzieher, 4664
phantasieren, 2614
Phosphor, 2287
Phthisis, 2820
Pigmentleber, 2997
Pille, 4258; 3522
Pincette, 3708
Pincettenschere, 3711
Pisse, 2279
(kalte) Pisse, 2280
pissen, 2278
plagen, 5232
platt, 2273
Platte, 947
Plattwürmer, 3540
plombieren, 5554
plötzlich, 1724
Pneumonia, 2827
pochen, 741
Pocken, 3135
pockenartig, 3137
Pockenfieber, 3138
Pockengift= Blatterngift, 3499
Pockenimpfung, 3150
Pockennarbe, 3141
pockennarbig, 3143
pökeln, 3977
Pökelfleisch, 3978
Polinurie, 2331
Polizei, 3814
Polster, 163
Polyklinik, 3870
Polyp, 2797
Pomeranze, 4051
Pomeranzenblätter, 4334
Portwein, 4093
practicieren, 3809
practisch, 3810
practischer Arzt, 3810
practicierender Arzt, 3810
Prellschuss, 3330
Preservationsmittel, 4209
pressen, 608
Pressschwamm, 4470
primär, 1617
probat, 5617
Profession, 4908
profus, 1976
Prostata, 1202
Provisor, 3852
puerperal, 2969
Puls, 663
Pulsader, 667
(die grosse) Pulsader, 671
Pulsadergeschwulst, 2847
Pulsation, 3022
pulsieren, 664
Pulsschlag, 666
Pult, 4644
Pulver, 4259
Pulverkörner, 3513

Punctiernadel, 3664
Punsch, 4102
Pupille, 1359
purgieren, 4193
Purgiermittel, 4196
Pustel, 1914
(mit) Pusteln bedeckt, 1915
Pustelflechte, 3193
putzen (ein Licht), 4615
Pyämie, 3073

## Q

Quacksalberei, 5628
quacksalbern, 5627
Quaddel, 1908
Qual, 1604
quälen, 1603
quälend, 1704
Qualm, 5164
Qualmbad, 3776
qualvoll, 1705
Quart, 5605
Quassiaholz, 4452
Quecksilber, 4130
Quecksilbercur, 4152
Quecksilberdampf, 5088
Quecksilberpflaster, 4431
quer, 625
Querbinde, 3629
Querbruch, 3413
Querlähmung, 2740
Querscheidewand, 626
Querspalte, 856
Querwunde, 3347
quetschen, 3327
Quetschung, 3328
Quetschwunde, 3329

## R

Race, 4
Rachen, 1061
(bösartige) Rachenbräune, 2914
Rachenhöhle, 1063
Rachenkatarrh, 2904
Rachenspiegel, 3754
radikal, 5293
Radikalmittel, 4232
Rahm, 4077
rahmähnlich, 2259
Rainfarrn, 4476
Rand, 131
rasch, 1989
rasen, 2691
rasend, 2523
Raserei, 2692
rastlos, 2346

Rastlosigkeit, 2348
Rath, 3838
(zu) Rathe ziehen, 4959
rathen, 4890
Rauch, 2318
rauchbad, 3795
rauchen, 4956
Rauchfleisch, 3976
rauchig, 2316
räudig, 1941
rauh, 1866
rauschen, 653
räuspern, 5104
Räuspern, 5105
Rautenblatt, 4342
Rebhuhn, 3986
Recept, 4189
recht, 426; 4792
Recht haben, 5625
Recidiv, 4110
recidive, 1619
Recipe, 4190
reflectiert, 1733
reflex, 1663
Reflexempfindung, 840
Regel, 3262
(in der) Regel, 5239
regelmässig, 2006
regeln, 5539
Regenbogen, 1355
Regenbogenhaut, 1356
Regenbogenhautentzündung, 5441
Regenwurm, 1118
regenwurmähnlich, 1119
regulieren, 4176
reichen, 5096; 5362
reichlich, 2089
reif, 1167
Reife, 3224
rein, 2125
reinigen, 3316
Reinigung, 3255
(monatliche) Reinigung, 3256
Reis, 4041
Reiseapotheke, 3864
reissen, 3341
reissend, 1701
reiswasserähnlich, 5161
reiswasserartig, 2254
Reiswasserstuhl, 2255
reizbar, 2510
Reizbarkeit, 827
reizen, 825
reizend, 1707
reizlos, 5111
Reizmittel, 4216
Reizung, 4243
Rettich (ig), 4483
Rhabarber, 4461
Rheinwein, 4086

rheumatisch, 2963
Rheumatismus, 3076
Ricinusöl, 4391
Riechbein, 240
riechen, 240
(aus dem Munde) riechen, 2044
Riechhaut, 1429
Riechkissen, 4571
Riechnerv, 1430
Riechzelle, 1431
Rinde, 964
Rindenmasse, 965
Rindensubstanz, 1157
Rinderbraten, 3968
Rindfleisch, 3956
Rindsbraten, 3968
Ring, 5429
Ringfinger, 446
Ringknorpel, 1466
Ringmesser, 3654
Ringmuskeln, 556
rinnen, 659
Rippe, 89
(die wahren) Rippen, 390
(die falschen) Rippen, 391
Rippenathmen, 2022
Rippenbruch, 5197
Rippenweh, 1804
Riss, 3342
rissig, 2133
Risswunde, 3343
Ritz, 1958
Ritze, 1473
ritzen, 1957
Ritzmesser, 3639
röcheln, 2041
Rock, 4681; 4721
Röhre, 96
Röhrenathmen, 2021
Röhrenknochen, 97
Rollbinde, 3628
rollen, 1316
Roller, 1317
Rollgelenk, 197
(der grosse) Rollhügel, 489
(der kleine) Rollhügel, 490
Rollmuskel, 1317
Rollmuskelnerv, 868
Rose, 3158
Rosmarin, 4339
Rosmarinöl, 4392
Rost, 4053
rostfarben, 2088
roth, 712
Röthe, 1764
Rötheln, 3133
röthen, 4241
Rotherhund, 3131
Rothlauf, 3159

Röthung, 4242
Rothwein, 4085
Rouleau, 4646
Rüben, 4037
(gelbe) Rüben, 4039
Rücken, 345
Rückenbein, 346
Rückenmark, 352
Rückenmarksfaden, 879
Rückenmarkshaut, 881
Rückenmarkskanal, 353
Rückenmarksliquor, 882
Rückenmarksnerv, 812
Rückenmarksrinde, 880
Rückenmarksschmerz, 1802
Rückenmarksnervensystem, 813
Rückenmarkszapfen, 878
Rückenmuskel, 605
Rückennerven, 891
Rückenschmerz, 1799
Rückenweh, 1800
Rückenwirbel, 349
Rückfall, 1524
rückfallen, 4109
Rückfallstyphus, 3125
Rückgrat, 348
Rückgratsgelenk, 351
Rückgratskanal, 877
Rückgratsschmerz, 1801
Rückgratswirbel, 350
Rücksicht, 5485
rücksichtlich, 5486
rufen, 4789
rufen lassen, 4790
Ruhe, 5067
ruhig, 2345
Ruhr, 2988
Rührei, 4018
rühren, 2765
rülpsen, 2203
Rülpsen, 2204
Rum, 4101
Rumpf, 340
Rumpfmuskel, 595
rund, 1469
rundlich, 1470
Rundwürmer, 3544
Runzel, 1899
runzeln, 1319
runzlich, 1864
Rupfzange, 3712
russartig, 2140
rüsten, 75
rüstig, 2352

## S

Sache, 4829
Sack, 918
Sackgeschwulst, 3226
Safran, 4475
Saft, 1595
(schlechte, böse, ungesunde) Säfte, 1596
Säge, 3650
sagen, 4837
Sahne, 4076
Salbe, 4265
Salbeiblätter, 4340
Salmiakgeist, 4416
Salpetersäure, 4301
Salpeterweingeist, 4439
Salz, 2284
(englisches) Salz, 4417
Salzbad, 3798
Samen, 1194
Samenbläschen, 1201
Samenstrang, 1197
Samenthierchen, 1196
Sand, 2297
Sandbad, 3794
sandig, 2298
sanft, 1860
Sardelle, 4007
Sarsaparille, 4360
Sassaparille, 4360
Sassafras, 4454
Satz, 2296
sauer, 1982
säuerlich, 5135
Sauerkraut, 4036
saufen, 3902
Säufer, 2708
Säuferleber, 3006
Säufernase, 3191
Säuferwahnsinn, 2709
Saugaderdrüsen, 976
Säugamme, 1299
saugen, 751
(an der Mutterbrust) saugen, 1294
säugen, 1295
Saugflasche, 1301
Saugeglass, 1302
Säule, 343
Säure, 2169
sausen, 2563
Schabeeisen, 3658
Schabemesser, 3657
schaben, 3699
Schachtel, 5615
Schädel, 586
Schädelbohrer, 3726
Schädelverletzung, 5026
schaden, 1597
Schaden, 1598
schadhaft, 5558
schädlich, 5165
schaffen, 13
Schaft, 961
Schale, 129
Schall, 656

schallen, 654
Scham, 460
Schambein, 461
Schamberg, 1205
Schambug, 5366
(sich) schämen, 459
Schamfuge, 462
Schamgeflechtsnerv, 900
Schamgegend, 1204
Schamhaar, 972
Schamlippe, 1208
Schamspalte, 1200
Schamtheile, die männlichen, 1183
Schamtheile, die weiblichen, 1184
Schamtheile, die äussern, 1203
Schanker, 3239
Schankergeschwür, 3240
Schar, 287
scharf, 1689
Scharlach, 3127
Scharlachfieber, 3128
Scharlachfriesel, 3129
Scharnier, 191
Scharniergelenk, 193
Schärpe, 3627
Schauer, 2113
Schauerbad, 3789
schaukeln, 4552
Schaukelstuhl, 4553
Schaum, 2310
schaumig, 5143
Scheibe, 492
Scheibenbinde, 3604
Scheide, 1206
scheiden, 282
Scheidenschmerz, 1844
Scheidewand, 283
Scheitel, 226
Schenkel, 478
Schenkelband, 481
Schenkelbein, 479
Schenkelbogen, 482
Schenkelgelenk, 480
Schenkelnerv, 897
Schenkelschmerz, 1850
Schere, 3643
scheren, 286
Sherry, 4095
scheuen, 5524
Schicht, 526
schichten, 525
schief, 3081
schielen, 2531
Schielen, 3476
Schielnadel, 3678
Schienbein, 496
Schiene, 495
Schierling, 4514
schiessen, 3325

(auf und ab) schiessend, 1736
Schiesspulver, 3511
Schiesswunde, 3326
Schiffsapotheke, 3866
Schild, 1033
Schilddrüse, 1034
Schildknorpel, 1465
Schildkröte, 4012
schinden, 3380
Schinken, 3980
Schirm, 4581
Schlaf, 2477
Schlafarzenei, 4102
Schläfe, 233
schlafen, 2476
schlafen gehen, 5236
Schlafengehen, 5236
Schläfenbein, 234
Schläfengrube, 339
schlaff, 1862
Schlafjacke, 4732
Schlaflosigkeit, 2589
Schlafmittel, 4201
Schlafmütze, 4714
Schläfrigkeit, 2581
Schlafrock, 4715; 4723
Schlafscheu, 2592
Schlafsessel, 4551
Schlafstube, 4546
Schlafstuhl, 4550
Schlafsucht, 2584
Schlaftrank, 4203
Schlafwandeln, 2610
Schlafzimmer, 4544
Schlag, 4946
(vom) Schlage gerührt, 2765
Schlagader, 668
(die grosse) Schlagader, 672
Schlaganfall, 2764
schlagen, 665
Schlagfluss, 2762
Schlange, 1110
schlangenförmig, 1111
Schlangengift, 3500
schlecht, 2630
schlechterdings, 5072
schleichend, 1637
Schleichfieber, 3113
Schleie, 4001
Schleife, 3625
Schleim, 543
Schleimbeutel, 545
Schleimfieber, 3110
Schleimhaut, 920
schleimig, 5160
Schleimnetz, 924
Schleimpapel, 3244
Schleimrasseln, 2051
Schleimscheide, 547

Schleimschicht, 940
schliessen, 405
Schliessmuskel, 557
schlimm, 3302
Schlingbeschwerde, 2782
Schlinge, 3626
schlingen, 3879
Schlitten, 3568
Schlitzmesser, 3645
schluchzen, 2047
Schluck, 4861
schlucken, 3880
Schlund, 1062
Schlundbräune, 2916
Schlundhöhle, 1064
Schlundkopf, 1066
Schlundkopfentzündung, 2915
Schlundkopfkrampf, 2917
Schlundkopflähmung, 2918
Schlundkopfpolyp, 2919
Schlundkopföffner, 3722
Schlundkopfspiegel, 3756
Schlundkopfstosser, 3724
Schlundkopfschwindsucht, 2920
Schlundsonde, 3671
Schlundstosser, 3723
schlürfen, 5260
Schlüssel, 406
Schlüsselbein, 407
Schlüsselblumenthee, 4350
schmal, 5299
schmale Kost, 5300
Schmalz, 1390
Schmarotzer, 3514
Schmarotzerpflanze, 3520
Schmarotzerschwamm, 3521
Schmarotzerthier, 3518
schmecken, 1433
(es) schmeckt mir, 4878
Schmeer, 2316
Schmeerbauch, 2397
Schmelz, 315
Schmerz, 1670
Schmerzanfall, 1750
Schmerz empfinden, 1671
Schmerz verursachen, 1672
Schmerz machen, 1673
schmerzen, 1553; 1669
schmerzfrei, 1682
schmerzhaft, 1680
Schmerzhaftigkeit, 1684
schmerzlich, 1679
schmerzlos, 1683
Schmerzlosigkeit, 1685
Schmerzparoxysmen, 1746

schmerzstillend, 1686
schmerzstillende Mittel, 4287
schmerzvoll, 1681
Schmiere, 214
schmieren, 213
Schmutzflechte, 3194
Schmutzgrind, 3195
schmutzig, 1886
schmutzigblau, 2312
schmutziggrau, 2094
Schnäpper, 3717
Schnapps, 4954
schnappsen, 3906
Schnecke, 1403
Schneckenfenster, 1404
Schneckennerv, 1417
Schneidemesser, 3647
schneiden, 208
schneidend, 1695
Schneidezahn, 299
schnell, 1629
schnellend, 1998
Schnelligkeit, 5214
schneuzen, 1490
Schnittmesser, 3640
Schnittwunde, 3320
schnüffeln, 1488
schnupfen, 5606
Schnupfen, 2790
Schnupfenfieber, 3104
schnüren, 4760
Schnürbrust, 4729
Schnürleib, 4728
Schnürmieder, 4727
Schnürriemen, 4730
schnurren, 2054
Schnürverband, 3615
schon, 5621
Schönheitsmittel, 4276
schöpfen, 987
Schorf, 1926
Schornstein, 4630
Schoten, 4030
Schramme, 1928
Schrank, 4637
Schreck, 5477
Schrecken, 5477
schrecklich, 1649
schreiben, 4900
Schreibtisch, 4645
Schrei, 2483
schreien, 2482
Schritt, 2353
schröpfen, 4250
Schröpfkopf, 3762
Schröpfschnepper, 3766
schrumpfen, 2468
Schrumpfung, 2469
Schrunde, 1930
schrunden, 1929
schrundig, 3340

Schublade, 4641
Schuh, 4698
Schuhband, 4699
Schuhriemen, 4700
Schuld, 5014
(sich zu) Schulden kommen lassen, 5015
schuldig sein, 5579
Schulter, 399
Schulterbein, 403
Schulterblattmuskeln, 612
Schulterknochen, 404
Schulterschmerz, 1798
Schuppe, 1931
Schuppen im Gesichte, 1932
Schuppen auf dem Kopfe, 1933
Schuppenausschlag, 3202
Schuppenflechte, 3203
Schuppengrind, 3204
schuppicht, 3205
schürfen, 1925
Schürze, 4733
Schüssel, 4676
schütteln, 5253
Schüttelfrost, 5184
Schüttelwehen, 3287
schützen, 3148
Schutzmittel, 4231
schwach, 1540
Schwäche, 1542
schwächen, 1543
schwächlich, 1544; 1545
Schwamm, 2486
Schwammhalter, 3682
schwammicht, 3234
schwanger, 1249
schwängern, 1251
Schwangerschaft, 1252
schwankend, 2000
Schwanz, 470
Schwanzbein, 471
Schwären, 3233
schwären, 2413
schwarz, 2263
Schwarzbrod, 3935
schwärzen, 5241
Schwefel, 2289
Schwefeläther, 4291
Schwefelbad, 3791
Schwefelholz, 4608
Schwefelsäure, 2290
Schwefelwasser, 4314
schweigen, 5328
Schweinefleisch, 3959
Schweiss, 937
Schweissbad, 3772
Schweissdrüse, 938
schweissfeucht, 1972
Schweissfieber, 3103

schweissig, 1980
Schweisskanal, 939
Schweissmittel, 4227
schwellen, 2398
Schwellung, 5062
schwer, 1610
Schwere, 1832
Schwermuth, 2761
schwermüthig, 4821
Schwiele, 949
schwielig, 950
schwierig, 5288
Schwierigkeit, 5115
schwimmen, 5447
Schwinde, 3201
Schwindel, 4798
schwinden, 2156
schwindlig, 4797
schwirrend, 2009
Schwitzbad, 3773
Schwitzcur, 5546
schwitzen, 1963
Schwitzen, 1964
schwitzend, 1979
Sclerotica, 1340
Sechswöchnerin, 1279
secundär, 1618
Seebad, 3799
Seekrankheit, 2945
segnen, 5502
(in) gesegneten Umständen, 5503
Sche, 1360
sehen, 1307
Sehkraft, 2527
Sehloch, 1358
Sehne, 208
Sehnenfaser, 209
Sehnenhäute, 542
Sehnerv, 865
sehnig, 584
sehnige Muskelbinden, 585
Sehorgan, 1308
Sehstörung, 2546
Sehwerkzeug, 1309
Seidlitzpulver, 4364
Seife, 4465
(medicinische) Seife, 4466
seifenartig, 2315
Seifenspiritus, 4436
seit, 4792
seitdem, 5080
Seite, 2342
Seitenkopfweh, 1782
Seitenstechen, 1803
Seitenstiche, 1803
Seitenstrang, 884
selbst, 5114
selten, 1651
Semmel, 3946
Senf, 4252

Senföl, 4393
Senfpflaster, 4252; 4432
Senfspiritus, 4437
Senfteig, 4455
Senfumschlag, 4253
senken, 3282
Senkung, 3282
Sennesblätter, 4341
Sensibilität, 831
Sensibilitätsstörung, 2657
sensitive Nerven, 835
serös, 700
Serum, 707
Serviette, 4668
Sessel, 3575
(sich) setzen, 5547
Seuche, 1643
seuchenartig, 1643
seufzen, 2060
Sichelnadel, 3663
sicher, 2364
sickern, 1921
Sieb, 237
Siebbein, 238
siech, 1567
Siechbett, 1561
siechen, 1568
Siechthum, 1570
Silbernitrat, 4315
simuliert, 1622
sinken, 2464
Sinn, 833
Sinnesnerv, 834
Sinnesstörungen, 2544
Sinnestäuschung, 2543
sinnlich, 1245
Sinnlichkeit, 1246
Sitz (der Schmerzen), 1751
Sitzbad, 3768
Sitzbein, 457
sitzen, 456
Skalpell, 3660
Skorbut, 3074
so wie, 5322
Sodawasser, 4072
Sodbrennen, 2208
Socken, 4705
sofort, 5251
sofortig, 5270
sogar, 4964
sogleich, 5098
Sohle, 513
Sommersprossen, 3207
sommersprossig, 3208
Somnambulismus, 26
Somnolenz, 2585
Sonde, 3665
sondieren, 3386
Sondiernadel, 3666
(hohle) Sondiernadel, 3667
Sonne, 909
Sonnenbad, 3793

Sonnengeflecht, 910
Sonnenhitze, 5380
Sonnenstich, 2766
sonor, 2050
sonst, 4810
sonstig, 5307
Soor, 2892
Sopha, 4643
soporös, 2588
Sorge, 4953
sorgen, 5136
Sorgfalt, 5565
sorgfältig, 4510
sorglos, 2515
Spalt, 850
spalten, 849
spanisch, 3251
Spanischfliegenpflaster, 4429
Spanischfliegensalbe, 4430
Spann, 498
spannend, 1698
Spannung, 1833
spärlich, 5186
sparsam, 2325
spät, 5100
special (or speciell), 3822
Specialarzt, 3823
specifisch, 2320
Speck, 3981
Speckleber, 3002
Speculum, 3752
Speiche, 418
Speichel, 981
speichelartig, 2269
Speichelcur, 4134
Speicheldrüse, 982
Speichelfluss, 5061
speien, 2080
Speise, 765
Speisebrei, 1086
Speisebreibildung, 1090
Speisegefässe, 769
speisen, 764
Speiseröhre, 1065
Speiseröhrenbräune, 2923
Speiseröhrenentartung, 2925
Speiseröhrenentzündung, 2924
Speiseröhrenkrampf, 2926
Speiseröhrenlähmung, 2927
Speisesaft, 766
Speisesaftgefässe, 768
Speiseschrank, 4650
Sperrpincette, 3710
spicken, 3000
Spickgans, 3991
Spiegel, 3751
Spinalganglion, 886

Spinne, 2549
Spinnensehen, 2553
Spinngewebe, 858
Spinnwebenhaut, 859
Spirituosen, 4955
Spital, 3871
spitz, 2497
Spitze, 274
Spitzpocken, 3155
Splitter, 3414
Splitterbruch, 3415
Splitterzange, 3709
Sprache, 1460
Sprachwerkzeug, 1461
sprechen, 1459
spreizen, 4780
(auseinander) spreizen, 4780
springen, 503
Springwürmer, 3548
Spritze, 3731
spritzen, 3729
Spritzenröhrchen, 3735
Sprudelbad, 3781
Spucke, 2082
spucken, 2081
Spucknapf, 4603
Spulwürmer, 3547
Staar, 3449
(den) Staar stechen, 3464
staarblind, 3463
Staarfell, 3453
Staaroperation, 3466
Staarstechen, 3465
Staat, 4905
Stadium, 1668
Stadien (pl.), 1668
Stahl, 4128
Stahlcur, 4129
stammen, 5291
Stand, 5222
(im) Stande sein, 5223
stark, 532
Stärke, 1559; 4477
stärken, 1558
Stärkung, 4163
Stärkungsmittel, 4165
starr, 2456
Starrheit, 2459
Starrkrampf, 2746
Starrsucht, 2768
starrsüchtig, 2769
stätig, 2120
stattfinden, 5027
Staub, 5123
Staubbad, 3783
staubig, 5149
Stearinkerze, 4611
Stechapfel, 4516
Stechbecken, 4602
stechen, 3321
stechend, 1094

Stecher, 3648
Stecknadelkopf, 5439
stehen, 2349
steif, 2362
Steifheit, 1759
Steigbügel, 1411
steigen, 675
steigern, 5341
Steigerung, 3394
Stein, 5332
Steinkolik, 3040
Steinkrankheit, 3038
Steinöl, 4389
Steinschmerzen, 1843
Steiss, 472
Steissbein 475
Steissbeinnerv, 901
Steissweh, 1847
Steisswirbel, 477
Steissbeinwirbel, 477
Stelle, 5087
Stellung, 4970
Steppdecke, 4574
steppen, 4573
sterben, 4915
stets, 5386
Stichlanzette, 3649
Stichwunde, 3322
sticken, 2074
Stickhusten, 2075
Stiefel, 4704
Stiefelzieher, 4647
Stier, 2534
still, 1654
stillen, 1297; 3384
stillstehend, 2004
Stimmband, 5071
Stimmbeschwerde, 2576
Stimme, 1458
stimmen, 2623
Stimmorgan, 1457
Stimmritze, 1474
Stimmritzband, 1475
Stimulans, 4293
Stimmung, 2625
Stinkasant, 4317
Stinknase, 2802
Stippchen, 1906
Stirn, 230
Stirnband, 3592
Stirnbein, 231
Stirnbinde, 3592
Stirnkopfschmerz, 1783
stochern, 5542
stockblind, 3468
Stockblindheit, 3469
stockend, 2005
stocktaub, 3179
Stockung, 5326
Stoff, 202
Stoffwechsel, 1055
stolpern, 2372

Stöpsel, 5611
stören, 2155
Störung, 2653
Stoss, 737
stossen, 736; 5395
Stosskrampf, 2747
straff, 219
Strahl, 1347; 5309
Strahlenband, 1348
Strahlenblättchen, 1371
Strahlenkörper, 1352
Strang, 883
Straussenmagen, 2929
Streckbett, 3581
strecken, 560
Strecker, 564
Streckstuhl, 3582
Streichhölzchen, 4607
streifen, 2100
Streifen, 2099
Streiffschuss, 3331
Strick, 5020
strict, 4879
Strictur, 2928
Strieme, 1959
strohartig, 1896
Strohhalm, 5262
Strohsack, 3580
Strom, 2334
Strömung, 5540
Strumpf, 4704
Strumpfbänder, 4739
Strychnin, 4527
Stube, 4543
Stück, 5117
Stuhl, 3574; 2225
Stuhldrang, 2247
Stuhlgang, 2224
Stuhlmangel, 2232
Stuhlverstopfung, 2233
Stuhlzäpfchen, 4269
Stuhlzwang, 5281
stumpf, 2519
Stumpfsichtigkeit, 3445
Stumpfsinn, 2618
Stumpfsinnigkeit, 2619
Stunde, 4809
stupid, 2526
Stupor, 2583
Sturzbad, 3784
Sturzbad von oben, 3785
Sturzbad von der Seite, 3786
Stütze, 5500
subcutan, 3736
Substanz, 80
suchen, 5110
summen, 2056
Sumpf, 5375
sumpfig, 5376
Suppe, 3961
Suppenfleisch, 3962

süss, 4455
Süssigkeit, 4456
Sympatheticus, 908
sympathetisch, 2962
sympathisch, 1657
Syntonin, 551
syphilitisch, 3005
Syrup, 4261

T

Tabak, 4343
Tabakqualm, 5164
Tag, 4983
tagelang, 5560
Tagesstunde, 4983
täglich, 5034
Talg, 935
Talgdrüse, 936
Tarantel, 4505
Tasche, 1480
Taschentuch, 5158
Tasse, 4674
Tastempfindung, 2573
tasten, 2572
taub, 1851
Taube, 3988
Taubgefühl, 2454
Taubheit, 4996
Taubsein, 1852
Taubstumme, 3482
Taubstummheit, 3483
tauchen, 5192
taumeln, 2366
Teig, 4254
teigartig, 1898
Teller, 4669
Temperatur, 2102
Terpentin, 4396
Terpentinöl, 4397
Terpentinölseife, 4468
Terpentinspiritus, 4398
Teufelsdreck, 4318
thätig, 5002
Thee, 4079
Theelöffel, 4672
Theer, 4312
theerartig, 2264
Theerwasser, 4313
Theil, 704
(zum) Theil, 4854
theilen, 5626
theilnahmlos, 2516
theilweise, 4995
Theriak, 4286
Thier, 1195
Thierarzeneikunde, 3834
Thierarzt, 3832
thierisch, 3516
thierische Schmarotzer, 3517

Thierkohle, 4320
thonartig, 2261
Thräne, 270
thränend, 2533
Thränenapparat, 1326
Thränenbein, 271
Thränendrüsen, 980
Thränenfistel, 3439
Thränenfluss, 3438
Thränengang, 1327
Thränenhügel, 1330
Thränenkanal, 1328
Thränenkarunkel, 1329
Thränensack, 1331
Thränensee, 1332
Thränenwärzchen, 1333
Thran, 4372
Thür, 5045
Thymianöl, 4410
tief, 2582
Tinctur, 4262
Tischtuch, 4667
Tischzeug, 4666
toben, 2696
Tobsucht, 2700
Todesschweiss, 1978
todtkrank, 1532
tödtlich, 1625
Tödtlichkeit, 3310
Tokaier, 4092
toll, 2697
Tolle, 2703
Tollheit, 2698
Tollkirsche, 4517
Tollwuth, 2699
Ton, 2055
tonisch, 2744
Topf, 4657
Torte, 4057
Toxication, 4537
Tracht, 4678
Traganth, 4472
Tragacanth, 4472
Tragbahre, 3571
Tragband, 3619
Tragbett, 3573
Tragbeutel, 3621
Tragbinde, 3620
träge, 1992
Trage, 3569
tragen, 360
Träger, 362
Trägheit, 2659
Tragstuhl, 3576
Trank, 4279
Tränkchen, 4280
Traube, 1366
Traubencur, 4125
Traubenhaut, 1367
Traufbad, 3780
träufeln, 5137
traumatisch, 1667

träumen, 2604
Träumen, 2605
treiben, 2401
Trepanschlüssel, 3727
Trichine, 3555
Trichter, 4661
Trieb, 1165
Triefauge, 3436
triefen, 1981
trinken, 3901
trinken über den Durst, 3908
Tripper, 3249
trocken, 1868
trocknen, 5607
Trommel, 1393
Trommelfell, 1395
Trommelhöhle, 1398
Trommelsucht, 3012
Trompete, 1221
Tropfbad, 3782
Tropfen, 2335; 4281
tropfenweise, 2236
Tropfenzähler, 5602
Tropfglas, 5608
trübe, 2253
Trübsinn, 2685
Trübung der Linse, 3448
Trunkenbold, 2705
Trunkenheit, 2704
Truthahn, 3989
Tuberkulose, 3017
tüchtig, 5570
turnen, 4169
Turnen, 4170
Turnübung, 4173
Tympanie, 3011

# U

Übel, 4942
übel, 2042; 2185
Übelbefinden, 1538
Übelkeit, 2186
übler Geruch, 2042
übelriechend, 2045
üben, 4171
Überanstrengung, 5003
(sich) überarbeiten, 4948
Überbinde, 3612
überbinden, 3610
übereinander legen, 4779
(sich) überessen, 3882
Überfluss, 2205
überfüllen, 5284
(sich) übergeben, 2183
Übergiessung, 3790
überhaupt, 5019
überknorpeln, 152

überladen, 1825
überlassen, 5629
überlaufen, 2118
(sich) überlegen, 4774
Übermass, 5490
übermässig, 1973
(sich) übermüden, 5379
Übermüdung, 5016
überraschen, 3788
Überraschungsbad, 3787
Überrock, 4682
Überschuh, 4703
Überschuss, 2210
übersichtig, 3473
Übersichtigkeit, 3474
übersteigen, 5407
überstrahlen, 837
Überstrahlung, 839
übertragbar, 1645
überziehen, 175
Überzug, 176; 2143
Übung, 4172
umbinden, 3611
umdrehen, 363
Umdreher, 364
Umfang, 2389
Umgang, 5352
umgeben, 2532
umherwandern, 5388
umhüllen, 913
Umhüllungsgewebe, 914
umkleiden, 4743
Umlauf, 723
umlaufen, 722
Umschlag, 4257
Umstand, 1250
(in andern) Umständen sein, 1250
(in gesegneten) Umständen sein, 5503
Umstrickung, 5329
umstülpen, 3470
Umstülpung, 3471
unangenehm, 5154
unaufhörlich, 2175
unbeachtet, 5267
unbeachtet lassen, 5268
unbedeutend, 5042
unbedingt, 5371
unbeeinflusst, 2542
Unbehagen, 1602
unbehaglich, 1601
Unbehaglichkeit, 1602
unbekümmert, 2634
unberührt, 5211
unbestimmt, 5000
unbeweglich, 141
unbewusst, 5385
unempfindlich, 1869
Unempfindlichkeit, 2458

unerlässlich, 5498
unerträglich, 1703
unfähig, 2355
Ungarwein, 4088
ungefähr, 4966
ungefährlich, 1616
ungegohren, 5331
ungeheuer, 5480
ungesund, 1533
ungewöhnlich, 4793
Ungeziefer, 3530
ungleich, 1996
unheilbar, 1624
unlöschbar, 5230
unmässig, 5046
unmittelbar, 5151
unmöglich, 2376
Unordnung, 5053
unpässlich, 1536
Unpässlichkeit, 1537
unregelmässig, 1723
unreif, 1168
Unreife, 1170
unrein, 4806
unrichtig, 3295
unrichtige Wochen, 5495
unruhig, 2346
unsicher, 2365
unstätt, 2520
unstillbar, 2176
unter, 395
Unterarm, 416
Unterband, 3613
Unterbauchschmerz, 1818
Unterbett, 4566
unterbinden, 3360
Unterbindung, 3361
Unterbrechung, 4856
unterbrochen, 2002
unterhalten, 5528
Unterhosen, 4706
Unterkiefer, 258
Unterkieferbein, 259
Unterkieferdrüse, 5083
Unterkleid, 4720
unterlaufen, 3347
(mit Blut) unterlaufen, 3338
Unterleib, 1817
unternehmen, 5458
Unterrock, 4722
Untersatz, 497
Unterschenkelmuskel, 618
untersuchen, 3312
(eine Wunde) untersuchen, 3313
Untersuchung, 5069
Untertasse, 4675
unterwerfen, 5367
unterworfen sein, 5368

(sich) unterziehen, 5301
unüberwindlich, 5236
ununterbrochen, 5314
unverändert, 2500
unverdaulich, 1050
unverfälscht, 5595
unverheirathet, 4910
unverhohlen, 4839
unvermischt, 5325
unvollkommen, 2385
unwillkührlich, 553
(die) unwillkührlichen Muskeln, 555
unwohl, 1534
(sich) unwohl fühlen, 1539
Unwohlsein, 1535
Unze, 5413
Unzeit, 1267
unzeitig, 1268
unzeitige Geburt, 1269
unzulänglich, 5283
üppig, 5290
Urin, 1137
uriniren, 1140
Ursache, 5033
urtheilen, 2650
Urtheilskraft, 2651

## V

Valerianöl, 4412
Vanille, 4446
variierend, 1738
Varioloid, 3156
variolös, 3157
vasomotorisch, 902
Veitstanz, 2774
Vene, 693
Venerie, 3238
venerisch, 3237
venös, 711
Venusbeule, 3247
Venuskrone, 3248
verabreichen, 5600
verändern, 4893
Veränderung, 4894
veranlassen, 4937
Veranlassung, 5032
(sich) verästeln, 1016
Verästelung, 1018
Verband, 3614
verbergen, 5465
verbessern, 4963
Verbesserung, 5081
verbinden, 99; 1760; 3315; 5591
Verbindung, 100
verbleiben, 5243
verbrauchen, 5543
verbreiten, 5039

Verbrennung, 3510
(sich) verbrühen, 3383
verdauen, 1047
verdaulich, 1049
Verdauung, 1048
Verdauungsapparat, 1051
Verdauungsbeschwerde, 2931
Verdauungsprocess, 1052
Verdauungsschwäche, 1827
verderben, 1823
(der) verdorbene Magen, 1824
verdrehen, 3425
verdreht, 2512
Verdrehung, 3426
verdriesslich, 5528
verdunkeln, 5420
verdünnen, 5363
Verdünstung, 5058
vereinen, 137
vereinigen, 137
Vereinigung, 138
verengen (oder ern), 1000
Verengerer, 1364
Verengerung, 1002
Verengung, 1002
verfallen, 5022
verfälschen, 5594
Verfettung, 2875
vergeben, 4491
vergiften, 3145
Vergiftung, 3505
vergrössern, 2730
vergrössert, 2138
Vergrösserung, 2423
(sich) verhalten, 4860
Verhältniss, 5401
(nach) Verhältniss, 5401
verharschen, 4156
Verharschung, 4157
verhärtet, 2142
Verhärtung, 2719
verheirathen, 4909
verhindern, 5059
verhungern, 5513
verhüten, 4208
Verhütungsmittel, 4211
verkleinern, 2731
verknöchern, 64
verkrümmen, 3429
Verkrümmung, 3430
verkrüppeln, 3431
Verkrüppelung, 3432
verlangen, 2167
Verlangen, 2168
verlarvt, 3113
verlassen, 4875
Verlauf, 5040
verletzen, 2713
Verletzung, 3308

verlieren, 2152
Verlust, 2603
vermehrt, 1741
Vermehrung, 2863
vermeiden, 5122
Verminderung, 5491
vermischen, 2213
vermittelst, 5261
Vermögen, 2643
vermuthen, 5090
vernarben, 3055
Vernarbung, 3356
verordnen, 4186
Verordnung, 4187
verpesten, 4540
Verpestung, 4541
verrenken, 3423
Verrenkung, 3424
verringern, 2678
verrücken, 5209
verrückt, 2695
(der) Verrückte, 2702
verschaffen, 5250
verschieden, 5085
Verschlechterung, 2679
verschliessen, 5150
verschlimmern, 1758
Verschlimmerung, 5180
verschlingen, 2978; 3886
verschlucken, 3887
(sich) verschlucken, 3888
verschreiben, 4808
Verschwörung, 5086
verschwinden, 5145
versichern, 4836
versprechen, 4870
verspüren, 4863
Verstand, 2662
verstärkt, 1739
verstauchen, 3427
Verstauchung, 3428
verstimmen, 2624
Verstimmung, 2626
Verstopfung, 2235
Verstummung, 2574
Versuch, 5084
versuchen, 4800
versüssen, 4438
vertheilen, 3221
Vertheilung, 3222
vertragen, 2164
vertrauen, 5630
Vertrauen, 5622
vertreiben, 1563
vertrocknen, 2821
Vertrocknung der Lunge, 2822
verursachen, 1671
verwachsen, 3353
Verwachsung, 3354
verwahren, 2407
Verwahrungsmittel, 4210

verwenden, 5566
verwirren, 2665
verwirrt, 2513
Verwirrung, 2666
Verworrenheit, 2667
verwunden, 3303
Verwundung, 3304
verzagen, 4830
verzeihen, 5582
verzerrt, 2511
verzögern, 5482
Verzückung, 2689
verzweifelt, 1613
(sich) verzweigen, 1017
Veterinärarzt, 3833
(zu) viel, 3907
(um) vieles, 4865
vielleicht, 4958
vierköpfig, 2623
vierköpfige Hauptbinde, 2624
Viper, 4504
Vitriolöl, 4302
voll, 1865
völlig, 5383
vollkommen, 5242
Vollsein, 2198
vollständig, 2602
Vomiermittel, 4198
vorangehen, 5009
vorausgehen, 5025
vorhergehen, 5128
vorbeugen, 2339
Vorcur, 4135
vorder, 221
Vorderarm, 416
Vorderarmmuskel, 615
Vorderfuss, 501
Vorderkopf, 223
Vorderzahn, 305
vorgeschützt, 1620
vorgestern, 4795
vorhanden, 4843
Vorhang, 4577
Vorhaut, 1189
Vorhautbündchen, 1190
Vorhof, 632; 1400
Vorhofkammerklappe, 636
Vorhofsfenster, 1402
Vorhofsnerv, 1418
Vorkammer, 631
Vorkammerklappe, 635
vorkommen, 5212
vornehmen, 4896
vornehmlich, 4977
vorschützen, 1621
Vorsicht, 5263
vorsichtig, 5306
vorstehen, 5075
vorübergehen, 4990
Vorverdauung, 1114

vorwärts, 2358
vorzeitig, 1272
vorzeitig niederkommen, 1273
vorziehen, 4969
vorzüglich, 5244

## W

Wachholderbeere, 4375
Wachholderbeeröl, 4376
Wachs, 4478
wachsähnlich, 1882
wachsartig, 1883
wachsen, 3161
Wachsleber, 2003
Wachsniere, 3033
Wachsröhrchen, 3573
Wachsschwamm, 4469
Wachssonde, 3672
Wachsstock, 4620
Wachstaffet, 3636
wackelig, 2368
wackeln, 2369
Wackeln, 2895
Wade, 499
Wadenbein, 500
Wadenmuskel, 619
Wagen, 3567
wählen, 5343
Wahnsinn, 2686
wahnsinnig, 2687
(der) Wahnsinnige, 2701
wahr, 155
wahrnehmen, 4939
währen, 5153
während, 5219
wahrscheinlich, 5445
Waizen, 4481
Waizenmehl, 4483
Waldluft, 5168
wallend, 2010
Wallnuss, 4053
Wand, 127
Wandermilz, 3013
Wanderniere, 3023
Wange, 251
Wangenbein, 252
wann, 4925
Wanze, 3534
warm, 1874
Warmbad, 3766
Wärmeempfindung, 5127
Wärmegefühl, 5127
Wärmegrad, 2104
Warze, regelförmige, 1445
(pilzförmige) Warze, 1448
(kelchförmige) Warze, 1450
Warzenhof, 1233

Warzenschützer, 5511
Waschbad, 3800
Waschbecken, 4589
waschen, 5532
Wäschschrank, 4639
Waschschwamm, 4590
Waschtisch, 4588
Waschtoilette, 4587
Waschung, 4266
Wasser, 705
Wasser lassen, 2277
Wasseransammlung, 5213
Wasserblase, 3183
wasserflicht, 5397
Wassergeschwulst, 2420
Wasserhaut, 916
wässerig, 708
Wasserkessel, 5537
Wasserkrebs, 2887
Wasserlefzen, 1227
Wasserpocken, 3154
Wasserscheu, 3498
Wassersucht, 2722
Wassersuppe, 4070
Watte, 4182
Wattenverband, 3616
weben, 84
Wechsel, 1054
Wechselfieber, 3100
wechseln, 1053
(den Sitz) wechselnd, 1735
weder noch, 5010
weg, 5471
wegbleiben, 5483
(zu) Wege bringen, 5472
wegnehmen, 3390
weh, 1284
wehe, 1284
weh thun, 1675
weh sein, 1676
weh um's Herz, 1678
Weh, 1285
Wehgefühl, 1747
Wehen, 1286
Wehen haben, 1288
weiblich, 376
weich, 385
Weiche, 386
Weichenband, 387
Weichengegend, 389
Wein, 4083
Weinsteinsäure, 4303
weise, 302
Weise, 4921
Weisheit, 303
Weisheitszahn, 304
weiss, 1880
Weissblütigkeit, 3071
Weissbrod, 3833
Weisse, 3272

(das) Weisse im Auge, 1338
weisser Fluss, 3270
weisslich, 2127
Weissling, 4002
weit, 999
weitsichtig, 1494
Weitsichtigkeit, 3447
Weizenbrod, 3934
werfen, 2344
(sich hin und her) werfen, 5533
Werg, 5354
Wermuth, 4450
wesentlich, 5384
Weste, 4683
Wickelband, 5204
widerlich, 2271
widerwärtig, 5523
Widerwillen, 5220
wie, 4928
Wiedereinrenkung, 3422
wiedereinrichten, 3419
Wiedereinrichtung, 3421
wiedereinsetzen, 3420
wiederherstellen, 4107
Wiederherstellung, 4108
wiederholen, 5201
wiederkehren, 5146
wiederkommen, 4817
Wieze, 4556
wild, 2517
wildes Fleisch, 3215
Wildpret, 3960
Wille, 2655
Willen, 2655
Willensstörung, 2656
willkührlich, 552
willkührliche Muskeln, 554
Wimper, 1322
Wind, 2195
Windel, 5520
winden, 846
Windgeschwulst, 2803
Windkolik, 2956
Windpocken, 3145
Windungen, 847
Wirbel, 225
Wirbel des Rückgrats, 341
Wirbelgelenk, 342
Wirbelsäule, 344
Wirbelschmerz, 1805
wirken, 567
wirklich, 4835
Wirkung, 5094
wissen, 2593
wissenschaftlich, 5632
Wittwe, 4912
Wittwer, 4911
Woche, 1274
(in) Wochen sein, 1276

(in) Wochen kommen, 1275
Wochenfieber, 3297
wochenlang, 5245
Wöchnerin, 1278
wodurch, 4934
wofern, 5360
woher, 4902
wohnen, 4902
Wohnung, 5166
wölben, 2428
Wölbung, 2429
Wolf, 3216
Wolke, 2319
wolkig, 2317
Wolle, 4487
Wollendecke, 4572
Wollust, 1247
wollüstig, 1248
worin, 5007
Wortgedächtniss, 2681
worüber, 4930
wuchern, 3371
Wucherung, 3372
Wundarzt, 3820
Wundbalsam, 4421
Wunde, 1675
Wundeisen, 3675
Wundenmal, 3146
Wunder, 4143
Wundercur, 4144
wunderlich, 1256
(der) wunderliche Appetit einer Schwangern, 1257
Wunderysipel, 3164
Wundfäden, 3693
Wundfieber, 3112
wundliegen, 3334
Wundliegen, 3332
Wundmittel, 4229
Wundpflaster, 3701
Wundpulver, 3702
Wundrand, 3349
Wundsein, 1748
würgen, 1794
Würgen, 2191; 1795
(sich) würgen, 2190
Wurm, 3164
Wurmfortsatz, 1122
Wurmmittel, 4230
Wurmsamen, 4440
Wurst, 3979
Wurstgift, 3501
Wurzel, 312
Wuth, 3495
Wuthgift, 3495

X

Xereswein, 4094

Z

Zacke, 133
zackig, 134
zackenförmig, 135
zäh, 2088
zählen, 5801
Zahn, 294
Zähne bekommen, 323
Zahnarzt, 3826
Zahnauszieher, 3685
Zahnbrecheisen, 3688
Zahnbrecher, 3686
Zahnbürste, 4592
Zahndurchbruch, 325
Zahneisen, 3687
zahnen, 321
Zahnen, 326
Zahnfäule, 2894
Zahnfäulniss, 3492
Zahnfeile, 3690
Zahnfistel, 3488
Zahnfleisch, 320
Zahnfleischentzündung, 2893
Zahnfleischgeschwür, 3487
Zahnfleischgewächs, 3485
Zahngeschwür, 3486
Zahngicht, 3489
Zahnglasur, 318
Zahnhals, 311
Zahnhöhle, 307
Zahnhusten, 2901
Zahninstrument, 3684
Zahnknirschen, 2481
Zahnkrampf, 2903
Zahnkrone, 309
Zahnlücke, 314
Zahnmittel, 4274
Zahnpilz, 5567
Zahnpulver, 5577
Zahnputzer, 3691
Zahnreihe, 5544
Zahnrose, 3490
Zahnruhr, 2902
Zahnschlüssel, 3689
Zahnschmelz, 317
Zahnschmerz, 1791
Zahnstein, 5572
Zahnung, 327
Zahnweh, 1792
Zahnweinstein, 2899
Zahnwurm, 3491
Zahnwurzel, 312
Zahnzange, 3716
Zauge, 3706
Zängelchen, 3707
Zangenentbindung, 3290
Zangengeburt, 3291
Zäpfchen, 5091
zart, 1859

Zauber, 4141
Zaubercur, 4142
zaubern, 4140
Zehe, 510
(die grosse) Zehe, 511
(die kleine) Zehe, 512
Zeigefinger, 444
zeigen, 443
(sich) zeigen, 5321
Zeit, 1266
(zur) Zeit, 5494
zeitweilig, 5574
zeitweise, 2680
Zelle, 911
Zellgewebe, 912
Zellschichte, 921
Zergliederungsmesser, 3646
zerreissen, 3376
Zerreissung, 3377
zerrütten, 2671
zerschlagen, v., 2449
zerschlagen, a., 2450
Zerschlagenheit, 2451
zerschmelzen, 5414
zerschmettern, 3378
Zerschmetterung, 3379
zersetzen, 2217
zersplittern, 3374
Zersplitterung, 3375
zerstören, 5443
Zerstreutheit, 2684
zertheilen, 3351
Zertheilung, 3352
Zeug, 5398
zeugen, 1172
Zeugung, 1174
zeugungsfähig, 1175
Zeugungsglied, 1178
Zeugungskraft, 1179
zeugungsunfähig, 1176
Zeugungsunfähigkeit, 1177
ziehen, 174
ziehend, 1699
Ziehpflaster, 4427
Zimmer, 4542
(das) Zimmer hüten, 1575
Zimmt, 4067
Zimmtöl, 4408
zischen, 2038
zittern, 2371

Zittern, 2479
Zitterwahnsinn, 2710
Zitwersamen, 4441
Zoll, 5433
Zopf, 963
zubereiten, 5234
zubinden, 4755
zubringen, 4853
zucken, 1777
Zucker, 2294
Zuckerwerk, 4063
Zuckung, 2381
zudecken, 5373
zuerst, 4932
Zufall (pl.), 5516
zufällig, 5112
Zug, 249
Zugbohrer, 3728
zugleich, 5249
Zugmittel, 3728
Zugpflaster, 4428
zuheilen, 4154
Zuheilung, 4155
zuknöpfen, 4759
zulassen, 5536
zumachen, 4757
Zunahme, 2393
Zündhölzchen, 4606
zunehmen, 2391
Zunehmen, 2392
Zunge, 328
Zungenband, 329
Zungenbändchen, 1451
Zungenbein, 330
Zungendrüse, 1456
Zungenentzündung, 2908
Zungenfleisch, 334
Zungenfleischner, 876
Zungengeschwulst, 2906
Zungengeschwür, 2907
Zungenhaut, 1439
Zungenkrampf, 2910
Zungenkrebs, 2909
Zungenlähmung, 2911
Zungenmuskel, 1436
Zungennerv, 1437
Zungenrand, 1438
Zungenrücken, 333
Zungenschaber, 3720
Zungenschlundkopfnerv, 873
Zungenschmerz, 1786

Zungenspitze, 332
Zungenvergrösserung, 2912
Zungenvorfall, 2913
Zungenwarze, 1440
Zungenwärzchen, 1441
Zungenwurzel, 331
zupfen, 3697
zurückbringen, 5304
zurückhalten, 5011
zurückkehren, 5001
zurückwerfen, 5534
zusammen, 568
zusammendrücken, 5564
zusammengezogen, 2541
Zusammenhang, 5478
zusammenpressen, 2732
zusammenschnüren, 4761
zusammenschnürend, 1700
zusammenwirken, 568
zusammenwirkend, 579
zusammenziehen, 3362
zusammenziehend, 581
Zusammenziehung des Herzens, 746
zuschnüren, 4763
zusetzen, 5568
Zustand, 1513
zuträglich, 5286
zuweilen, 4796
zuziehen, 4943
(sich eine Krankheit) zuziehen, 4943
Zweck, 5514
zweckmässig, 5515
Zweifel, 5551
Zweig, 1014
zwerch, 602
Zwerchfell, 603
zwickend, 1719
Zwieback, 3953
Zwiebel, 958
Zwielicht, 5424
Zwilling, 3293
zwischen, 161
Zwischenknorpel, 162
Zwischenraum, 5036
Zwischenrippennerv, 893
Zwölffingerdarm, 1046

# ALPHABETICAL INDEX

OF THE

# ENGLISH WORDS AND TERMS.

## A

abandoned (to vice), 5235
abate, v., 3115; 5298
abatement, 5338
abdomen, 340; 378
abdominal bandage for hernia, 3609
abdominal enlargement, 3009
abdominal pressure, 609
aberration of mind, 2670
ability, 32; 2645
able, 44
able to, to be —, 5223
abnormal, 3437
abortion, 1269; 1270
abortion, to cause an —, 3296
about (to be), 2124
above, 394
abscess, 2416; 3233
abscess lancet, 3642
abscess of the brain, 2721
abscess of the kidney, 3035
absolute, 5371
absolute blindness, 3469
absolutely, 5541
absorb, v., 751; 5356
abstain from, 4886
accelerate, v., 2028
accident, 2777
accidental, 5112
accipiter, 3594
accompany, v., 4992
accoucher, 3827; 3828; 3829
accumulation of water, 5213
accustomed (to be), 5034
accurate, 4905
acetabulum, 463
acetate of copper, 4526
acetic acid, 4296
acetic ether, 5569
acetous, 4523
ache, 1670
ache, v., 1553; 1675
aching feeling, 1747
acid, 1982
acidulated, 5135
acne, 3184; 3185
acne mentagra, 3192
acne rosacea, 3189
acneform dermatosis, 3187
act, v., 567
act together, 568
action, 5094
active, 5002
acupuncture needle, 3664
acute, 1627; 1628
Adam's apple, 1463
add, v., 5191; 5311; 5568
adder, 4503
Addison's disease, 3068
adhere, v., 5176
adipose, 926
adipose degeneration, 2874
adipsia, 2179
administer, v., 5600
admit, v., 5536
adulterate, v., 5594
adverse, 5523
advise, v., 4890
affect, v., 4999
affect the head, 1788
afflict, v., 1605
afford, v., 5217
affusion, 5396
afraid, to be afraid, 4785; 5587
afterbirth, 1293
after-digestion, 1115
after-pains, 3288
after-treatment, 4136
against, 569
age, 1242
age of puberty, 1243
aggravate, v., 1758
agitate, v., 139; 2640; 3117
agitation of mind, 5132
agree with, to, 4862
agrypnia, 2591
ague, 3118
ague fits, to have —, 3098
aguish, 1633
ail, v., 1553
ailing, 1525
ailment, 1549
ailment, long —, 1550
aim, n., 5514
aim at, to —, 5199
air, 1005
air, v., 5124
air-bed, 3585
air-cells, 1045
air passages, 1006
air-pillow, 4570
air-poisoning, 4589
air vesicle, 1043
alarm, v., 4844
alarmed, to be —, 5129
albinism, 3214
albugo, 3460
alcohol, 4305
alcoholic liquors, 4955
ale, 4098
alienation of mind, 2661
alike, 2384
alive, 46

alleviate, v., 1673; 753
alleviation, 4864
all-healing remedy, 4284
allopathic cure, 4116
allopathic doctor, 3836
allow, v., 5349
almond, 1454
almond emulsion, 4370
aloes, 4306
along with, 5249
alter, v., 4820; 4893
alteration, 4894
alternate, v., 2115
alternating, 1666
althea syrup, 4329; 4330
alum, 4307
alveolar abscess, 3493
always, 5386
amaurosis, 3451
amber, 4394
ambliopia amaurotica, 3452
ambulance, 3587; 3875
ambulatory clinic, 3870
amenorrhœa, 3365
amiable, 5280
amniotic fluid, 1291
amount, 5400
amount to, to —, 5400
amuse, v., 5527; 5528
anæsthesia, 2620; 2321; 2622
anal region, 1134
analogous, 5493
anasarca, 2620
anchovy, 4007
aneurism, 2846; 2847
angina maligna, 2914
angina pectoris, 2857; 2858
angle, 191; 192
angle of the jaw, 5078
anger, 2854; 2855
angry, 5522
anguish, 2639
animal, 1195; 3516
animalcule, 1195
ankle, 508
annular knife, 3654
ano lyne, 1686; 4287
anodynes, 4214
anæmia, 3065; 3066; 5051
anæsthetic, 4289
answering a purpose, 5515
ant, 1767
antagonists, 583
anterior, 221
antidote, 4285; 4286; 4495
anti-febrile, 4356
antispasmodic powder, 4369
anus, 1133
anvil, 1409

anxiety, 5139
anxious, 2506
any, 4938
apathetic, 2636
aperient roots, 4352
aphex pulmonis, 1041
aphthæ, 2891; 2892
apophysis, 1121
apoplectic seizure, 2764
apoplexy, 2762; 2763
apoplexy, fit of, 2764
apothecary, 3849
a.'s apprentice, 3851
a.'s assistant, 3850
a.'s shop, 3859
apparatus for deformed feet, 3584
appear, v., 1920; 2033; 2322; 4849; 5183; 5321
appearance, 5103
appease, v., 3889
appetite, 3885
a., canine —, 2161
a., loss of —, 2154
a., to get an —, 2149
a., to have a hearty —, 3897
a., to provoke the —, 2150
a., to stay the —, 3890; 3891; 3892
a., to whet the —, 2150
a., vitiated —, 1257
a., want of —, 2153
apple, 4044
apple of the eye, 1336
apply, v., 4961; 5438
(to) apply to the breast, 5509
applying, 5055
apprehend, v., 5587
apricot, 4048
apron, 4733
arachnoid membrane, 859
arch, v., 2428
arcus cruralis, 482
ardor urinæ, 3059
arena, 2300; 2301
areola mammæ, 1233
arise, v., 2374; 4907; 5549
arm, 410
arm, upper part of the —, 411
arm-chair, 4548; 4549
arm-pit, 408; 409
arrange, v., 5505
arrest, v., 3184
arrest (progress), v., 5297
(to) arrive at, 2119
arsenic, 4520
arsenic, white —, 4521
art, 3801
arterial, 714

artery, 667; 668; 669
arthalgia, 1854
arthritic pains, 1856
arthritis, 3078
arthrodia, 189; 190
articular, capsule —, 199; 217
a. cartilage —, 168
a. surface —, 183
a. trochlea —, 415
articulation, 167
a., defective —, 2575
a. of the bones, 101-104
a., trochoid —, 197; 198
artificial, 5584
ascaris lumbricoides, 3547
asleep, to be —, 2478
athwart, 602; 625
atlas, 361
atmosphere, 1005
atrium, 631-636
attack, n., 2212; 2777
attack of indigestion, 2982
attack of pain, a violent —, 1750
attacked, to be —, 5545
attain, v., 2119; 5199
attempt, 5084
attempt, v., 4800
attend, v., 5587
attention, 5530
attentive, 4935
attenuate, v., 5363
attire, 4680
asafœtida, 4317; 4318
ascarides, 3544
ascaris, 3550; 3551; 3552
ascend, v., 675-677
ascending aorta, 679
ascites, 3010
as for, 5484
ashamed, to be —, 459
ash-pan, 4625
ask counsel of, 4959
aspect, 1589
assert, v., 4836
assistance, 4786
associates, 582
assuaging, 4213
assure, v., 4836
asthma, 2817
as to, 5484
astragalus, 504
asunder, 2729
aorta, 670-674
aorta, arch of the —, 680
aorta, descending —, 678
audible, 2037
audition, 1383
auditory canal, obstruction of the —, 3483

**auditory** duct, external —, 1389
a. organ, 1381
a. ossicle, 1407
auricle, 631; 632; 650; 651; 1400
auricular valve, 635
aurist, 3825
aversion, 5220
avoid, v., 5122; 5215; 5524
awake, v., 4804; 5405
awaking, 4805
aware of (to be), 4863
awry, 3081
axilla, 408; 409
axungia, 215

## B

back, n., 222; 345
backbone, 348
bacon, 3981
bad, 2042; 2680; 3302; 4834; 5358
bake, v., 3939
ball, 187
ball extractor, 3714
ball gimlet, 3725
balm, 4282
balm-mint, 4336; 4337
balsam of Peru, 4420
band, 105
b., navel —, 3599
b., plaited —, 1351
b., swathing —, 5204
b., transverse —, 3629
bandage, 105; 3588; 3314
b., abdominal —, 3597
b. after bleeding, 3634-35
b., body —, 3598
b., circular —, 3604
b., compressive —, 3615
b. for the chest, 3602
b. for the chin, 3595-96
b. for the head, 3622
b. for the foot, 3606
b. for a fracture or hernia, 3608
b. for the upper arm, 3601
b., four-tailed (or four healed) — of the head, 3624
b., frontal —, 3592
b., handkerchief —, 3589
b., hanging —, 3626
b., head —, 3592
b., knee-cap —, 3605
b., laced —, 3615
b., navel —, 3599
b., nose —, 3594
b., roller —, 3628

**bandage**, shaving —, 3590
b., star —, 3632
b., suspensory —, 3619-21
b., under —, 3613
b., upper —, 3612
b. to tie up a vein with, 3593
b., to apply a —, 3615
bandaging, 3661
banquet, 3921
barber, 3843-45
bare, 5342
barely, 5273
bark, v., 2076
barley, 4484
barley meal, 5285
basin, 449
basis, 3391
basket, 4659
bath, 3764
b., air —, 3792
b. for a limb, 3769
b. of medicinal herbs, 3796
b. taken for cleanliness, 3800
b., douche —, 3777-79
b., foot —, 3771
b., hand —, 3770
b., hip —, 3768; 5333
b., hot-air —, 3772-73
b., medical —, 3765
b., medicated —, 3797
b., mineral —, 3765
b., plunge —, 3787
b., salt-water —, 3798
b., sand —, 3794
b., sea —, 3799
b., shower —, 3780-5
b., shower — (from above), 3786
b., shower — (from the side), 3787
b., slipper —, 3768
b., steam —, 3772-73
b., sulphur —, 3791
b., sun —, 3793
b., sweating —, 3772-73
b., tepid —, 3767
b., vapor —, 3774-76 3795
be, v., 1502
beans, 4031
bear, v., 360; 2164; 5202
bearable, 5196
beard, 969
bearer, 362
beard-hair, 970
beat, v., 665
beating, 1692
beat in pieces (to), 2449
bed, 3572
bed, portable —, 3573

bed-clothes, 4557
bed-cover, 4560
bed-curtains, 4579-80
bedding, 4557
bed-hangings, 4579-80
bed-linen, 4558
bed-pan, 4602
bed-ridden, 1574
bed-ridden (to be), 1573
bed-room, 4544-46
bedstead, 4554-55
bed, go to —, 5236
bed, keep one's —, 1573
bed, put to —, 1258
beef, 3956
befall, v., 5545
before, 4989
before (not), 5099
beget, v., 1172-73
beget, able to —, 1175
begetting, 1174
begin, v., 4927
beginning, in the —, 5097
behave one's self, to —, 4860
belch, v., 2201-3
belching, 2202-4
believe, v., 4831
belladonna, 4517
belly, 378
b., paunch —, 2397
belly-ache, 1813-14-15
below, 395
belt, 4738
bench, 4635
bend, v., 419; 559; 2338
bend forward, to —, 2339
bend over, to —, 2356
bend, cause to —, 559
beneficial, 5286
benign, 1646
bent, 2357
benumb, v., 2452
benumbed, 1851; 2456
b., to be —, 2478
benumb the head, to —, 1780
besides, 4810
best part, 121
bestow upon, to —, 5217; 5566
better, is getting —, 1560
between, 161
beverage, 4279-80; 2177
beware of (to), 4815
bile, 2215
bilious, 2136
bifurcation, 1021
bile, 1099
biliary calculus, 5255
biliary duct, 1102-3
bilious, 2958

bind, v., 98
bind over, to —, 3610
bind round, to —, 3611
bind together, to —, 99
bind up, to —, 3360
birth, 1264
birth to, to give —, 1263
biscuit, 3953-54
bistoury, 3640; 3645
bit (morsel) of bread, 3938
bite, v., 291
bitter, 2180
bitter as gall (as), 1100
black, 2263
blacken, v., 5241
bladder, 915; 1158
bladder, neck of —, 1159
blanket, 4564-72
bleed, v., 684
bleed a person, to —, 4251
bleeding, 3357
bleeding from the nose, 2800
bleeding instruments, 3761
blennorrhœa, 3249
blepharospasm, 3444
bless, v., 5502
blind, 1116; 4646
b., absolutely —, 3468
b. from cataract, 3463
b., to make —, 1353
blindness, 3467
blister, 1924; 3403
blister, v., 1911
blister, small —, 1909
blistering fly, 4319
blistering plaster, 4426
blisters, to get —, 1911
blood, 683
b., arterial —, 717
b. coagulum, 693
b. colored, 690
b. corpuscle, 701-3
b., dark —, 715
b., effusion of —, 2091
b., flow of —, 3261
b. globule, 701-3
b., light-red —, 718
b., mass of —, 695
b. passage, greater —, 730
b. passage, lesser —, 731
o., red —, 691
b., venous —, 716
b. -vessels, 694
b. -water, 706-7
bloody, 685
bloody flux, 5476
bloodless, 686-7
blotch, 3184-5
blow, 4946
blow, v., 2404

blow up, to —, 2405
blubber, 4372
blue, 1889
blue, dirty —, 2312
blunt, v., 5390
board, 3578; 3811
board of health, 3812-13
bodice, 4724-6
bodily, 19; 20
bodily strength, 23
body, 8; 21
b., extent of —, 2427
b., frame of —, 12
b. heat, 2105
b. linen, 5347
b., part of the —, 5227
b., upper part of the —, 5173-4
boil, n., 2415-17; 3233
boil, v., 3963
boil up, to —, 5604
bolster, 3633
bon-bon, 4060
bone, 61; 91
b. affection, 4979
b. -black, 4320
b. capsula, 117
b.s, cheek —, 268-9
b., chine —, 466-8
b., collar —, 355; 407
b., cylindrical —, 97
b., edge of the —, 132
b. edge with serrations, 136
b., ethmoid —, 238-9
b., facial —, 242
b. -file, 3719
b. -forceps, 3652; 3715
b., frontal —, 231
b., hollow —, 97
b., hyoid —, 330
b., inferior turbinated —, 281-2
b., lachrymal —, 271
b., like —, 67
b. -like, 72; 94-5
b., made of —, 65; 93
b., malar —, 252
b., maxillary —, 255-63
b., metacarpal —, 436
b., nucleus —, 124
b., occipital —, 229
b., palatal —, 265
b., parietal —, 225-6
b., pelvic —, 450
b., petrous —, 236
b., rocky —, 236
b. -saw, 3651
b. scraper, 3721
b., sesamoid —, 519
b., shin —, 496
b., small —, 62

bone, splenoid —, 232
b. structure, 74
b. substance, 83
b., superior maxillary —, 257
b., temporal —, 234
b., zygomatic —, 268-9
bony, 66
bony substance, 81
boot, 4701
bootjack, 4647
borborygmus, 2230
Bordeaux, 4089
border, 905
boring, 1696
born, to be —, 1271
bosom, 1234
both, 2386
bottle, 1300
bottom, 3391; 5344
bough, 794; 1014
bougie, 3679
boundary, 905
bounding, 1998
bow, 420
bow, v., 419; 2153
bowed, 2357
bowels, 919
b., action of the —, 2228
b., open —, 4973
b., openness of the —, 5137
b., sluggish action of the —, 2233
bowl, 4676
brachial muscles, 614
brachial plexus, 890
brain, 807-8
b., cavity of the —, 841
b., central —, 844
b., concussion of the —, 2725
b. disease, 2711
b., dropsy of the —, 2723
b., inflammation of the —, 2715
b., inflammation of the membranes of the —, 2753-4
b., injury of the —, 2712
b., lobe of the —, 855
b., softening of the —, 2718
b., surface of the —, 861
brains, 807-8
bran, 3196
branch, 794; 1014-15
branch out, to —, 1016-17
brandy, 4099; 4954
bread, 3932
b. and butter, 3947
b., brown —, 3935

304

**bread**, dry —, 3943
b., fresh —, 3942
b., home-baked —, 3941
b., stale —, 3943
b., wheat —, 3934
b., white —, 3933
**break**, *v.*, 324
b. open, to —, 3223
b. out, to —, 5183
b. small, to —, 5395
b. through, to —, 324
breakfast, *v.*, 3884
breakfast, *n.*, 3922
breast, 365; 1231-4
breast bone, 367 9
breast glass, 3746
breast to a child, to give the —, 1296 7
**breath**, 55; 985; 992
b., to draw —, 988-9
b., drawing of —, 992
b., to fetch —, 988-9
b., fetch a long, deep —, 990
b., to have a foul —, 2044
b., hold one's —, 995
b., out of —, 994
b., to recover —, 2015
b., shortness of —, 2014-25
b., to take —, 988-9
b. of life, 56
**breathe**, *v.*, 54; 984-9
b. hard, to —, 991
b. into, to —, 4239
b. out, to —, 997
**breathing**, 4151
b., difficulty in —, 5065
b., enfeebled —, 2026
breathless, 994
breech, 458
breeches, 4684-5
breed, 4
bridge, 845
bridge of the nose, 1425
Bright's disease, acute —, 3028
**bring**, *v.*, 4853
b. back, to —, 5304
b. into, to —, 5317
b. near, to —, 3393
b. on, to —, 5041; 5224
b. through, to —, 5116
b. to bed, to —, 1258
b. up, to —, 1303
b. up a child by hand, to —, 1304
b. upon one's self, to —, 4943
brisk, 1303
broad, 3243
broken, to be —, 2447
bronchial dilatation, 2805

bronchial stem, 1020
bronchial trunk, 1020
bronchitis, acute —, 2803
bronchus, 1010; 1019
bronzed skin, 3209
broom, 46.9
broth, 1085; 3965
brothers and sisters, 4916
brought to bed, to be —, 1260
brought to bed of a son, to be —, 1281
brow, 230
brown, 1892
brownish-red, 1893
brows, to knit the —, 1319
bruise, 3328
bruise, *v.*, 3327
bruised all over, 2450
brush, 4591
brush, *v.*, 5570
bubbling, large —, 2049
bubo, 3247; 5348; 5353
bucket, 4655
bud, *v.*, 954
bug, 3534
build, *v.*, 10
building, 73
bulb (esp. onion), 958
bulging, 2473
bulk, 82; 2387-9; 2427
bullet forceps, 3714
bulkiness, 2388
bulky, 2389
bunch, 422
bundle, 206
Burgundy, 4090
Burgundy nose, 3190-1
burn, 3382 95; 3510
burn (coffee), 3940
burning, 1691
burn one's self, to —, 3509
bursa mucosa, 547
burst, *v.*, 3223; 5083
burst out with, to —, 2484
business, 4907
butter, 1334
buttock, 458
buttocks, 472-3; 1133
button, 4090
button hole, 4691
button up, to —, 4757-9
buzz, to —, 2054-6

**C**

cabbage, 4033-5
cachexia, 2461-2
cadaverous, 2493
cæcal pouch, 1080
cæcal sack, 1080

cæcum, 1117
cæcum, inflammation of the —, 2969
Cæsarean section, 3292
cake, 698; 3952
calamus, 4402
calculus, 5332
calf of the leg, 499
calix, 1149
call, *v.*, 4789
call again, to —, 4817
called, to be —, 4897
call for, to —, 2167
call forth, to —, 5041
callosity, 949; 1962
callous, 950
calm, 5504
camphor, 4404
camphor liniment, 4405
canal, 353; 622
canal, intestinal —, 1092
canal of the spinal marrow, 353
canals, semi-circular —, 1412
**cancer**,
c. of the bladder, 3044
c. of the chest, 2813
c. of the eye, 3441
c. of the intestines, 2972
c. in the kidneys, 3027
c. of the liver, 3004
c. of the mouth, 2890
c., pancreatic —, 3015
c. of the stomach, 2935
c. of the tongue, 2909
c. of the womb, 3277
candidly, 4839
candle, 4609-10
candlestick, 4617
canker, *v.*, 2897
cantharides, 4319
cap, 4694-5; 4735
capacity, 32
capability, 2643
capable of, 44
capillary, 771-2
capillary bronchitis, 2804
capillary vessels, 771-2
caprice, 2631
capricious, 2159; 2632
capriciousness, 2633
capsula, 116
carbolic acid, 4297
carbonic acid, 4298
carbuncle, 2490
carcinoma, 2796
carcinoma cutis, 3218
cardalgia, 1809
**cardiac** affection, 2850
c. dropsy, 2884; 3062
c. region, 645

cardiac softening, 2869
c. sound, 657-8
cardialgia, 2930
cardiectasis, 2866
carditis, 2859
care, 4841; 4953; 5565
care for, to —, 2635; 3385
careful, 5410
carefully, 5306
careless, 2636
carelessness, 5269
care much or little, to —, 4846
caries, 2894
carnal, 5351
carminative species, 4353
carp, 3998
carpet, 4576
carriage, 3567
carrier, 362
carrots, 4038-9
carry, v., 360
carry out, to —, 3385
cartilage, 147
c., annular —, 165
c., arytenoid —, 1468
c., costal —, 370
c., cover with —, to, 152
c., cricoid —, 1466
c., false —, 158
c., interarticular —, 162
c., laryngeal —, 1464
c., lateral —, 1426
c., like —, 149-51
c., osseous —, 154
c. substance, 153
c., roundish —, 1471
c., tarsal —, 1324
c., thyroid —, 1465
c., true —, 157
cartilaginous, 148
c. coating, 177
c. covering, 177
c. cushion, 177
c. membrane, 160
caruncula lachrymalis, 1329-30
case, 5156
case, in —, 5156
case, in any —, 4859
cask, 4660
casualty, 5516
catalepsy, 2767-8
cataleptic, 2769
cataplasm, 4256; 4283
cataracta punctata, 3458
cataracta spuria, 3462
cataract, 3449-50
c., capsular —, 3453-4
c., false —, 3462
c., hard —, 3456
c., lenticular —, 3455

cataract, membranous —, 3461
c., soft —, 3457
cataractous, 3463
catarrh, 2798
c., intestinal —, 2966
catarrhal, 2828
catch, v., 4794; 5529
catching, 1638
catheter, 3672-4
cauda, 470
cauliflower, 4034
cause, 5032-3
cause, v., 1671; 5472
caustic, 4228
caustic-case, 3680-81
cauterize, v., 3314
cautery iron, 3364
cavern, 121
cavity, 121
c., articular —, 463
c., bone —, 122
c., cervical —, 357
c., laryngeal —, 1472
c., mediastinal —, 1032
c., oral —, 336
c., pharyngeal —, 1063-4
c., thoracic —, 372
c. of the uterus, 1214
cease, v., 3115; 5238
ceasing to circulate, 2005
cell, 911
cellar, 4654
cellula, 911
cellular layer, 921
cellular tissue, 912
central, 1659
central part, 123
cerebellum, 843
cerebral anæmia, 2726
c. fluid, 862
c. nerves, system of the —, 811
c. tumor, 2778
c. symptoms, 2613
cerebrum, 807-8; 842
cerumen auris, 1301-2
cerumen, indurated —, 3483
cervical articulation, 356
cervical plexus, 889
chain saw, 3656
chair, 2225; 3575
chalice, 1149
chalk, 4490
chalky, 5345
chamber, 629; 4542
ch., anterior —, 1362
ch., posterior —, 1363
ch. pot, 4599, 4600
ch. -robe, 4715
Champagne, 4091

chancre, hard —, 3239
chancre, soft —, 3240
chancroid, 3240-41
change, 1054; 4894
change, v., 1053; 4820; 4893
ch. of air, 4895
ch. of matter, 1055
ch. of position, 2884
chap, v., 3339-40; 3381
chapped, 2133
charcoal, 4321
charcoal, animal —, 4320
charpie, 3692-4
chasm, 313
cheek, 251; 267
cheeked, hollow —, 2499
cheer, to be of good —, 4824
cheerful, 4822
cheese, 2129
chemise, 4734
chemist, 3849
cherry, 4047
chest, 365; 373-4
ch. disease, 2810
ch., oppression of the —, 2811
ch., prominence of the —, 2430
ch. -wall, 2466; 2470
ch. of drawers, 4640
chew, v., 593
choroid, inflammation of the —, 3433
chronic, 1630-1
chicken-pox, 3145; 3154-5
chief, 227; 5244
chiefly, 5259
child, 4923
child, big with —, 1249
child, quick with —, 1253
childbed, to be in —, 1276
chilblain, 3507-8
chill, 2112-14; 2117
chilliness, 2112
chilly, to be —, 2116
chimney, 4630
chin, 260
chinpiece, 3595-6
chiragra, 3084
chiropodist, 3842
chisel, 3323
chlorine water, 4311
chlorosis, 3067
chocolate, 4080
choke, 2074
choke, v., 1794
choking, 1795
choler, 1099
cholera morbus, 2990
chop, v., 1929; 3318
chopping blade, 3647

choroid, 1343-6
chyme, 1086 8
chymification, 1090-1
cicatrix, 3139 40
cicatrization, 3356-7
cicatrize, v., 3355; 4156
cider, 4096
ciliary body, 1352
ciliary ligament, 1348
ciliary motions, 2556
cingulum, 3179-81
cinnamon, 4067
circle, 720
circle about, to —, 719
circular saw, 3655
circulation, 723; 2844
c., derangement of —, 2845
c., greater (systemic) —, 728
c., lesser (pulmonary) —, 729
c., portal —, 778
c. of the blood, 725-7
c. of the body, 735
circumvallate papillæ, 1450
circumscribe, v., 5004
circumstance, 1250
cirrhosis, 3006
citric acid, 4299
clammy, 1897; 2126
Claret, 4089
clavicle, 355
clavus pedum, 3560
claw, 2487
claw, like a —, 2488
clayey, 2261
clean, 2125
clean, v., 3316
clear, 2087
clear the throat, to —, 5104
cleave, v., 849; 3363
clever, 38 5
climate, 4148
clinic, 3869
clitoris, 1229
cloak, 4686
clonic, 2745
close, 998
close, v., 405; 1000; 3393; 4154; 4156; 5150
close-stool, 4601
close up, to —, 4757
clot, v., 5118
cloth, 5308
c'othe, v., 4741
clo'hes, 4558
clothes press, 4638-9
clothing, 4677
cloud, 2319

cloudy, 2317
clove, 4406
clump, 697
clyster, 3739
c., to apply a —, 3738
c.-pipe, 3740-1
c.-pump, 3743
coagulate, v., 660
coagulum, 699-700
coal, 4628
coalesce, v., 3353
coalescence, 3354
coarctation of the prepuce, 3249
coarse, 5206
coat, 792; 4681
coat of the stomach, 1075
coat of the tongue, 2143
coated, 2126
coating, 176
coccygeal region, 472-3
coccyx, 471; 475-6
cochlea, 1403
cochlear window, 1404
coffee, 4078
coffee-grounds, 2218
coffee-ground like, 2219
coil, v., 846
coils, 847
coin, 2097
coition, 1181
cold, 1878; 2112
cold, to catch —, 4957
cold, catching —, 4813
cold, have a —, 5466
cold (in the head), 2799
colic, 1828-9; 2951-3
c., acid —, 1101
c., bilious —, 2957
c., copper —, 2961
c., flatulent —, 2956
c., lead —, 2960
c., metallic —, 2959
c., nephritic —, 1840; 3040-41
c., rheumatic or inflammatory —, 2963
c., simple —, 2954
c., uterine —, 3267
c.-like, 1816
collar, 3250
collection, 5469
collyrium, 4270-71
colon, 1125
color, 243
color, v., 5221
colored, 689
colorless, 2251
colpalgia, 1844
column, 343
coma, 2586
comb, 4594

comb, v., 4595
combination, 138
combustion, 3510
come before one's eyes, to —, 5212
c. from, to —, 5549
c. off, to —, 5487
c. on or forth, to —, 2033
c. out, to —, 3472; 5406
c. to, to —, 5400
c. to see, to —, 4851
comedone, 3188
comfortable, 1600; 5289
commendable, 5167
commission, 5590
commissure, 108
commit, v., 4994; 5015
commode, 4640
common, 866; 1650
common rue, 4342
comatous, 5023
companion, 582
complain of, v., 4931
complain, v., 1546
complaint, 1537; 1547
complaint, bodily —, 1582
complaint, to make —, 1546
complete, 2602; 5242; 5383
complexion, 244-5
complicated, 3344
component, 2282
compose, v., 5589
composed, 5504
compose one's self, to —, 4827
comport one's self, to —, 4860
comprehend, v., 2123
compress, 3633; 3369
compress, v., 2732
compressio cerebri, 2716
compress of lint, 3370
conception, 2663
concern, v., 5484
concern one's self, to —, 2635
concha, 280
concha inferior, 281-2
concussion, 5327
condition, 1513; 1755; 5222
condition, natural —, 15
condition, to be in a —, 5223
conductor, 1224
condyle, 487
condyloma, 3242
cone, 1444
confectionery, 4064
confine, v., 5004
confine one's self, to —, 4881-2

confined, to be —, 1277
confinement, 1261-2
confound, 2665
confuse, v., 2665
confused, 2513
confusion, 2666-7; 2670; 2673
confusions of the conceptions, 2669
conical, 1445
conical papillæ, 1446
conjecture, 5090
conjunctiva, 3433
conjunctivitis, 3433
connect, v., 99; 102
connection, 100; 5478
connection, immovable —, 143
conscious, 2594
consciousness, 2596
consciousness, complete loss of —, 2603
consequence, 4871; 5066
consider as, to —, 5559
considerable, 5473
consist of, to —, 1917
constant, 1725; 2165; 2341
constantly, 5386
constipation, 2232
constipation, habitual —, 2987
constipated, to be —, 2227
constituted, 14
constitution, bodily —, 16; 22
constrict, v., 3362
constricting, 1700
constriction, 5329
constrictor, 1364
constrictors, 581
construct, v., 10
construction, 11
consult, v., 1985; 4959
consume, v., 3881; 5543
**consumption**, 2437
c., fall into a —, 2437
c., galloping —, 2823
c., hepatic —, 3007
c. of the lungs, 2818-20
consumptive, 2824
contagious, 1639
contagious, non —, 1640
contagium, 3503; 4532-4
contain, v., 817; 2123; 5240
continual, 4980
continue, v., 48; 1120; 4941; 5153; 5392
continuing, 1726
continuous, 5314
contortion, 2689
contract, v., 1000; 3362
contracted, 2540

contracted, closely —, 2541
contraction, 746
contrary, on the —, 5403
contuse, v., 3327
contusion, 3328
contusion, gun-shot —, 3330
convolutions, 847
convulsion, 1765; 2381; 2689
**convulsive**, 1710-12
c. attacks, 2748
c. diseases, 2743
c. moment, 2381
c. twitches, to feel —, 1777
cook, v., 3963
cool, 1877
cool, v., 4217; 4224
cooperate, v., 568
cophasis, 3481
copper, 4501
copper-snake, 4502
copulate, v., 1180
copulation, 1181
copulative organs, 1182
cord, 883
coriander, 4444
cork, 4663
corkscrew, 4664
corn, 3560
corn-cutter, 3842
corn-doctor, 3842
cornea, 1341-2
corneous, 946
corneous tunic, 1341-2
corner, 191-2
corporeal, 19-20
corpse, like that of a —, 2518
corrode, v., 2896
corrugate, v., 1319
corrugator, 1320
corrupt, v., 1823
corset, v., 4762-3
corsets, 4724-5
cortex, 964
cortical substance, 965; 1156
coryza, 2799
cosmetic, 4276
cost, v., 5586
costive, 2238
costive, to be —, 2237
costiveness, 2234; 2236
costume, 4678
cotyla, 172
cotyle, 113; 172
cotton, 4489
couch, v., 3464

couching, 3465-66
**cough**, v., 2067
cough, 2069
c. a little, to —, 2068
c., choking —, 2075
c. during teething, 2901
c., have a —, to, 2067
c. up, to —, 2078; 5126
c., whooping —, 2072-3
coughing, a fit of —, 2070
councillor (counsellor), 3838
count, v., 5601
countenance, half of —, 248
countenance, expression of —, 2502
counteract, v., 570; 5095; 5573
country, 4891
couple, v., 1180
courage, 4823
course, 724; 5040
cover, 1477; 4564
cover, v., 175; 1476; 1915; 5373
cover with, to —, 3948
cover for the feet, 4576
covering, 176; 529
coverlet, 4564; 4575
crab, 3529
crack, v., 1929; 3381
crackle, v., 2052; 2057
crackling sound, 2058
cradle, 4556
cradle (for a broken foot), 3583
cramp, 1765
cranium, 586
crassamentum, 699; 700
crave, v., 2167
craving, 2160; 2168
craw fish, 4009
craziness, 2670
crazy, 2695
creak, v., 5044
cream, 4076-77
creamy, 2259
create, v., 13
creep, v., 1768
crepitate, v., 2052
cribri-form, 239
cripple, to make a —, 3431
crippled, to be —, 3431
crippling, 3432
critical, 1612; 2248
crook, v., 3429
crooked, 1107
cross, 464
cross (one's arms), to —, 4778-9
cross-wise, 602; 625

croup, false —, 2790
croup, true —, 2791
crout, 4036
crow, v., 2059
crow-foot (plant), 4511
crown, 225-6; 308; 1350
crucial, 3631
crumb, 3936
crush, v., 3378; 5395
crust, 176; 964; 1926; 1937; 3175; 3937
crusta lactea, 3176-7
crusta lamellosa, 3204
cry, 2483
cry, v., 2482
cubit, 417
cuckoo, 474
cucumbers, 4023
cucurbita, 3762
cup, 1149; 4665; 4674
cup, v., 4250
cupboard, 4637
curable, 1623
curdle, v., 5118
curd-like, 2130
cure, v., 4104; 4111
cure, 4105; 4112; 4167-68
c., air —, 4153
c. by emaciation, 4146-7
c. by fasting, 4145
c. by grapes, 4125
c., homœopathic —, 4117
c., hunger —, 4145
c., hydropathic —, 4118
c. by iron, 4127; 4129
c. by means of herbs, 4150
c. by mercury, 4132-3
c., climate —, 4149
c., cold water —, 4121
c., magic —, 4142
c., magnetic —, 4137
c., milk —, 4122
c., miraculous —, 4144
c., respiratory —, 4152
c., sweating —, 5546
c., sympathetic —, 4139
c., whey —, 4124
curled mint, 4337
current, 5540
current of air, 5043
curtain, 4577-8
curvature, 3430
curve, v., 559; 3429
cushion, 163; 4568
cushion, medicated —, 4183
cut, v., 298; 3318
cut by a blow, 3319
cut in, to —, 1023
cutaneous irritation, 1955
cuticula, 925
cuticular, 2792

cutis anserina, 1950
cutis vera, 933
cutting, 1695
cutting knife, 3647
cyanosis, 2880
cyanosis, cardiac —, 2881
cynanche tonsillaris, 2786
cyst, 918
cyst, vesicle —, 915
cysticercus, 3556
cysticercus cellulosæ, 3557-8
cystitis, acute —, 3034
cystodynia, 1842

## D

daily, 5035
damage, v., 1597
damp, 1972
dampness, 1373
dandruff, 1939; 1948; 3197-9; 3200
danger, 4832
dangerous, 1614
dangerous, not —, 1616
dark, 710
dark-brown, 2135
dark colored, 2491
dark hue, of a —, 2313
dark-red, 1884
darken, v., 5420
date, v., 5497
day, 4983
day before yesterday, 4795
days, for —, 5560
dazzle, v., 1353
dazzling, 5431
deaden, v., 5390
deadly, 1625
deaf (numb), 1851
deaf, absolutely —, 3479
deaf and dumbness, 3483
deaf and dumb person (a), 3482
deafness, 3481; 4996
deal, 704
deathlike, 2495
death-like paleness, 1594
death-sweat, 1978
debilitate, v., 1543; 1556
debility, 2463
deceptions of the senses, 2543-5
decidedly, 2095
deciduous, 1584
declare, v., 4968
decline, 2121
decline, v., 2688
decompose, v., 2217

decrease, n., 2426-7
decrease, v., 2107; 2424
decrepit, 1577; 1584
decrepitude, 1584
deep, 2582
defect, 1579
defect, bodily —, 1580
deficient, 2157
deficient, to be —, 1552; 4852
definite, 4984
deformity, bodily —, 1581; 1599
degenerate, v., 4842; 5295
degree, 2103
degree of heat, 2104
degrees, by —, 2108
dejection, 2685
dejection of spirits, 2761
delay, v., 5482
delicate, 1544; 1859
delirious, to be —, 2614-15
delirium, 2616
delirium tremens, 2709-10
deliver (a woman), to —, 1258
delivered, to be, 1259-60; 1263; 1275
delivered prematurely, to be —, 1273; 3294
delivery, 1261-2
delivery by Cæsarean section, 3291
delivery, forceps —, 3290-91
demand, 5266
demean one's self, to —, 4860
dental instruments, 3684
dentifrice, 4274
dentist, 3826
dentition, 326
depilatory, 4275
deposit, 931; 2291; 2296
deposit, v., 930; 4368
depraved, 2158
depress, v., 1314; 5021
depressed, 2471; 2504-5
deprive, v., 2378
deprived of a thing, to be —, 2379
derma, 933
deranged, 2668; 2695
deranged man, 2702-3
deranged woman, 2702-3
derangement, 2653; 2670
descend, v., 675-6
describe, v., 4936
description, to give a —, 4936
desire, 2163; 2188
desire, v., 1254; 2167

desk, 4644-5
desperate, 1613
despond, v., 4830
desquamation, 1934
destroy, v., 2671; 5443
detach, v., 4766
deteriorate, v., 4842
deterioration, 2679; 5180
develop, v., 5340
devil's dung, 4317-18
devoted, 5235
devote to, to —, 5566
devour, v., 3878-9; 3886
diabetes, 3046
diabetes insipidus, 3051
diabetes mellitus, 3047-9; 3050
dialysis, 1586
diaphoretic, 4227
diaphragm, 603-4
**diarrhœa**, 2239-41; 2243
d. from teething, 2902
d., to have the —, 2242
d., tendency to —, 2242
diastole, 745
diathesis, 16
die, v., 4915; 5474
diet, 3930; 4177
diet-drink, 4263-4
diet, indiscretion in —, 4993
diet, low —, 5300
different, 5085
difficult, 5288
difficulty, 2375; 5115
difficulty in respiration, 2032
diffuse, 1731
diffuse, v., 795
dig, v., 337
**digest**, v., 1047
d. everything, able to —, 3900
d. little, able to —, 3899
d. much, able to —, 3898
digested, half —, 5141
digestible, 1049
digestion, 1048
digestion, fore —, 1114
digestive apparatus, 1051
digitalis, 4518
dilatation of the œsophagus, 2928
dilatation of the stomach, 2939
dilate, v., 1001
dilated, 2539
dilator, 1365
dilator oris, 3753
diluent, 4222
dilute, v., 5363

dim, 5423
dim, to grow —, 5420
diminish, v., 2424; 2678; 2731; 4986
diminution, 2425-6; 5491
diminution in bulk, 2430
dinner, 3924-6
dip, v., 5192
direct, v., 5364
direction, 5365
directly, 5098
dirty, 1886
dirty-grayish, 2094
disagreeable, 5154
disappear, v., 2156; 5145
disc, 492
discernibly, 4962
discharge, v., 3387-8; 5578
**discharge** from the bowels, 2249
d. from the ear, 3478
d. from the nostrils, 3484
d. of matter, 5355
d. like matter, saliva, 2269
d. of pus, 5355
discharged, to be —, 5487
discomfort, 1602; 5231
discontinue, v., 5436
discover, v., 5359
discuss, 492
discuss, v., 3351
disease, 1521
disease of the bones, 4979
disease peculiar to children, 1522
diseased, 3437
disfigure, v., 5583
disgust, 2162
dish, 4676
dishes, 3917
dislocate, v., 3423; 5209
dislocation, 3424
disorder, 2653; 2666; 5053
disorder, v., 2671
disorder of mind, 2672
dispensary, field —, 3853
dispersion, 3222
displace, v., 5209
displacement, 2884
dispose in strata, to —, 523
disposed, 4819; 4975
disposition, 2625
disposition, natural —, 32
disregard, v., 5268
dissecting knife, 3646
dissipate, v., 3351
dissolute, 5162
dissolute, to be —, 5162
dissolve, v., 4224; 4770
distance, 5425

distance, in or at a —, 5426
distend, v., 2402; 4780
distention, 2198
distill, v., 4308
distinct, to grow —, 5421
distinguish, v., 2422
distort, v., 3425
distorted, 2511-12; 3081
distortion, 3426
distracted, 2668
distressed, 2029; 2506
distribute, v., 795; 3221
distributed, to be —, 5039
disturb, v., 2155
disturbance, 2653
ditch, 338
diuretic, 5382
dive, v., 5192
divide, v., 283; 3221; 3351
divest one's self of clothes, to —, 4751-2
dizziness, 1781; 4798
dizzy, 4797
do, v., 5271
do good, to —, 4862
doctor, 3805
doctor's advice, to ask a —, 3810
doctor, to consult a —, 3810
dog, 3496
done, 5323
done, to be —, 5312
done, to have —, 5324
door, 5045
dorsal stitches, 1799; 1800
dorsum, 345
double, 5428
double (one's arms), to —, 4778-79
double beating, 1995
doubt, 5551
douche, 3777-9; 3790
dough, 4254
doughy, 1898
dove, 3988
down, 2446; 5334
downward, 5334
drain, v., 1585
draught, 4861
draught, cooling —, 4263-4
draught, narcotic —, 4203
**draw**, v., 174
d. away, to —, 561
d. down, to —, 1314
d. in, to —, 2338
d. off, to —, 561; 5313
d. on (towards one's self), to —, 502
d. over, to —, 175
drawer, 4641

drawers, 4706
drawer, secret —, 4642
drawing, 1699
dream, v., 2604
dreaming, 2605
dregs, 5320
drench, v., 5205
dress, 4679-80; 4718
dress, v., 3315
dress anew, to —, 4743
dress, mode of —, 4678
dress one's self, to —, 4747
dressing, 3614
dressing-gown, 4715
dressing, padded —, 3616
drink, 2178
drink, v., 3901-2
d. beyond the thirst, 3908
d. drams, 3908
d. hard, 3902
d. like a beast, 3901
d. to excess, 3908
drip, v., 1981
drive, v., 2401
drive up, 2402
drop, 2335
drop, v., 659; 5437
drops, 4281
drop by drop, 2336
drop-counter, 5502
dropsical swelling, 5213
**dropsy**, 2722
d., hepatic —, 3064
d. in the chest, 2814
drowse, v., 4199
drowsiness, 2581
drugs, 3854-5
druggist, 3849; 3853
druggist's shop, 3856; 3860-61
drum, 1393-4
drunk, to be —, 3912-3
drunk, to get —, 3909 11
drunkard, 2705; 2708
drunkard, madness of —, 2710
drunkenness, 2704
dry, 1868
dry, v., 5409
dry up, 2821
drying up, 2822
duck, 3984
dull, 1688; 2519; 2525; 5391; 5423
dull, v., 5390
dullness, 2619
dullness of intellect, 2674
dullness of vision, 3445
dull-witted, 2617
duodenum, 1096
dura mater, 857

duration, 49
duration of life, 50
duration of time, 5479
during, 5219
dusky-red, 2492
dust, 5123
dusty, 5149
dwell, v., 4902
dwelling, 5166
dysentery, 2988 9
dysmenorrhœa, 3268; 3273
dyspepsia, 1827; 2931
dysphagia, 2782
dyspnœa, 2023
dysuria, 3057-8

# E

**ear**, 1385
ear-ache, 1789; 1790
e., buzzing in the —, 2565-9
e., cavity of the —, 1368
e., drum of the —, 1395-6
e., external —, 1385
e., lobe of the —, 1387-8
e., middle —, 1397; 5468
e., noise in the —, 2584
e., pierce the —, to, 1486
ear wax, 1391 2
early, 4816
earnestly, 5442
earth-colored, 1888
earthworm, 1118
earthworm-like, 1119
earthy, 1888
easy, 5289
easy-chair, 4550-51
easier, to make —, 4847
eat, v., 3877
eat away, to —, 2896
eat greedily, 3878
eat (of animals), to — 2896
eat up, to —, 3881
eating, 3883
eating tetter, 3216
echinococcus, 3559
ecthyma, 3193
eczema, 1947; 3170
eczema rubrum, 3173
edge, 131; 347
edge of the tumor, 1438
eel, 4008
effect, 4871; 5094
effervescent, 4362
efflorescence, 1902-3
effort, 4985
effuse, v., 2090
effusion, 5492

egg, 1211
egg globule, 1218
egg, white of an —, 2258
eggs, boiled —, 4014
eggs, buttered —, 4017
eggs, hard boiled —, 4013
eggs on butter, 4019
eggs, poached —, 4018
ejaculate, v., 1198
ejaculation, 1199
ejaculatory duct, 1200
elbow, 421
elbow, bend of the —, 424
elder flowers, 4331-2
electric shocks, 1737
elephantiasis Arabum, 3220
elephantiasis Græcorum, 3219
elevate, v., 1312
elevated, 1916
elevation, curved —, 2429
elevator, 1313
else, 4810
emaciate, 2438-9
emaciate, v., 2433
emaciated, 2498
emaciation, 2435-6
emanate, v., 5549
embarrassed, 2030
embolus, 3359
emetic, 4197-8
emit, v., 1983
emollient, 4220
emotion, 5132
emphysema, 2407-8; 2808-7
emphysematous, 1871
employ, v., 4113
emptiness, 2166
empty, 688
empty, v., 5155
empyema, 2812
enamel, 315
enamel of the teeth, 317-18
encase, v., 5374
encompass, v., 5190
encysted tumor, 3225-6
end, 185
end, v., 184
endemic, 1642
endure, v., 48; 5202
enema, 3739
enema apparatus, 3743
enervate, v., 1556; 2380
enervation, 1557
eneuresis, 2332
enfeeble, v., 1556
enfeeblement, 1557
engaged in, to be —, 2124
enjoy, v., 3894

enjoy (food or drink), to —, 4879
enlarge, v., 2730
enlarged, 2138
enlargement, 2423
enormous, 5480
enter, v., 5210
enteritis, acute —, 2965
entertain, v., 5528
entozoa (living within the body), 3537-9
enuresis, 3017; 3061
envelop, 529
envelop, v., 528; 913
envy, 3564
ephelid, 3207
epidemic, 1641-3
epidemical, 1644
epididymis, 1193
epigastric tumor, 3013
epiglottis, 1478-9
epiglottitis, 2793
epilepsy, 2771-2
epileptic, 2773
ep'phora, 3438
epistaxis, 2800
epithelioma, 3218
epithelium, 925
Epsom salts, 4417
epulis, 3485
equal, 2384
eradicate, v., 3391
erect, 2340
ergot, 4474
errand, 5590
eructate, v., 2201; 2203
eructation, 2202; 2204
e., sulphurous —, 2205
e., sour —, 2205
eruption, 1900
e., miliary —, 3130
e. of a tooth, 325
e., ulcerated —, 1946
erysipelas, 3159
e. bullosum, 3163
e., facial —, 3161
e. of the head, 3160
e. odontalgicum, 3490
e. vesiculosum seu bullosum, 3162
e., wound —, 3164
erythemia, 2475
especially, 4799
essential, 5384
establish, v., 31; 5404
establishment, 3867
ether, 4290
Eustachian tube, 1405-6
evacuant, 4194-6
evacuate, v., 4191-3; 5155
evacuation, 2244
evaporate, v., 1965

evaporation, 1966; 5058
even, 180; 4964; 5114
events, at all —, 4859
ever since, 5080
everywhere, 5531
evil, 2042; 4834; 4942
evoke, v., 5041
eye, 296
e., aqueous humor of the —, 1374
e., chamber of the —, 1361
e., coat of the —, 1337
e., globe of the —, 1336
e., or'it (socket) of the —, 1311
e., tunic of the —, 1337
e., white of the —, 1338
e., watering of the —, 3438
e.-ball, 1336
e.-ball, protrusion of the —, 3472
e.-brow, 1318
e.-brows, to lower the —, 1483
e.-brows, to raise the —, 1482
e.-lash, 1322-3
e.-lid, 1321
e.-lids, eversion of the —, 3471
e.-lids, inflammation of the —, 3433
e.-lids, turning over of the —, 3471
e.-powder, 4273
e.-salve, 4272
e.-tooth, 297
exacer'ation, 3394; 5180
exact, 4965
exaltation, 2641
examination, 5069
examine, v., 2302-4; 3312; 3386
exanthem, 1902-3
excavate, v., 119; 120
exceed, v., 5407
exceedingly, 5296
excellent, 3896
exception, 5461
excess, 2210; 2205; 5490
excess, v., 3907
excite, v., 2231
excited, 2509
excitement, 2641; 5131
exclusively, 5254
excrement, 2220
excrescence, 3389; 5070
excruciating, 1704-5; 5248
excoriate, v., 3334-5
excoriation, 1927; 3332-3

excurvation, 3430
exercise, 4172
exercise, v., 4171
execute, v., 5271
exert one's self, to —, 4950
exertion, 4985
exhale, v., 54
exhalation, 2013
exhaust, v., 1585
exhausted, 1565
exhausting, 1714
exhaustion, 1586; 2442
exist, v., 5029; 5031
existence, in —, 4843
expand, v., 1001
expansion, 1003
expectorant, 4222
expectorate, v., 2078
expectoration, 2079
expedient, 3918
expel, v., 2078; 3296
expel a disease, to —, 1563
experience, v., 5077
experiment, 5084
expire, v., 997
explicit, 5593
exploring needle, 3668
exposure to the atmosphere, 5792
expression, 1589; 2638
exsanguine, 686-7
extend, v., 743; 2337; 5335; 5418
extension chair, 3582
extent, 2389
external, 922; 1721
extinguish a lamp, to —, 4616
extinguished, to be —, 2577
extirpate, v., 3391
extract, 3683
extreme, 2174
extremities, 398
extremity, 185
extremity, articular —, 186
extuberance, 3372
exudate, 1971
exudation, 1968-9; 2131
exude, v., 1967

**F**

face, 241
fact, in —, 5489
faculty, 2645
f. of thought, 2647
f. of volition, impairment of —, 2656

fæces, 2220
fail, v., 1552; 2156; 4852
faint, 1565
faint, v., 2601
faint, feel, to —, 1677
fainting, 2598
faintness, 1566
fair, 4885
fall, 4944
fall, v., 1583; 5022; 5298
fall asleep, to —, 2478
f. away, to —, 2434
f. down, to —, 2467
f., to get a —, 4945
f. in, to —, 2467
f. into, to —, 2688; 5052
falling, 2121
falling away, 2436
falling sickness, 2771-2
Fallopian ligament, 387
Fallopian tube, 1222-3
fallow, 1887
false, 156
family, 4924
family way, in the 1249; 5503
famish, v., 5513
far advanced, 1253
far, by —, 4865
fare, 3930
farinaceous food, 4884
far-sighted, to be —, 1494
far-sightedness, 3447
fasciculus, 206; 883
fasciculus lateralis, 884
fat, 926-7; 1127; 1390; 2394; 2396
fatal, 1625
fat deposit, 931-2
fat membrane, 928
fatigue, 1016
fatigue one's self, to —, 4950
fatigue, over —, 5003
fatigued, 2443
fatten, 1128
fauces, 1007
fault, 5014
favus, 1447; 3522-3
fear, v., 5557
feather-bed, 4565
features, 250
febrifuge, 4356
febrile attack, 3088
febris urticata, 3168-9
fee, 5580
feeble, 1540; 1544; 2949
feeble-minded, 2755
feebleness, 1578
feed, v., 748
feed with mast, to —, 1128

feel, v., 828; 1984; 2123; 4863; 5302
feeling, 1500
f., loss of —, 5339
f. of soreness, 1748
feign, v., 1621
feigned. 1620; 1622
female, 376
f. attire, 4717
f. sex, 1163
femur, 478-79; 486
fenestra vestibuli, 1402
ferment, v., 2214
fermentation, 2215
fester, v., 2412-13
fetch, v., 986
fetor, 2093
fetter, v., 5529
fetus, 1282
fever, 3087
f., abatement of the —, 3116
f., acute puerperal —, 2909
f., arthritic —, 3086
f. attended with vomiting, 3106
f., autumnal —, 3108
f., to be in a —, 3094
f., bilious —, 3109
f., catarrhal —, 3102; 3104; 3105
f., cough —, 3102
f., double quotidian —, 3091
f., free from —, 5381
f., to have —, 3094
f., hay —, 3101
f., hectic —, 3113
f., inflammatory —, 3111
f., intermittent —, 3100
f., lingering —, 3113
f., masked —, 3113
f., miliary —, 3130
f., milk —, 3298
f., mucous —, 3110
f., nervous —, 3120
f., pernicious malarial —, 3113
f., puerperal —, 3297
f., quartan —, 3089
f., quotidian — 3089
f., relapsing —, 3125
f., scarlet —, 3127-9
f., simple quotidian —, 3090
f., slow —, 3113
f., spring —, 3107
f., sweating —, 3103
f., syphilitic —, 3113
f., tertian —, 3092
f., traumatic —, 3112

f., typhoid —, 3119
f., typhus —, 3121-2
f., variolous —, 3138
f., yellow —, 3113
feverish, 1633; 2011; 4802
f. odor, to emit a —, 3096
f., to feel —, 3095
few, a —, 4873
few, some —, 4873
fibre, 201
f., axis —, 890-91
f., sarcous —, 539
f., sensitive —, 832
f. of the motory nerve, 885
f.s, bundle of —, 207
f. of the cerbral nerves, 863
f., sympathetic nerve —, 887
f., tendinous —, 209
fibrin, 203
fibrinous, 204
fibro-cartilage, 159; 548
fibrous, 204
fibrous ligament, 205
fibula, 500
field-dispensary, 3863
fierce, 1655
fig, 4056
figure, 25-6; 37
figure to, give —, 24
filament, 3545
filaria, 3546
file, 3718
fill to excess, 5284
fill up, 5555
film, 3443
find out, to —, 5077; 5359
fine, 5508
finely powdered, 5576
finger, 437
f. -articulation, 440
f., ear —, 447-8
f., fore —, 444
f., little —, 447-8
f., middle —, 445
f., ring —, 446
f., sore —, 3229
finished, 5323
finished, to have —, 5324
fire-hearth, 4627
fire-wood, 4626
firm, 2354
first, 5101
first, at —, 4932
fish, 3997
fish-skin disease, 3206
fissure, 850; 1930
fissured, 2134
fissure, longitudinal —, 851

fissure, transverse —, 856
fistula, 3230-1
fistula lachrymalis, 3439; 3940
fistula of the gum, 3488
fistulous ulcer, 3232
fit, 2212; 2381; 2777
fit, v., 5449
fits, 5517
fit for, 44
fit together, to —, 102
fit out, to —, 75
fit, go into a —, 2601
fit well, to —, 5208
fitting, 5450
fitting together, 103
fix, v., 5404
fixed, 2456; 4984
flabby, 1862
flaccid, 1862
flake, 2550
flank, 2342
flannel, 5203
flap, 854
flash (of water), 5309
flask, 4662
flat, 178; 2273
flatness, 179
flatten, v., 2467
flattened, 2472
flatulence, 2195-6
flatulent, 2194
flatulency, cause of —, 2193
flax-seed, 4379
flay, v., 3380
flea, 3532
flea-bite, 3533
fleam, 3717
fleeting, 2683
**flesh** 319
f., dead —, 3235
f., to gain —, 2430
f., to lose —, 2430
fleshless, 537
fleshly, 5351
fleshy, 536
fleshy substance, 538
float, v., 5447
flocculus, 2550
floor, 5344
floor-cloth, 4576
flour, 4479
flow, 2063; 3269
flow, v., 210; 659
fluid, 211-12
fluidity, 1970
flushed, 2501
flux, 2063; 3269
flux albus, 3270-1
fly, 2548

fly blister, 4429-30
fly-net, 4585-6
flying 1708
foam, 2310
focus of suppuration, 3311
fœtus, 1282
fog, 5430
fold, 1349
fold, v., 5207
fold across (one's arms), 4778-9
follicle, 544
follicle, mucous —, 545
follow, v., 4998
folly, 2698-9
foment, v., 3317
fomentation, 4233
fond of, 5599
food, 1057; 1059; 1126; 3930
foot, 451
foot-stool, 4636
foot, upper part of —, 514
foramen, 278
**forceps**, 3687
f., crutch-shaped —, 3713
f., delivery —, 3290-1
f. scissors, 3711
f., scraping —, 3713
f. with a catch, 3710
fore, 221
fore-arm, 416-7
foreboding, 5024
fore-digestion, 11
fore-foot, 501
forehead, 230
foreign, 3367
fore-skin, 1189
fore-skin band, 1190
fork, 1021
form, 25; 36; 2272
form, v., 24; 27; 35
form, bodily —, 26
formation, 28
formerly, 4919
formication, 1769-70
formidable, 1649
forth, 2359
forthwith, 5251
forward, 221; 2358-9
foul, 2199; 4806
found, v., 51
foundation, 3391
four-headed, 3623
fowl, 3985
foxglove, 4518
**fracture**, 3407-8
f., comminuted —, 3415
f., longitudinal —, 3409· 3412
f., oblique —, 3411

f. of a rib, 5197
f., reduce a —, 3416
f., setting of a —, 3417
f., transverse —, 3410; 3413
frænulum, 329
frænum, 1190
frail, 1577
frailty, 1578-9
frame, 25
frame of mind, 2625; 2628
framework, 76
framework, bony —, 89
frank, 4839
frantic, 2697
freckle, 3207
freckled, 3208
freckle-faced, 3208
free, 2343
free, v., 5358
freeze, v., 3506
frenzy, 2692; 2700
frequency, 2027
frequent, 1990; 2329
frequently, 4880
French roll (milk bread), 3945
freshen, v., 4217
fright, 5477
fritter, 4016
front, 221; 230
frothy, 5143
fruit, 1219; 4042
fruitage, 4043
fugitive, 2683
full, 1865; 5383
full-blooded, 2501
fulness, 2198
fumigate, v., 4234
fumigation, 4235
function, 2652
functional, 1660
fūngiform, 1448
fungiform papillæ, 1449
fungous, 2485; 3234
fungus, 1447; 2486; 3235
funiculus lateralis, 884
f. marginalis, 906
funnel, 4661
fur, 2131; 4688
fur collar, 4740
furnish one's self with, to —, 5603
furniture, 4633-4
furred, 2132
furred on one side, 2139
furrow, 848
furthermore, 5275
furuncle, 2417-19
fury, 1123; 3495

# G

gait, 39
galactischesis, 3301
galactorrhœa, 3299
gall, 1099
gall-bladder, 1104
galls, 4447
gall-stone, 5255
game, 3960
ganglion, 755; 801; 814-15
ganglion chain, 907
ganglion globules, 824
ganglionic system, 816
gangrene, v., 2897; 3406
**gangrene, 3395**
g., acute —, 3397
g., dry —, 3400
g., hospital —, 3405
g., moist —, 3399
gangrenous eschar, 3365
g. sanies, 3402
g. stomatitis, 2887
gap, 313
gape, v., 3350
garden sage, 4340
gargle, 4268
garment, 3527; 4679
garters, 4759
gas lamp, 4012
gasp, v., 2071
gasps, 2018
gastric juice, 1089
g. membrane, 1075
g. nerve, 1078
g. paralysis, 2942
g. plexus, 1072
g. region, 1068
g. tumor, 2934
g. ulcer, 2936
g. vessel, 1071
g. wall, 1069
gastritis, 2938
gastrodynia, 1819-20; 2211
gate, 775
gelatin, 4448
generally, 5019; 5239
generate, v., 1172-3
generation, 1174
genital member, 1179
genital parts, 1164
gentle, 1757
germ, 955
germinal bed, 1217
germinate, v., 954
get asleep, to —, 2478
get into, to —, 5052
get on clothes, to— 4746
get up, to —, 4872
ghastliness, 1594
ghastly, 1592
gherkins, 4024

giblets, 3995
giddy, 4797
ginger, 4068
girdle, 3178
give, v., 5362
give way, 5107
given to, 5235
glad, to be —, 4922; 5082
**gland, 1188**
g., bronchial —, 1039
g., cardiac —, 643
g., gastric —, 1074
g., lachrymal —, 980
g., lactiferous —, 1230
g., lingual —, 1456
g., lymphatic —, 976
g., mammary —, 983
g., mesenteric —, 1113
g., prostate —, 1202
g., pulmonary —, 1039
g., salivary —, 982
g., sebaceous —, 936
g.s, secretory —, 975
g., shield-shaped —, 1034
g., submaxillary —, 5093
g., sweat —, 938
g., thoracic —, 1035
g., thyroid —, 1034
g.s, vascular —, 973; 978-9
glans penis, 1188
glaring, 5431
glass, 3762; 3915
glassy look, 2536
Glauber's salts, 4418
glaucoma, 3459
glazing, 316
glene, 113
globule, 187
glossalgia, 1786
glossitis, 2908
glottis, 1009; 1474
gluttonize, v., 3878
glycerine, 4449
gnash, v., 2480
gnat, 2547
gnathus, 262
gnawing, 1702
go, v., 88; 5408
go about, 4896
go down, 2107
goblet, 4665
gold leaf, 5556
gonagra, 8085
gonorrhœa, 3249
gone, to be —, 5471
**goose, 3983**
g., breast of a —, 3993
g., roasted —, 3992
g. skin, 1950
g., smoked —, 3991
gory, 685
**gout, 1854; 3082**

g. in the foot, 3083
g. in the hand, 3084
gradual, 1653
gradually, 2108
grain, 3512
grains of gunpowder, 3513
granular, 5457
granulations, 3373
grape, 1366
grasp, v., 2122
grate, 4658
grate, v., 1942; 2480
gravel, 2299; 2300
gravity, 2321
gravy, 3971
graze, v., 1957
grease, 214; 927; 1390; 2396
grease, v., 213
greenish-yellow, 2309
greens, 4027
grief, 4952
gripe, v., 1124; 2122
gripes, 1815; 1828-9; 2951-3
griping, 1718
gritty, 1858; 2298
groan, v., 2060
groin, 3861; 5366
grooved probe, 3667
grooved sound, 3669-70
grooved staff, 3669-70
gross, 5206
ground, 3391
grow, v., 40
grow exuberantly, 3371
grow into, 3562
growth, 41; 5070
gruel, 4071
guard, v., 3148; 5538
guard against, 4815
guess, v., 5090
guess, by a rough —, 4966
guide, 1224
guilty of, to be —, 5015
gullet, 1065
gulp, 4861
gulp down, to —, 3886; 4783
gum, 320
gum-boil, 3486-7
gums, 320
gums, inflammation of —, 2893
gunpowder, 3511
gustatory papillæ, 1441-2
gut, 454
gutta rosacea, 3190-1
gymnastic, 4174
gymnastic art, 4170
gymnastic exercises, 4173
gymnastics, 4175
gynglimus, 193-4

# H

habitual, 2987
hæmaturia, 5029
hæmoptysis, 2834-5
hair, axillary —, 971
h. -brush, 4593
h. bulb, 959
h. -germ, 956
h. of the head, 968
h. of the pubes, 972
h. -root, 960
hair shaft, 962
hairy, 957
hale and sound, 1509
half, n., 247
half, a., 246
h. -boot, 4702
h. digested, 5141
h., left —, 628
h., open —, 2530
h., right —, 627
hallucinations, 2543-5
halt, v., 2363; 2383
ham, 494; 3980
hammer, 1408
hammering, 1693
hand, v., 5362
hand, 425
h., at —, 4843
h., back of the —, 429
hand-barrow, 3569; 3570
h., hollow of the —, 430-1
h., left —, 428
h., palm of the —, 430-1
h., right —, 428
hangnail, 3565-6
hanker after, to —, 1254
hanker for, to —, 1254
hankering, 1255
happen, v., 5312
hard, 1610; 1858
hard, grow, to —, 4156
hardening, 2719
hardness, 1762
harm, v., 1597
harsh, 1866
hartshorn, 4413
hasten, v., 2028
hat, 4693
haunch, 382
hawk, v., 5104
hawking, 5105
hay asthma, 3001
hazel nut, 4054
head, 125; 227
h. ache, 1778-9
h. ache, frontal —, 1783
h. ache, hysterical —, 1781
h. -band, 3591
h. -clerk, 3852

head-nodder, 598
h. of a bone, 126
h., back of the —, 224
h., forepart of the —, 223
h., heaviness of the —, 2727
h., hind part of the —, 224; 228
h., top of the —, 225-6
heal, v., 4104
heal over, to —, 3355
heal up, to —, 4154
healing, 4105
healing up, 4155
health, 1511; 4166
h., to be in —, 1506
h., in good —, 1512
h., state of —, 1513
h., weak —, 1545
healthy, 1507
hear, v., 1382
hearing, 1383
h., to be dull of —, 1496
h., illusions of the sense of —, 2561
h. (person) hard of —, 3480
h., to be quick of —, 1495
heart, 624
heart ache, 1809
heart aches, my —, 1678
h. -beat, 739
h. -burn, 2207-8; 2856
h. -cavity, 646
h. -clot, 661
h. disease, 5050
heart's impulse, 740
h. motion, 648
h. tumor, 2877
h., abscess of the —, 2874
h., apex of the —, 649
h., atrophy of —, 2872-3
h., chamber of the —, 630
h., coarctation of the —, 2868
h., defect of the —, 2849
h., dilatation of the —, 2866
h., disease of the —, 2848
h., displacement of the —, 2884
h., expansion of the —, 774
h., fatty —, 2876
h., hypertrophy of the —, 2864-5
h., induration of the —, 2870
h., irritable —, 2860
h., narrowing of the —, 2868

heart, oppression of the —, 1810
h., ossification of the —, 2871
h., palpitation of the —, 742; 2851
h., paralysis of the —, 2853
h., point of the —, 649
h., polypus of the —, 2879
h., spasm of the —, 2857-8
h., valve of the —, 634
hearth, 4651
hearty, 3896
heat, 2206
heat, v., 5225
heat of the sun, 5380
heat, vital —, 53
heating, 5131
heave the gorge, to —, 2948
heaviness, 1781; 1832; 2321
heavings, 2018
hectic-red, 1885
heed, 4841
heel, 505-6
heel-bone, 507
height, 2110
helminth, 5279
helminthic, 4230
helminths, 3536
help, v., 5318
help to, to —, 5250
hematine, 691
hemicrania, 1782
hemiplagia, 2739
hemisphere, 852-3
hemlock, 4511; 4514
hemorrhage, 5476
hemorrhage from the lungs, 2836
hemorrhoids, 1132; 2991
hemorrhoidal vessel, 1131
hemorrhoidal ulcers, 5287
hemp seed, 4443
henbane leaves, 4335
hepatalgia, 1831
hepatic dropsy, 3064
hepatic lobe, 1008
hepatic obstruction, 2995
hepaticum, 3543
hepatitis, acute —, 2996
herb medicine, 4184
hereditary, 1648
hernia, 3407
hernia cæcalis, 2970
herpes, 1945; 3172
herpes tonsurans, 3216
herpes zoster, 3179; 3180-1
herring, 4008
hew, v., 3318

hiccough, v., 2047
hide, 110; 601
hide from, to —, 5465
high, 1988
high-colored, 2314
hill, 488
hind, 222
hinder, 222
hinder, v., 5059
hinge, 191
hip, 382
hip-gout, 1878; 3085
hip-joint, 480
hiss, v., 2038; 2054
hitherto, 4960
hoarse, 2064
hoarse, to get —, 2066
hoarseness, 2065
Hock, 4086
hæmidrosis, 1975
Hoffman's drops, 4419
hold, v., 817
hold back, to —, 5011
hold fast, to —, 5176
hold of, to lay —, 5369
hole, 121; 278
holes in, make —, 1485
hollow, 115
hollow out, to —, 119-20
holy, 465
homœopathic doctor, 3837
homœopathist, 3837
honey, 4458
hoped, it is to be —, 4818
horizontal, 1754
horn, 942
horn, of —, 943
horn-like, 945
horny, 943 4
horny tunic, 1341-2
horse-doctor, 3832-3
horse radish, 4486
hose, 3740·41
hospital, 3404; 3871-74
hospital, field —, 3876
hot, 1875
hot, to make —, 5225
hough, 494
hour, 4809
hour, of half an —, 5252
how? 4928
hum, v., 2054; 2056
**human**, 7
h. being, 1
h. body, 9
h. kind, 3
humanity, 6
humerus, 403-4
humerus, head of —, 412
humidity, 1971
humor, 212; 1373; 2631

**humors**, bad, 1596
h., unhealthy —, 1596
hump, 442
hunger, stay the —, 3890-2
hungry, to be —, 2151
hungry, to get —, 4876
hurt, n., 1598; 3307·9; 4942
hurt, v., 2713; 3305-6
hurting, 3304
hushed, to be —, 5328
hyaloid, 1380
hyaloid membrane, 916
hydrocephalus, 2723
hydrocyanic acid, 4522
hydronephrosis, 3036
hydro-pericardium, 2882
hydrophobia, 3498
hydro-pneumonia, 2840
hydrothorax, 2814
hymen, 1226
hyoscyamus leaves, 4335
hypermetropia, 3474
hypermetropic, 3473
hypnophobia, 2592
hypnotic, 4202
hypodermic, 3736
hypogastralgia, 1818
hypogastric, 3021
hypostasis, 2291
hysteralgia, 1845; 3281
hysteria, 2776; 3274-75
hysterical, 1665
hysterics, 2776
hysterocele, 3283-84
hysterotomy, 3292

# I

ice, 4061-2
ice-cream, 4061-2
Iceland moss, 4355
ichthyosis, 3206
icy, 1879
idea, 2663
idiocy, 2779
idiopathic, 1617
if, 5156
ileus volvulus, 2986
ilium, 455; 1107
ill, 1519; 2185; 2630
ill, to be —, 1527
ill, to fall —, 1529
ill, feel one's self —, 1528
ill, I feel —, 1677
ill, to be taken —, 1529
iliac region, 3020
illness, 1521; 1535; 4942
illusions, 2543-45
imbecile, 2755 57
imbecility, 2618
imbecility of mind, 2779

immediate, 5151
immediately, 5098; 5251
immense, 5480
immovable, 141
impairment, 2677
impassive, 2516
impede, v., 2031
impeded, 2034
impediment, 5113
imperceptible, 1997
imperfect, 2385
impetigo, 1949; 3176-7; 3193
impetus, 1165
importance, 5066
important, 1611
impossible, 2376
impotence, 1177
impotent, 1176
impregnate, v., 1251
improve, v., 4963
improvement, 5081
impuberty, 1170
impulse, 737; 2245
inch, 5433
incise, v., 1023
incision, 1024
incisor, 299; 305
inclination, 2188
inclinations, to indulge one's —, 4828
inclose, v., 5190
inclovation, 146
increase, v., 1756; 2391; 5341
increase, 2392-93; 2863; 3394
incubus, 2606-7
incunneation, 146
incur, v., 4943
incurable, 1624
indent, v., 2489
index, 444
indicate, v., 443
indication, 5024
indifferent to, 2636
indigestion, 1824
indigestible, 1050
indispensable, 5498
indisposition, 1535; 1537; 1538
indisposition after a drunken debauch, 2933
indisposed, 1534
indolence, 2659
indulge one's self (to), 5280
indurated, 2142
induration, 2719
inebriated, 2706-7
infect, v., 3345; 3502; 4540
infection, 3504; 4537; 4541

infectious, 1638
infirm, 1525; 1540
infirmary, 3872-73
infirmity, 1526; 1542
infirmity, bodily —, 1580
inflamed, 2137
inflammation, 2714
inflammatory, 1627; 1636
inflate, v., 2193; 2405; 5470
inflation, 1761; 2400
inflect, v., 2338
influence, 5226
influence, v., 5337
influenza, 3104-5
inguinal region, 389
inhale, v., 996; 5159
inject, v., 1198; 3729-30; 4244
injected, 2528
injecting, 4245-47
injection, 3737; 4236
injection-pipe, 3735
injure, v., 2031; 2713; 3305-6
injured, 5558
injury, 1598; 3307-9
inoculate, v., 3149
insalivation, 1084
insane, 2524; 2687
insane, the — 2758
insanity, 2686; 2700; 2759; 3497
insect, noxious —, 3530
insensibility, 2597
insensible, 1869
insidious, 1637
insolation, 2766
insomnia, 2589 90
inspect, v., 2302-4
inspire, v., 996
instantaneous, 5270
instep, 498
instil, v., 4237
instillation, 4238
instinct, 1165
institution, 3867
instruction, 5365
instrument for drawing teeth, 3688
instrument for vaccinating, 3677
insufficient, 5283
insufflation, 4240
insuperable, 5256
integumentum commune, 923
intellect, 2662
intellectual, 2654
intellectual indolence, 2660
intemperate, 5046

intense, 1713
intensity, 5005
intercalate, v., 2976-77
intercostal nerves, 893
intercourse, 5352
intercurrent, 2000; 2003
intergrowth, 3354
interior, 5460
intermission, 5310
intermit, v., 2115
intermittent, 1664; 2002-3
intermittingly, 5310
internal, 1720
interrupt, v., 2155
interruption, 4856
interstice, 5036
intertwine, v., 2978
interval, 2693; 5036
intestine, 454
**intestines**, 919
i., induration of the —, 2984
i., intussusception of the —, 2979-82
i., invagination of the —, 1112
i., large —, 1112
i., small —, 1094
i., perforation of the —, 2964
i., softening of the —, 2973
intestinal obstruction, 2985
intestinal phthisis, 2974
intestinal tumor, 2971
intolerable, almost —, 1703
intoxicated, 2706; 2707
intoxicated, to be —, 3912
introduce, v., 5076; 5317
introvert, v., 5432
intumescence, 1761
inveterate, 1635
invigorate, v., 1555
invigoration, 4162-3
invincible, 5256
involuntary, 553
inward, 571
inwards, 571
iris, 1354; 1356-7
iritis, 5441
irksome, 2029
iron, 4126
irradiate, 837
irradiation, 836
irregular, 1724
irritability, 827
irritable, 2508; 2510
irritant, 4292
irritate, v., 825

irritating, 1707
irritating, not —, 5111
irritation, 4243
ischium, 382 84; 453; 457
ischuretic, 5382
ischuria, 2326-8; 3057-8
itch, 1952
itch, v., 1951
itch for (to), 1254
itching, 1255; 1706; 1952
itch-tick, 3525
itchy, 1876; 1943

**J**

jacket, 4731
jag, v., 2489
jag-formed, 135
jag-shaped, 135
jar, v., 5044
jaundice, 2992
jaw, 253; 261
j., lower —, 258-9
j., upper —, 256
jaws, 1061
jejune, 5228
jejunum, 1106
jelly, 4022
jelly, like —, 2267
jerk, v., 1777
jerking, 1999
Jesuits' powder, 2361
jet, 5309
join, v., 102
join together, 137
joining, 138
**joint**, 103; 167
j. articulation, 171
j., comminution of the —, 3379
j.s. crushing of the —, 3379
j., ligament of the —, 169
j., maxillary —, 254
j., quite out of —, 2455
j., rigid —, 220
j., ring —, 170
j., rotary —, 197-8
j., tight —, 220
j., turning —, 351
j., vertebral —, 351
judge, v., 2650
judgment, 2651
jug, 4656
juice, 1595
juice, intestinal —, 1093
juncture, 171
juncture of bones, 104
juniper, 4375
just, 4971

## K

keep, *v.*, 817; 5181
keep away, to —, 5483
kernel, 123
kettle, 4658
kettle of water, 5537
key, 406; 3689
**kidney**, 1146
k., disease of the —, 5049
k., fatty —, 3032
k., movable —, 3033
k., pelvis of the — 1151
k., waxy —, 3033
kind, 70
kind of, of the —, 71
kitchen, 4648
kitchen closet, 4649
knee, 491
knee-pan, 493
kneel down, 4777
knife, 3637
knob, 422; 488
knock, *v.*, 2728
knot, 755
know, *v.*, 2421; 2593
knowledge, 3802
known, 5518
known, to make —, 2347
knuckle, 63; 413; 439
krout, 4036

## L

labor, difficult —, 1265
labor, in (to be), 1238; 1290
labor pains, 1846
l. pains, tremulous —, 3287
laboratory, 3802
labored, 2039
laborious, 2034; 2370
lace, *v.*, 4760; 4762-3
lace one's self, to —, 4760
lace together, to —, 4761
laces, 4730
lacerate, *v.*, 3376
laceration, 3342-3
laceration of vessels, 3377
**lachrymal** apparatus, 1326
l. canal, 1328
l. duct, 1327
l. lake, 1332
l. papilla, 1333
l. sack, 1331
lachrymalis, caruncula —, 1329-30
l., lacus —, 1332
lacteal vessel, 1238
lactic acid, 4300
lactiferous canal, 1237
l. duct, 1235
lame, 2382
lame, *v.*, 2733
lameness, 2734
lamp, 4604
lancet, 3638-9; 3645
lancinating, 1697
language, 1460
languid, 1565
languish, *v.*, 1568
languor, 2637
lantern, 4618
lard, *v.*, 3990
larder, 4649
large, 1987
**laryngeal** cover, 1478-9
l. phthisis, 2794-5
l. pocket, 1481
laryngitis, 2783-4
laryngoscope, 3755
larynx, 1462
lassitude, 1566; 2442; 2445; 5389
l., extreme —, 2451
last, *v.*, 40; 4941; 5153
lasting, 40
late, 5100
lately, 5422; 5481
laugh, *v.*, 2749
laugh, convulsive —, 2751
laughter, 2750
lavement, 3739
lavender, 4377
lavender flowers, 4442
lax, 1862
laxness, 5137
lay, *v.*, 30
l. hold of, to —, 2122; 4.5; 5369
l. off clothes, 4750
l. on, 31
l. one's self over, to —, 4774
l. over, 3948
l. up, 5181
layer, 526
layer, mucous —, 940
lazaretto, 3874
lazaretto, flying —, 3875
laziness, 2659
lead, *v.*, 5235
lead, 4422
lead salve, 4423
leaf, 366
lean, 2431
lean or thin, to become —, 2433
leanness, 2432
learn, *v.*, 5077
leather, 932
leave, *v.*, 3114; 4875
leech, 5435
left, 427
leg, 91; 497
leg of mutton, 3972
legumes, 4028
lemonade, 4073
**lens**, 1875
l., capsula of the —, 1376
l., opacity of the crystalline —, 3448
lonitive, *a.*, 4213
lenitive, *n.*, 4212
lenitives, 4214
lenticulus (of the stapes), 1410
lentils, 4032
lepra, 3217; 3219
lesion, 3.07-9
lessen, *v.*, 4986
let, *v.*, 3114
let down, to —, 3282; 4772 73
lethargy, 2459-60; 2582-85
letter, 5427
lettuce, 4512
lettuce, strong-scented —, 4512
leucocythæmia, 3070-71
leucoma, 3460
leucorrhœa, 3270-71
level, 178-9; 2273
liable to, 4975
liberty, at —, 2343
licentious, 5162
licentious, to be —, 5162
licorice, juice of —, 4471
licorice root, 4456
lid or eyelid, 1321
lids, secretion from —, 1335
lie, *v.*, 2192
l. close to, 5208
l. down, 2447; 4775; 4981; 5298
l. in, to —, 1259; 1277
l. near to, 5208
l. one's self sore, 3336
life, 43
life, manner of —, 4949
lift, *v.*, 1312; 2373
lift up, to —, 4771
**ligament**, 105
l., capsular —, 200
l., crural —, 481
l., digital —, 438
ligamental juncture, 106
ligamentum ciliare, 1348
ligamentum glottidis, 1475
ligate, *v.*, 3360
ligation, 3361
ligature, 105; 3361; 3630

light, 1609 ; 4609-10
light a lamp, 4614
light-red, 713
light-screen, 4619
light, very —, 2307
lighter, to make —, 4847
lightning-like, 1709
like, 68
like, v., 5415 ; 5599
like it, I —, 4847
liking, 5416
likewise, 5488
limb, 392
limbs, 393
lime, 4309
lime water, 4310
limit, n., 803
limit, v., 5004 ; 5037
limited, to be —, 5038
limp, v., 2363
linden flowers, 4333
line, 249
line, v., 175
lineament, 249
lineaments, 250
linen, 5057 ; 5347
linen cloth, 5057
linen-press, 4639
lingering, 1637
lingual frenulum, 1451
l. membrane, 1439
l. parenchyma, 334
lining, 176 ; 4692
lint, 3369 ; 3693
lip, 1210
lippitudo, 3431-6
liquid, 212
liquor, 4100
liquor, intestinal —, 1093
listless, 2514-15 ; 2634
listless, to be —, 2635
litter, 3569 ; 3571 ; 3573 ; 3577
live, v., 42
**liver**, 1097
l. spots, 3210
l. worm, 3542-3
l., abscess of the —, 3000
l., acute congestion of the —, 2993-4
l., chronic congestion of the —, 2998
l., disease of the —, 5048
l., fatty —, 3001
l., hob-nail —, 3006
l., hypertrophy of the —, 2999
l., softening of the —, 3008
l., waxy —, 3002-3
livid, 1890
living, 46

loathing, 2162
lobe, 854 ; 1027
lobster, 4010
lobule, 1025
local, 1661 ; 5292
localised, 1728
lochia, 3279-80
lock up, to —, 405
loin, 379
loins, 464
long, a., 4904
longing, 2168
look, v., 2322 ; 3914
look at, to —, 5562
look for, to —, 5110
look ill, to —, 1520
look on, to —, 2302
look to, to —, 5136
look unhealthy, 1590
look well, to —, 1518
loop, 3625
loose, 1861
loose, to make —, 4767-9
loosen, v., 4221 ; 4756 ; 4767-70
looseness, 2243
lord, 4833
lose, v., 2152
loss of feeling, 5339
lotion, 4266 ; 5372
loud, 2037
louse, 3526
louse, felt —, 3529
love, v., 5599
low-diet, 5300
lower, 395
lower, v., 3282
low regimen, 4179
lozenge, medicated —, 4278
lucid intervals, 2694
lukewarm, 5246
lull (to sleep), 4199
lumbago, 1836 ; 3080
lumbar region, 3019
lumbi, 464
lump, 697
lunacy, 2698-9
lunar caustic, 4316
lunatic, 2690 ; 2702-3 ; 2758
lunch, 3923
lung, 782
lungs, lobe of the —, 1023-9
l., paralysis of the —, 2831
lung-sick, 5177
lupus, 3216
l. vorax, 3216
lust, 1247
lustful, 1248
luxurious, 5290
lying in, to be —, 1276

lymph, 752
lymphatic absorbents, 753-4
l. vessels, 753-4

# M

macaroni, 4058
macaroon, 4059
mackerel, 4006
macroglossia, 2912
macula, 1369 ; 3212
mad, 2687 ; 2697
mad, to run or go —, 2688 ; 2695
madman, 2701
madness, 2686 ; 2698-9 ; 3495 ; 3497
madness of drunkards, 2710
mad woman, 2701
made up, 14
maggot, 3549
magic, 4141
magnesia, 4366
mainly, 4977
main point, 5108
maintain, v., 5535
**make**, v., 13 ; 27 ; 1672 ; 5589
m. easier, to —, 4847
m. little, to —, 4846
m. much, to —, 4846
m. one, to —, 137
m. use of, to —, 5172
malady, 1547-82
malar-bones, 268-9
male, 375
male sex, 1162
malign, 1647
malignant, 1647
malleolus, 508
malt, 4457
mamma, 1231
mammæ, areola —, 1233
maniacal, 2523
man, 1
manage, v., 3385
manager, 3852
mandible, the —, 253
mangy, 1941
maniac, 2701
manifest, v., 5182
manifestation of the disease, 5102
mankind, 3
manner, 4921
manner of life, 4949
marasmus, 2430
margin, 131 ; 1210
mark, 5211

marked, 1715
marriage, 1240
marriageable, 1241
marrow, 114
marrow, bone —, 115
marry, v., 1239; 4909
marsh, 5375
marshmallow root, 4328
marshy, 5376
mass, 82; 697
masseters, 594
mast (for fattening), 1196
master, 4833
masticate, v., 593
match, 4606 8
material, 202
matter, 202; 1922; 4829
matters not, it —, 4869
mattery, 2096
mattress, 3579
maturity, 3224
maxilla, 253; 261-2
m., inferior —, 258-9
m., superior —, 253
meagre, 2431
meagreness, 2432
meal, 3883; 3920; 4479; 5216
mean, v., 4868
means, 3918
m., by all —, 5072; 5541
m., by no —, 4878; 5109
m., by what —, 4934
m. of, by —, 5261
m. of subsistence, 3919
measles, 3132
meat, 319
meat, boiled —, 3962; 3964
mediastinum, 1031
medical, 3804; 3806
m. establishment, 3868
m. science, 3803
m. student, 3816
medicament, 4277
medication, 4112
medicinal herbs, 3858
medicinal soap, 4466
medicine, 3857; 4185
m., course of —, 4112
m. (science of), 3803
m., family —, 4188
m. chest, a family —, 3865
m. chest, a ship's —, 3866
m. chest, traveling —, 3864
medium, 3918
medullary cortex, 880
m. matter, 1157
m. sheath, 822
m. substance, 966; 1157
medullated, 818
medullated, non —, 819

melancholia, 2761
melancholy, 2685; 4821
melt, v., 4221
melt away, v., 5414
members, 398
m., anterior (fore) —, 396
m., hinder —, 397
m., upper —, 396
membranes covering the brain, 3752
memory, 2649
memory, verbal —, 2681
meninges, 3752
menses, 3258-59; 3262-3
menstruate, v., 3253
menstruation, 3254; 3260
m., derangement of —, 3264
m., vicarious —, 3266
mental, 2629
m. alienation, 2759
m. depression, 2629
m. emotion, 2642
m. faculties, 5561
m. wandering, 2682
mention, v., 4940
mercurial plaster, 4431
mercury, 4131
mesentery, 1095; 1108
mesenteric, 3018
m. intestines, 1109
mesocardium, 647
mesogastrium, 1076
metacarpus, 435
metallic, 2077
metatarsus, 509
miasma, 4531
miasmal, 5377
miasmatic, 5377
middle, 434
midge, 2547
midwife, 3830-31
midwife, man-, 3827-9
miliary, 1919
military surgeon, 3818
milk, 300
milk-ducts, inflammation of the —, 3300
millet, 1918
mind, 2627
m., absence of —, 2684
m., diseased in —, 2757
m., frame of —, 2625; 2628
mineral water, 4103
minute, 1986
miracle, 4113
mirror, 3751-2
miscarriage, 1269-70; 3289
miscarriage, cause a —, 3296

miscarry, v., 1267; 5495
misery, 1289
missing, to be —, 1552
mist, 5430
mite, 3549
mitigate, v., 1673
mitigation, 4864
mix, v., 2213; 5125
moiety, 247
moist, 1372; 1972
moisten, v., 3171; 5056
moisture, 1373; 1971
molar, 306
mole, 3211
mollify, v., 4219; 5274
moment, 2693
momentary, 1658
money-like, 2098
monomania, 2760
monster, 3289
mons veneris, 1205
month, 4892
monthly, 3256
monthly flow, 3257
mood, 2631; 4823
moonstruck, 2690
morbid, 3437
morbili, 3132
morning, 4987
morning-cap, 4737
morning-dress, 4719
morning, in the —, 4988
morphine, 4530
mortal, 1615
mortification, 3395
mortify, v., 3406
Moselle, 4087
mosquito-bar, 4585-6
moss, 4354
mostly, 5387
mother-spot, 3212
motion, 523
motionless, 141
mount, v., 675
mouth, 335; 633; 1061
mouth, roof of the —, 264
movable, 140
move, v., 139; 5408
m. back and forth, to —, 3117
m. in a vehicle, to —, 5408
m. rapidly, to —, 721
m. round rapidly, to —, 722
movement, 523
moving round, 723
Mr., 4833
much, by —, 4865
much, too —, 3007
muco-enteritis, 2066
mucous, 5160
m. membrane, 920

mucous membrane, nasal —, 1428
m. patch, 3244
m. rattle, 2051
m. reticulum, 924
mucus, 543
m., blood-streaked —, 2265
m., glairy —, 2087
muguet, 2891-2
mumble, v., 5399
murmur, v., 653
muscle, 515
m., abdominal —, 606
m. of abdominal wall, 607
m., abductor —, 565
m.s buccinator —, 591
m.s, cervical —, 596-7
m.s, circular —, 556
m.s, cooperative —, 579
m.s, counteracting —, 580
m.s, cranial —, 587
m., deltoid —, 613
m., depressor —, 1315
m.s, dorsal —, 605
m. of the eye, 589
m., extensor —, 564
m.s, facial —, 588
m.s, femoral —, 617
m., flexor —, 563
m.s, fore-arm —, 615
m.s of the hand, 616
m., hollow —, 558
m.s, involuntary —, 555
m. of the leg, 618
m., lingual —, 1436
m. of the mouth, 592
m.s, pelvic —, 610
m.s, peroneal —, 619
m.s, respiratory —, 1004
m.s, scapular —, 612
m.s, thoracic , 600
m.s, trunk , 595
m.s, vascular — 611
m. of the ventricles of the heart, 652
m.s, voluntary —, 554
muscae volitantes, 2551
muscular, 516-17; 533
m. band, 521
m. envelope, 530
m. exercise, 4175
m. fibre, 519
m. fibrin, 550-1
m. jerking, 2474
m. layer, 527
m. ligament, 522
m. ligaments, sinewy —, 585
m. membrane, 520-1
m. motion, 524

muscular power, 531
m. rigidity, 2770
m. stratum, 527
m. strength, 531 ; 534
m. structure, 535
musk, 4459
must, 5315
mustard poultice, 4253
mutilate, v., 3431
mutilating, 3432
mutter, v., 5399
mutton, 3958
mutton-chop, 3973-4
myelalgia, 1802
myocarditis, 2861-2
myodesopsia, 2551
myodynia, 1774
myopia, 3446
myrrh, 4460

# N

nævus, 3212
nævus vasculosus, 3213
nail, 951
n., body of the —, 952
n., root of the —, 953
nails growing into the flesh, 3563
naked, 5342
name, 4899
nape, 358-9
nape of the neck, 358-9
naphtha, 4389
napkin, 4668
narcosis, 2587
narcotic, 4200 ; 4206
narcotize, v., 4205
nares, 279
narrow, 998 ; 4178 ; 5299
narrow, v., 5133
narrow, to make —, 1000
nasal bone, 1423
n. bones, 273
n. cartilage, 290
n. cavity, 1008
n. mucus, 1427
n. muscles, 590
n. polypus, 2801
n. roof, 277
n. syringe, 3747
nature of, of the —, 71
natural parts of men, 1183
natural parts of women, 1184
nausea, 2188 ; 2947 ; 5140
navel string, 1292
near it, 5548
nearly, 4966
nearness, 5598

near-sighted, to be —, 1493
nebula, 5430
necessary, 4787 ; 5188
neck, 310-11
neckerchief, 4697
necktie, 4696
necrosis, 3396
necrotic, 3401
need, 1289 ; 4951
need, v., 4788 ; 5357
needful, 4787
needle, 3661
needle probe, 3667
negligence, 5269
neighborhood, 5597-8
neither—nor, 5010
neoplasm, 3215
nephralgia, 1839
nephritis, 3026
nerve, 779
n., accessory —, 875
n., auditory —, 872; 1416
n. -bundle, 787
n. cell, 823
n., cervical —, 888
n., coccygeal —, 901
n., cochlear —, 1417
n. corpuscle, 802
n., cranial cerebral —, 809-10
n., crural —, 897
n.s, dorsal —, 891
n. envelope, 792
n., facial —, 871
n. fibre, 786
n., gustatory —, 1443
n. irritation, 826
n. layer, 788
n. -like, 806
n., lingual —, 1437
n., lumbar —, 894
n. of the lumbar vertebra, 895
n. medulla, 789
n., olfactory —, 864 ; 1430
n., optic —, 865 ; 5444
n., pathetic —, 868
n. of the pudic plexus, 900
n. plexus, 785
n. pulp, 789
n. root, 798
n., sacral —, 898
n., sciatic —, 899
n.s of sensation, 835
n., spinal —, 812
n. substance, 780-1
n., sympathetic —, 908
n. termination, 797
n.s, thoracic —, 892
n. -tissue, 782

**nerve,** trochlearis —, 868
n. trunk, 793
n. tubercle, 814 15
n., vestibular —, 1418
n., limited to one —, 1730
nerveless, 805
**nervous,** 584; 803-4; 1541
n. distribution, 796
n. fluid, 790
n. pain, 1775-6
n. papilla, 799-800
n. plexus, 783
**nervus** abducens, 869
n. glossopharyngeus, 873
n. hypoglossus, 876
n. motor ocule, 867
n. pneumogastricus, 874
n. trigeminus, 870
net, 784
nettle, 3165
nettle-rash, 3166-7
**neuralgia,** 1775-6
n., facial —. 2780
n. facialis, 1784
n. of the kidney, 3026
neuralgic, 1656
neurasthenic, 1541
nevertheless, 4855
new-born, 1283
newly, 5481
nicotine, 4528
**night,** 4982
n. -cap, 4712-14; 4736
n. -dress, 4700
n. -gown, 4715
n. jacket, 4711; 4732
n. -lamp, 4605
n. -shade, 4515
n. -shade, deadly —, 4517
n. -shirt, 4710
n. sweat, 1977
nightmare, 2306-7
night's rest, 5018
nippers, 3707
nipping, 1717
nipple, 1234
nipple glass, 3746
n., artificial —, 5511
nit (a), 3531
nitrate of silver, 4315-16
nitric acid, 4301
nixus, 609
nod, v., 597
node, 755
nodule, 1904-6
nævus flammeus, 3213
noise, 655-6
noiseless, 2046
noisy, 2035
noma, 2887
nook, 192

noose, 3625
normal standard, 2109
nose, 272
n., blow one's —, 1490
n., bridge of the —, 289; 1445
n., Burgundy —, 3190-1
n., root of the —, 1424
n., tip of the —, 275
n., wing of the —, 1426
nostril, 279
not at all, 4791
notch, v., 2489
notice, v., 4939-40
nourish, v., 748; 1056
nourishing, 1058
nourishment, 1057
now, 5121
noways, 4878
noxious, 5165
nubile, 1241
nubility, 1244
nude, 5342
numb, 1851; 1869
numbness, 1852; 2453-4
nummular, 2098
nurse, 1298; 3846-8
nurse, v., 1295
nursing bottle, 1301-2
nut, 4052
nutmeg, 4089
nutriment, 1059
nutrition, 749
nutritious fluid, 750
nutritive matter, 1060
nymphæ, 1227
nympho-mania, 3235-6

## O

oakum, 5354
oat meal, 4482
oats, 4180
obedience, 5272
obese, 2394
obesity, 2395
object, 4991
oblige one, to —, 5591
oblique, 3081
observe, v., 4811; 4939
obstetrician, 3327-29
obstinate, 1631; 1634
obtuse, 2519
occasion, 5032
occasion, v., 4943
occasional, 5474-5
occiput, 223
occupy, v., 5030
occur, v., 5212
oculist, 3324
odd, 1256

odious, 5523
odontagra, 3489
odontalgia, 1791-2
odontitis, 2900
odor, 1420
œlema, 2420
œdema of the glottis, 2787-8
œdematous, 1873
œsophagitis, 2923-4
œsophagus, 1065
œ., degeneration of the —, 2925
œ., paralysis of the —, 2927
œ., stricture of the —, 2928
offense, 5014
offensive, 2045; 2271
offensive smell, 2093
offer, v., 5501; 5600
office, 3311
officinal herbs, 3858
often, 4880
oil, v., 213; 5305
**oil,** 985
o., anise-seed —, 4400
o. of almonds, 4399
o. of amber, 4395
o. of bergamot, 4401
o., camphorated —, 4405
o. of cassia, 4408
o., castor —, 4391
o. of cloves, 4407
o., cod-liver —, 4373
o., croton —, 4409
o. of curled mint, 4383
o. of juniper, 4376
o. of lavender, 4378
o., linseed —, 4380
o. of marjoram, 4381-2
o. of mustard, 4393
o., neats-foot —, 4374
o. of nutmeg, 4384
o., poppy —, 4388
o., resin —, 4300
o. of rosemary, 4392
o., sweet —, 4385
o. of sweet flag, 4403
o. of thyme, 4410
o., train —, 4372
o. of turpentine, 4397-8
o. of valerian, 4411-12
o. of vitriol, 4302
o., olive —, 4385-86
oiled silk, 3636
ointment, 214; 4267
old, 4901
olecranon, 423
**olfactory** cell, 1431
o. membrane, 1429
o. organ, 1422

omalgia, 1798
omelet, 4015
once, 4920
one another, 3392
one, to make —, 137
one-sided, 1729
only, 5273
on no account, 5109
onward, 2359
ooze, v., 1921
open, v., 4756; 4780; 5455-6
open air, 5346
open one's legs, to —, 4781
operate, v., 587
operate on, to —, 5337
ophthalmalgia, 1787-8
opiate, 4288
opinion, 4838
opium, 4529
opponents, 583
oppress, v., 5185
oppression, 2639; 5139; 5144
oppressive, 1690
orange, 4050-1
orange leaves, 4334
order, 3273; 4187
order, v., 4186
orderly, 5489
organ of hearing, 1384
o. of perception, 833
o. of sight, 1310
o. of smell, 1419
o. of speech, 1461
o. of taste, 1432
o. of vision, 1308-9
orgeat, 4371
orifice, 633
originate, v., 5291
oscillating, 2537
os coxæ, 383-4; 453
os pubis, 461
os sacrum, 466-8
osseous, 65; 69
o. envelope, 130
o. shell, 130
o. substance, 83
o. wall, 128
ossicle, 62
ossification, 90
ossify, v., 64
osteocopus, 1771-3
osteo-genesis, 90
osteoid, 69; 72
osteotome, 36-9
otalgia, 1789-90
other, 4917; 5307
otitis, 3477
otolith, 1413-15
ounce, 5113
outcry, raise an —, 5519

outer, 922
outlet, 5330
outside, 1763
outwards, 572
ovarian tumors, 3024-5
ovarium, 1211
ovary, 1211
oven, 4652
over, 594
overcoat, 4682
overcrowd, v., 5284
over-eat one's self, 3882
over-fatigue, v., 5379
overshoe, 4703
overspread, v., 2118
overtake, v., 3788
over-tire one's self, 5379
overweariness, 5016
overwork one's self, 4948
oviduct, 1222-3
owe, v., 5579
oxalates, 2293
oxyuris vermicularis, 3548
oyster, 4011
ozœna, 2802

## P

pachydermia, 3220
pad (little), 3339
p. of straw, 3580
pail, 4655
pain, v., 1669-70
pain, n., 1285; 1547; 1608
pains, 1286
painful, 1679-81; 1284
p. sensation, 1538
painfulness, 1684
pain in bones, 1771-3
p. in the bowels, 1811-12; 1828-29
p. in the buttock, 1847
p. in the chest, 1803-7
p. in the eyes, 1787-8
p. of the foot, 1857
p.s of the gout, 1835
p. in the head, 1778-9
p. in the hip-joint, 1837
p. in the left hypochondrium, 1817
p. in the knee, 1853
p. in the loins, 1835
p. in the neck, 1793; 1797
p. in the stomach, 1819-20
p. in the thigh, 1850
p. in the throat, 1793
p., allaying —, 1686
p., assuaging —, 1686
p., paroxysmal —, 1740
p., to cause —, 1305; 1669

p., to feel —, 1669; 1676
p., to give —, 1287
p., to give severe —, 2950
p., seat of the —, 1751
palate, 264
palate sail, 1452
palatine arch, 1453
p. swelling, 2905
p. tumor, 2905
pale, 1591; 1881
p. as death, 1592
paleness, 1593
palliative, 4212
pallid, 2494
pallor, 1593
palm, 435
palpebral conjunctiva, 1325
palpitate, v., 741
palsy, v., 2733
palsy, n., 2463
p. occurring on both sides, 2739
p., unilateral —, 2738
pan, 112; 586
panacea, 4284
pancake, 4016
pancreas, 1083
pancreatic juice, 1082
p. tumor, 3014
pane (window), 4631
pang, 1285; 1608
pant, v., 2071
pantaloons, 4684-5
pantry, 4650
pap, 1085; 4260
papilla, 2146
pappy, 5276
papula, 1904-6
papule, 1904-6
paralysis, 2734
p., general —, 2735
p., partial —, 2736
paralytic disease, 2742
paralyze, v., 2733; 5453
paraplegia, 2739-41
paraphimosis, 3252
parasites, 3514-15
parasites, animal —, 3517
parasitic animal, 3518
p. fungus, 3521
p. plant, 3520
pardon, v., 5582
parents, 4914
paronychia, 3565-66
parrot-beak, 3700
parsley seed, 4445
part, 704
partial, 4995
particular, 4984
particularly, 4799
partition wall, 284

partition-wall, transverse —, 626
partly, 4854
partridge, 3986
parulis, 3486-7
pass (over), v., 5408
pass sleeping, 4857
pass through, 5454
passing, 4990
past remedy, 1613
paste, 4254
pastile, 4278
pastry, 3950-1; 4021
patch, 1936
patent lint, 3695-6
path, 724
patience, 5063
patient, n.,
paunch, 2397
p. belly, 2397
pay, v., 5578
p. a fee, 5581
peach, 4049
peak, 274
pear, 4045
pearly, 2535
peas, 4029
p., green —, 4030
**pectoral**, 4226
p. drops, 4324-5
p. powder, 4324-5
p. syrup, 4326-7
peculiar, 5089; 5258
p. nature, 70
pediculus pubis, 3529
p. vestimenti, 3528
peel, v., 1935
peep, v., 3914
peevish, 2503; 2632; 5522; 5526
peevishness of temper, 2630
pelvis, 449
pemphigus, 3182
penetrate, v., 5210
penis, 1185-7; 5350
pepper, 4065
peppermint, 4338
peptic, 2937
p. ulcer, 2937
perceive, v., 4863; 4939
perception of warmth, 5127
perceptible, scarcely —, 1997
perceptibly, 4962
perch, 3999
perfect, 2602; 5242
perforate, v., 1485; 5467
perforation, 2964
perforator, 3726
perform, v., 4811; 5271

p. gymnastic exercises, 4169
perfume cushion, 4571
perhaps, 4958
pericarditis, 2860
pericardium, 638-42
perichondrium, 160
perihepatitis, 2996
perilous, 1615
period, 3262-3
peripheral, 1662
periosteum, 111
peritonitis, 2967-8
permit, v., 5349
pernicious, 3069; 5165
pernio, 3507 8
perone, 500
persistent, 1727; 4980
perspiration, 937
perspiration, profuse —, 1976
perspire, v., 1963; 1965
perspiring, 1964; 1979-80
purulent, 5496
Peruvian bark, 4358-9
perversion of the sense of vision, 2546
perverted, 2158
pest, 3126
pestilence, 3126
petroleum, 4389
petticoat, 4722
phagedenic, 3241
phalanx, 441
phantasms of the sense of smelling, 2570
ph. of the sense of taste, 2571
**pharyngeal** catarrh, 2904
ph. catheter, 3671
ph. cavity, 1063-4
ph. mirror, 3754
ph. paralysis, 2918
ph. phthisis, 2920
ph. spasm, 2917
ph. speculum, 3754
pharyngeotome, 3722
pharyngitis, 2915 16
pharyngo-oral cavity, 1097
pharyngoplegia, 2918
pharyngoscope, 3756
pharynx, 1061-2; 1066
pheasant, 3987
phimosis, 3249
phlegm, 543
phlegmon, 2714
phonopathia, 2576
phosphorus, 2287
photopsia, 2551
phthisical, 2824

phthisis pulmonalis, 2823-4
physician, 3805; 3816
pia mater, 860
pica, 1257
pick, v., 5542
picker, 3648
pickle, v., 3977
pickled cucumbers, 4042
picked lint, 3698
pickles, 4026
pie, 4021
piece, 5117
pieces of ice, 5237
p. of fæcal matter, 5282
pierce, v., 3321
pierce, 3648
pigeon, 3988
pigment, 941
pigmented liver, 2997
pike, 4000
piles, 2991; 5287
pill, 4258
pillow, 4569
p. -case, 4560
pimple, 1923; 3184-5
pimpled, 3186
pincers, 3707
pincette, 3708
pinch, v., 5185; 5278
pinched, 2498
pinching, 1717
pine away, to —, 1569
pine-apple, 4055
pinhead, 5439
pin-worm, 3550
pipe, 96
pit, 338
pitcher, 4656
pityriasis, 1931; 3197-9; 3200
place, v., 30
place, n., 4967; 5087
plague, 3126
plain, 178
p. fare, 2931
p. surface, 179
plank, 3578
**plaster**, 3692; 3701
p., cantharides —, 4429-30
p., compound litharge —, 4433
p., court —, 3705
p., drawing —, 4427-8
p., healing —, 3704
p., lead —, 4424
p., mercurial —, 4431
p., mustard —, 4252; 4255; 4432
p., pitch —, 4425
p., sticking —, 3703; 4424
plate, 947; 4669

plate, corneous —, 948
pleasant, 1600
please, 4971; 5106
p., as you —, 5416
p., if you —, 5106
pleased, to be —, 5415
pleasure, 5416
p., at —, 5416
pledget, 163
plenty, 2256
pleura, 1030
pleurisy, acute —, 2842
pleurodynia, 1804
plexus, lumbar —, 896
p. of veins, 682
plica, 963
plowshare, 288
pluck, v., 3697
plug, v., 5554
plug, n., 3358
plum, 4046
pneumonia, 2826-7
p., acute —, 2829
pneumothorax, 2843
pocket, 918; 1480
pocket handkerchief, 5158
pock-mark, 3141-2
pock-marked, 3143-4
pock-pit, 3141-2
pock-pitted, 3143-4
podagra, 3083
point, 274
p. of, in —, 5484
point out, to —, 5364
p., main —, 5108
pointed, 2497
poison, v., 3345; 4491
**poison**, n., 3494; 4490
p., cadaveric —, 4536
p., pestilential —, 4535
p., sausage, 3501
p., septic —, 4536
p., void of —, 4493
poisoned flour, 4521
poisoning, 3505; 4537
p. of blood, 4538
**poisonous**, 4492
p. cresses, 4513
p. minerals, 4519
p. matter, 4494; 4525
p. mushrooms, 4509-10
p. vapors, 4531
poke, v., 5542
police, 3814
polypus, 2797
p. of the pharynx, 2919
polyuria, 2831
pomum Adami, 1463
pomphus, 1908
pons, 845
poorly, 1536

poppy (plant), 4387
p. heads, 4463
p. seed, 4464
pork, 3959
porridge, 4260
porrigo larvalis, 3174
port, 4093
porta, 775
portable bed, 3573
p. chair, 3576
portal blood, 777
p. vein, 776
p. v., inflammation of the —, 3006
position, 1755; 4970
possible, 4883
posture, 1755; 4970
pot, 4657
potato, 4040
potent, 1175
potion, 4279-80; 4845
poultice, 4256; 4283
poultry, 3982
pour, v., 5157
powder, 4259
power, 2643
powers of locomotion, 2360
practical, 3810
practice, v., 3809
p. magic, to —, 4140
p. witchcraft, to —, 4140
practising, 3810
p. physician, 3810
practitioner, 3810
pray, 4803; 4971
precaution, 5263
precede, v., 5009; 5025; 5128
pregnancy, 1252
pregnant, 1249
prefer, v., 4969
preference, in —, 5277
premature, 1272
p. birth, 3294
preparation, 5589
preparatory treatment, 4135
prepare, v., 75; 5234; 5592
prepared for, 5504
p. sponge, 4469
prepuce, 1189
p., coarctation of the —, 3249
presbyopic, 1494
prescribe, v., 4186; 4808
prescription, 4187
present, v., 5257
present, 2323; 4926; 4972
present, for the —, 4926
preservative, 4209-11

preserve, 4025
preserve, v., 5181
p. from, to —, 4207
press, 4637
press, v., 608; 2246; 2274
p. forward, 5303
p. together, 5564
pressing, 1690
pressure, 2245
pressure, perception of —, 1749
presume, v., 5090
pretend, v., 1621
pretended, 1620
pretty, 5508
prevent, v., 4208; 5059; 5573
previously, 4919
prick, v., 3321
primary, 1617
principal, 5244
p. lobe, 1026
p. thing, 5108
p. trunk, 906
principally, 4977
principle of life, 58
privy, 4597-8
p. councillor, 3839
p. parts, 460
probable, 5445
probang, 3723-4
probe, v., 3386
p. a wound, to —, 3313
probe, 3665; 3675
proceed, v., 1120
process, 1121
p. of digestion, 1052
procreative power, 1179
proctagra, 1849
proctalgia, 1848
procure, v., 5250; 5603
produce, v., 18; 1910
profession, 4908
project, v., 2145; 5075
projection, 488
prolapse of the womb, 3283-4
prolapsus linguæ, 2913
proliferate, v., 3371
proliferation, 3372
prolific power, 1179
prominent, 2473
p., to be —, 2145
promise, v., 4870
promotor, 573-5
prong, 133
proper, 3896; 5163; 5450; 5452
p., to be —, 5451
proportion, 5401
prostata, 1202
prostration, 1584; 2448

protect, v., 3148
protrude, v., 3472; 5303
protuberance, 413; 3389
p., articular —, 414
proud flesh, 3235
provided that, 5360
provisions, 3919
provisor, 3852
pruriency, 1954
pruriform, 5496
pruritus, 1953
p. cutaneus, 1952
prussic acid, 4522
psoriasis, 3203; 3524
p. nummularis, 3204
ptisan, 4263-4
puberty, 1169
pubic region, 1204
p. symphysis, 462
pudenda, 460
p., lip of the —, 1208
pudendal fissure, 1209
puerperal fever, 3297
p. peritonitis, 2969
puffed up, to be —, 2406
puffiness, 2407-8
puffy, 1872
pull, v., 174; 3341
p. out, to —, 3683
pulling, 1699
pulmo, 732
pulmonalgia, 1830
**pulmonary** abscess, 2832
p. apoplexy, 2833
p. atrophy, 2838
p. cells, 1044
p. circulation, 733-4
p. emphysema, 2809
p. gangrene, 2825
p. lobe, 1013
p. œdema, 2830
p. plexus, 1037
p. tissue, 1036
p. ventricle, 1040
p. vesicle, 1042
p. vessel, 1038
p. ulcer, 2841
pulp, 1085; 4260
pulsate, v., 664
p. violently, 2728
pulsating, 1692
pulsation, 666; 3022
pulse, 663
p. of the heart, 738
pulse-beat, 666
punch, 4102
punctured cataract, 3458
p. wound, 3322
pungent, 1694
pupil, 1358-60
purgative, 4194-6

purgative, effervescent —, 4365
purge, v., 4191-3
purging, 3255
purpose, 5514
purpura, 3075
purulent, 2096
pus, 1922
push, v., 736
p. in, to —, 2976-7
p. off, to —, 2144
p. out, to —, 2484
pustular, 1915
pustule, 1912-4; 1923
put, v., 30
p. clothes on, 4742
p. other clothes on, 4743
p. far off, 3366
p. in a seton, 4249
p. out a lamp, 4616
p. out of humor, 2624
p. out of tune, 2624
putrefy, v., 2217
putrescent, 2205
putrid, 2205; 2199
pyæmia, 3072-3
pyelitis, 3036
pyloric valve, 1031
pylorus, 1077-9
pyroligneous vinegar, 4294
pyrosis, 2207-8
pyuria, 3030

## Q

qualified, 3835
quality, 15
qualm, 2191; 2947
qualmish, 2185
qu., to feel —, 2187
quantity, 82; 2256
quart, 5605
quassia, 4451-2
quench the thirst, 3303-4
quick, 1988
quickness, 5214
quicksilver, 4130
quiet, 1654; 2345
quilt, v., 4573
quilt, n., 4574
quinine, 4357-9
quinsy, 2785; 2921-2
quiver, v., 1777
quite, 4792

## R

**race**, 2; 4
r. of men, 5

**rachidian** cone, 878
r. filament, 879
r. liquor, 882
r. meninge or membrane, 881
racking, 1716
radiating on all sides, 1732
radical, 5293
radish, 4485
rage, v., 2691; 2696
rage, n., 1123; 3495
rain bow, 1355
raise, v., 1312; 1756; 2373; 4771; 5341
raise one's self, 2374
r. an outcry, 5519
raised, 1916
rale, 2048; 2051
rales, fine —, 2052
r., sibilant —, 2054
ramble, v., 5388
ramification, 1018
ramify, v., 1016-17
rapid, 1628-9; 1989
r. decline, 2823
rare, 1651
rash, 1900-1
raspatory, 8657-8; 3690
raspberry syrup, 4074-5
r vinegar, 4295
rather, 5277
rattle in the throat, 2041
rattle-snake, 4500
rave, v., 2614-5; 2691; 2696
ravenous, 2160
r. desire for food, 2161
ray (of light), 1347
ring rays, 1351
reach, v., 2119; 5096; 5362
r. into, 5210
react, v., 570
read, v., 5419
ready for, 5504
r. to fall down, 1584
really, 4835
reappear, v., 5146
reason, for that —, 5596
receive, v., 322
receptacle for keeping anything, 1012
recipe, 4189-90
recognize, v., 2422
recollect, v., 4933
recommend, v., 5247
recover, v., 1516; 4107
r. from illness, 1576
recovery, 1562; 4105; 4106; 4108
recreate, v., 4158
recreation, 4159

recruit, v., 4107
rectal flexus, 1130
rectum, 1129; 1135
recumbent, 1754
r., to be —, 2192
recur, v., 5001
red, 712
redden, v., 4241
redness, 1764
reduce, v., 3419-20
reduction, 3417-18; 5338
r. of a dislocation, 3421-2
reek, n., 2318
reel, v., 2336
reeling, 2455
reference, 5485
reflex, 1663; 1733
r. irradiation, 838-9
r. sensations, 840
refresh, v., 4158; 4224
refreshment, 4159
refrigerant, 4218; 4225
refrigerate, v., 2224; 4217
regard, 5178
r. to, with —, 5485
regarding, 5486
regimen, 3930
region, 388
regular, 2006
regulate, v., 4176; 5539
rein, 379
reins, 464
rejoice, v., 4922
relapse, v., 4109
relapse, 1524; 4110
relapsing, 1619
relation, 5178
relax, v., 2380
relaxation, 2243; 2263; 5389
relief, 4864; 5195
relieve, v., 1753
remain, 4812; 5243
remark, v., 4940
remedy, v., 5412
**remedy**, n., 4167-8; 4185; 4277
r., old woman's —, 4188
r., prophylactic —, 4231
r., refreshing, 4160-61
r., strengthening, 4164-5
remember, v., 4933
remove, v., 3366; 5417; 5446
r. a disease, 1563
renal calculus, 3038
r. calix, 1150
r. conductor, 1152
r. dropsy, 3063
r. pyramid, 1148
r. radix, 1147
rend, v., 5376

render a service, 5553
renew, v., 5194
repast, 3883
repeat, v., 5194; 5201
replace, v., 5304
repugnance, 2163
require, v., 5357
requisite, 5163
resemble, v., 2257
resembling, 68
r. rice water, 2254
reservoir, 1012
reset, v., 3419-20
resetting, 3421-2
reside, v., 4902
resin oil, 4390
resolution, 3352
r. of a tumor, 3222
respect, 5178; 5485
**respiration**, 985; 992; 2012; 4151
r., costal —, 2022
r., difficulty in —, 2032
r., nasal —, 2020
r., oral —, 2019
r. through the mouth, 2019
r. through the nose, 2020
r. through a tube, 2021
respiratory, 1046
r. process, 993
rest one's self, to —, 2377
rest, 5067
restless, 2346
restlessness, intense —, 2348
restoration, 4108
restore, v., 4107
r. (sick persons), to —, 1561
restored to health, to be —, 1517; 4889
restrict one's self, 4881-2
retain, v., 1011; 5011; 5200
retard, v., 5482
retch, v., 2184
retching, 2191
retina, 791; 1368
r., inflammation of the choroid and of the —, 3433
retract, v., 2338
return, v., 5001; 5073; 5146
revolve, v., 195; 719
r., cause to —, 196
rhachialgia, 1801
rhagades, 1930
**rheumatism** (acute and chronic), 3076
rh., cerebral —, 3077
rh., muscular —, 3079
rhinorrhagia, 2800

rhinoscope, 3757
rhonchus, 2048; 2051
rhubarb, 4461
rib, 87
ribs, false —, 391
r., true —, 390
rice, 4041
rice-water stool, 2255
rid of, to —, 5358
r. one's self of, 5175
ridge, 347
rift, 1473
right, 428
r., not —, 3295
rigid, 219; 2456
r. joint, 220
rigor, 2114; 5184
rigors, 3118
rim, 131
r. of the bone, 132
rima glottidis, 1473
r. pudendi, 1209
rind, 964
ring forth, to —, 654
**ring**, 5429
r. knife, 3654
r. rays, 1351
r. worm, 3216; 3235
rinse out, to —, 5563
ripe, 1167
ripeness, 3224
rise, v., 2111; 2374
rise from, 4997
r. out of, 4997
r. up, 2201
risus, 2750
river, 3269
rivet, v., 5529
road, 724
roar, v., 2562
roast, v., 3966
r. (coffee), 3940
**roast**, 3967
r. -beef, 3968
r. meat, 3967
r. veal, 3969
roasted duck, 3994
rob, v., 2378
rock, v., 4552
rock, 235
rocking-chair, 4553
roll, v., 195; 1316; 2229
roll, 3946
roller, 5204
roller, articular —, 415
roof, 276
room, 4542-3
r., to keep one's —, 1575
root, 312
rope, 5020
rose, 3158
rose rash, 3133

rosemary, 4339
roseola, 3131; 3134
rotate, v., 195; 363
rotator, 364; 1317
rotten, 2199; 5558
round, 1469
r. -worms, 3544
roundish, 1470
rouse, v., 5405
rub, v., 4848
rubber bag, 5394
rubification, 4242
rue, common —, 4342
rule, as a general —, 5239
rum, 4001
rumble, v., 2229
rump, 472-3
run, v., 210; 721; 3387-8
r. into, 2688
r. out, 5073
r. over, 2118
r. under (the skin), 3337
rupia, 3194-5
rushing, 1708
rust-colored, 2088
rustle, v., 653
rustling, 655

## S

sac, 544
sack, 918
sacred, 465
sacrum, the —, 466-8
safe, 4650
s. and sound, 1510
s. and well, 1510
saffron, 4475
salep, 4473
saliva, 981
salivary glands, 982
salivation, 4134; 5061
salmon, 4004
salop, 4473
salt, v., 3977
salt, n., 2284
s., English —, 4417
s. of hartshorn, 4414
salve, 4265
sand, 2297; 2299
s. -heat, 3794
sandy, 2298
sanitary establishment, 3808
s. police, 3815
sandwich, 3949
sap, 1595
sarcous fibre, 539
sardonic laugh, 2751
sarsaparilla, 4360

sash, 4738
sassafras, 4453-4
satisfy, v., 3889
s. one's self, 3893
s. one's hunger, 3893
sauce, 3970
saucer, 4675
sausage, 3979
s. poison, 3501
saw, 3650
say, v., 4837; 4866
scab, 1937-8; 3175
scabbed, 1940
scabies, 1944
scabious, 1943
scaffold, 76
scald, n., 3382
s. one's self, 3383
scale, 1931
scalpel, 3641; 3660
scaly, 3205
scanty, 2325; 5186; 5299
scapula, 401-2
scar, 1928; 3139-40
scarificator, 3639; 3763
scarlatina, 3127-9
scent, 1420
s. -bag, 4571
science, 3802
scirrhophthalmia, 3441
scissors, 3643
sclerotic, 1339-40
scoop, v., 987
scour, v., 5575
scrape, v., 1925; 1942; 3699
scraped lint, 3700
scraper, 3691
scratch, v., 1925; 1942; 1957
scratch, n., 1926; 1956-7
s., light —, 1928
scream, v., 2482
scream, n., 2483
screen, 4581; 4622
scrotum, 1192
scum, 2310
scurf, 1926; 1933; 1939; 3175
s. like, 1940
scurfy, 1940
scurvy, 3074
sea-sickness, 2945
search out, to —, 5359
season, 5170
seat one's self, to —, 5547
seat of disease, 1564
seat of the pain, 1751
sebum, 935; 2396
secondary disease, 1523
secrete, v., 974; 2275
secundines, 1293

sediment, 2291; 2296; 5319
see, v., 1307
s. to, 5136
seeing but half of an object, 2559
seek, v., 5110
Seidlitz's powder, 4364
seize, v., 5369
seized, to be —, 5370; 5545
seizure, 2212
select, v., 5343
semen, 1194
seminal vesicle, 1201
send for, 4790
senna leaves, 4341
sensation, 1500; 2644; 2658
s., impairment of —, 2657
sense, 833
s. of pressure, 1749
s. of hearing, 5092
s. of seeing, 241
s. of smelling, 1421
s. of smell, destitute of the —, 1499
s. of touch, 2573
s. of warmth, 5127
senseless, 2452; 2675
sensible, 829
s. of, to be —, 828
sensibility, 830-1
sensitive, 829
s. fibre, 832
sensitiveness in the pit of the stomach, 1822
sensual, 1245
sensuality, 1246
separate, v., 283
septum, 284
s. narium, 285
serious, 1612
serous, 708-9
s. membrane, 917
serpent, 1111
serpentine, 1111
serrated, 134
serration, 133; 136
scrum, 706-7
serve, v., 4784
service, 5552
set apart, to —, 2275
s. at rest, 3889
s. to, 5568
seton, 4248
settle, n., 3575
severe, 1610
s. chill, 5184
severity, 5147
sew, v., 107
sex, 2
sexual, 5012
s. instinct, 1166

s. organ, 1161
shade, 4619
shaft, 961
shake, v., 2724; 3117; 5253
s. with the ague, 3097
shaking, 5327
shaky, 2368
shame, 460
shape, v., 27; 35
shape, n., 28
shape, to give —, 24
share of a plow, 287
sharp, 1689
sharpened, 2498
shatter, v., 2671
shave off, to —, 5393
shear, v., 286; 5393
shears, 3643
sheath, 529-30; 546; 792
shield, 1033
shin, 495
shingles, 3178
shirt, 4708
shiver, v., 2116-17
s. with cold, 3099
shiver, n., 3414
shivering, 2113
shock, 737
s., violent —, 5327
shoe, 4698
shoe lace, 4699; 4700
shoestring, 4699; 4700
shoot, v., 3125
shooting up and down, 736
short, 1991
shortsighted, to be —, 1493
shoulder, 399; 400
s.-blade, 401-2
show, v., 443; 2347; 5182; 5364
s. forth, 5257
s. one's self, 5321
shriek, v., 2482
shriek, n., 2483
shriek, to give a —, 5519
shrine, 4637
shrink, v., 2468
shrinking, 2469
shrivel, v., 2468
shuddering, 2113
shun, v., 5524
shut, v., 405
sibilant rales, 2054
sick, 1519; 1567
sick, fatally —, 1532
sick, to fall —, 1529
s., to feel —, 1677; 2184
s., to grow —, 1529
sick-bed, 1571-2
sick-nurse, 3846-8

sick-room, 4545; 4547
sickle needle, 3663
sickliness, 1526
sickly, 1525; 1533; 1544; 1567
s., to be —, 1568
sickly look, 1590
sickness, 1521; 1570
side, 2342
siesta, 4204
sieve, 237
sigh, v., 2060
sight, 241
sight impaired, the —, 2527
sight or eyesight, to have a good —, 1492
sign, n., 5024
signify, v., 4868
s., it does not —, 4869
silent, to be —, 5328
silly, 2522
similar, 68; 4850
simple, 1652
simply, 5273
sinapism, 4252
sinew, 208; 541
sinewy, 584; 803
single, 4910
sink, v., 2464; 5022
s. in, to —, 2465
s., cause to —, 3282
sinking, 1994
sip, v., 5260
sir, 4833
sit, v., 456
s. upright, to —, 4776
situated above, 394
size, 41; 2387; 2389; 2427
skeleton, 77; 88
skillful, 3835
skim (over), v., 5408
skin, v., 3380
skin, n., 110; 601
s., external —, 923
s., hard —, 949
s., burning sensation of the —, 1960-61
s.-color, 244-5
skinned, 2792
skirt, 4721
skull, 586
s., injury of the —, 5026
s.-saw, 3653
slack, 1861
slash, 3319
sleep, v., 2476
sleep, n., 2477
s. all night long, 4857
sleeplessness, 2589; 2500
sleeve, 4689
sleigh, 3568

slight, 1609; 1687
slime, 543
slimy, 5160
sling, 3617-21; 3626-7
slippers, 4716
slow, 1630; 1632; 1637; 1992
small, 1988; 5299
small-pox, 3135-6
smart, 1670
smarting, 1679-80
smear, v., 213
smell, v., 240
smell, n., 1420
s., bad —, 2043
s., without —, 5253
smoke, v., 4956
smoke, n., 2318
smoked beef, 3976
smoky, 2316
smooth, 180; 1757
s. surface, 181
smother, v., 2074
snake, 1110
snake-formed, 1111
sneeze, v., 2061
sneezing, frequent —, 2062
snore, v., 2036
snuff, v., 5606
snuffle, v., 1488-9
so-so, 5507
soak, v., 5205
soap, 4465
s., hard —, 4467
soapy-looking, 2315
sob, v., 2047
socks, 4705
soda, 4363
soda-water, 4072
sofa, 4663
soft, 385; 1757; 1860; 2485
soften, v., 1753; 4219; 5274
softening, 2717
s. of the urinary bladder, 3043
soil, 5344
sojourn, v., 5378
sojourn, n., 5168
solar plexus, 910
sole (of the foot), the —, 513
solicitude, 5565
solid, 2354
some, 4938
somnambulism, 2610-12
somnolence, 2582-5
sonorous, 2050
soon as, as —, 5322
sooner than, 5277
sooty, 2140
soporific, 4200-1

soporous state, 2588
sore, 1284; 3302
sorrow, 4953
sorry for, to be —, 5588
sort, 70
soul, 57; 2627
sound, v., 654; 2579; 3386
sound, 656; 1507; 2055; 3365
s., thoroughly —, 1508
soundness, 4166
sounds, blowing —, 2053
soup, 3361
sour, 1982
Spaniard, 325
Spanish fly, 4319
spark, 59
s. of life, 60
spasm, 1765
s., clonic —, 2747
s., facial —, 2775
s., tonic —, 2746
s., uterine —, 3276
s. in the chest, 2815
s. of the lung, 2815
s. of the œsophagus, 2926
s. of the tongue, 2910
s. with rigidity, 2748
spasmodic, 1710-12
s. attacks, 2748
speak, v., 1459
s. through the nose, 1487
special, 3822; 5244
specialist, 3823
species, 2
specific, 2320; 4232
specify, v., 4968
speck, 3443; 5440
spectacles, 5464
speculum, 3751-2
s. auris, 3758-9
s. oris, 3758
s., vaginal —, 3760
speech, 1460
s., loss of —, 2574
spell, v., 4898
spell, n., 4141
spermatic cord, 1197
spermatozoon, 1196
sphacelus, 3398
sphere, 187
sphincters, 557
spider, 2549
spider's web, 858
spinal column, 344
s. cone, 878
s. filament, 879
s. ganglion, 883
s. liquor, 882
s. marrow, 352
spinal meninge or membrane, 881

s. nervous system, 813
spine, 346; 348
spirit, 57
s. of mustard, 4437
s. of nitre, 4439
s. of life, 58
spirits, 4097; 4955
s. of hartshorn, 4415-16
spit, v., 2080-1
s. blood, to —, 2085
s. out, to —, 2083-4
s. up, to —, 2181
spitting, 2079; 2086
spittle, 2082
spittoon, 4603
spleen, 1105
s., inflammation of the —, 3012
s., movable —, 3013
splenalgia, 1834
splenoid bones, 232
splinter, v., 3374
splinter, n., 3414
splinter forceps, 3709
splintering of bones, 3375
split, v., 849; 1929; 3374
spoil, v., 1823
spoil in bending, to —, 3420
spoke, 418
spondylalgia, 1805
sponge, v., 5308
sponge, n., 2486; 4590
s. tents, 4470
s., prepared —, 4469
s. -holder for cauterizing the pharynx, 3682
spongy, 2485; 3234
spoon, 4670
spot, 1399; 2557
sprain, v., 3427
spraining, 3423
spread, v., 795; 5039
s. out, 4780
s. over, 2118
spread, n., 4257
spring, v., 503
spring, warm —, 3763
spring-lancet, 3717; 3763
spring mattress, 4567
sprout, v., 3561
sprung, 2133
spurt out, 5142
sputum, 2079
squama, 1931
squamose, 3205
squamous dermatosis, 3202
square, 5431
squeamishness, 2188
squeeze, v., 603; 2274
squeeze in, 2975

squint, v., 3475
squinting, 2531; 3476
squirt, 3742
stage, 1668
staggering, 2367
stagnating, 2004
stain, v., 5221
stain. n., 1369
stand, v., 2349
s. out, 5075
standing still, 2001
stapes, 1411
starch, 4477
staring, 2534
starting up out of one's sleep, 2308-9
starve, 5313
starving system, 4145
state, v., 4968
state, n., 1513; 4905; 5222
s. of health, 1514
staunch, v., 3384
staunching of the blood, 3334
stay, v., 4812
s. away, 5483
s. out, 4976
stay (progress), to —, 5297
stays, 4727-9
steady, 2334
steadily, 2120
steam, 4531
stearine candle, 4611
steel, 4128
step, v., 675
step, n., 2353; 2341
step out, to —, 3472
stercoraceous matter, 2221
sterno-cleido mastoideus, 593
sternodynia, 1803
sternum, 367-9; 687
sternutatory, 4228
stew, v., 5411
stick, v., 3363; 5176
stiff, 2332
stiffness, 1759
stifle, v., 2074
still, v., 1296-7; 3384
stimulant, 4216; 4293
stimulate, v., 2231; 4215
sting, v., 3321
stitch, v., 4573
stitches in one's side, 1803
stocking, 4704
stomacace, 2885-6
stomach, 1037
st. -ache, 2211; 2930
st. of an ostrich, 2929

stomach, acidity of —, 2209
st., atrophy of the —, 2943
st., catarrh of the —. 2932
st., an empty —, 5228
st., frigidity of the —, 1826
st., fundus of —, 1070
st., induration of —, 2941
st., orifice of the —, 1077-9
st. out of order, 1824
st., overload with food one's —, 1825
st., pain in the —, 1819-20
st., pit of the —, 644
st., press on one's —, 2192
st. powder, 4322-3
st. pump, 3744-5
st., rising of the —, 2186-7
st., rupture of the —, 2944
st., softening of the —, 2940
st., a spoiled —, 1824
st., weakness of —, 1826
stomalgia, 1785
stomatoscope, 3753
stomatitis, 2888-9
**stone**, 5332
st. in the bladder, 3038
st. -blind, 3468
st. -colic, 1843
**stool**, 2224-5; 2244
st., to go to —, 2226
st., continuous desire to go to —, 2246
st.s, fatty —, 2268
st.s, incontinence of —, 2259
st., rice-water —, 2255
stop, v., 3384; 5436
stop (progress), 5297
stoppage, 5526
stopper, 3358
stopping, 2005
stove, 4621
stove-door, 4623
strabismus, 3476
strabismus needle, 3678
straight, 5463
straighten, v., 5133
strain, v., 3427
straining, 609; 1698; 4985
stramonium, 4516
strange, 1256; 2580; 3367
strangle, v., 1794
strangling, 1795
strangling sensation (a), 1766
strangury, 2330; 3059
stratum, 526
straw-bed, 3580

straw-colored, 1896
straw halm, 5262
streak, 1959; 2099
streak, bloody —, 2101
streaks, to form —, 2100
streaked with yellow, 2092
stream, 2334
stream forth, 1983
strength, 17; 1559
strength of the body, 18
strengthen, v., 1555; 1558
strengthening, 4162 3
stretch, v., 560; 5335
stretch out, 2337
stretch chair, 3582
stretcher, 3577; 3581
stretcher-bearer, 3577
strew over, 4257
strike, v., 736; 4947
s. down, 4368
s. up, 2765
stricture, 2928
string, 883
stripe, 1959; 2099
stripes, to form —, 2100
stroke, 4946
stroma, 1217
strong, 532; 1554
strong, only moderately —, 1994
structure, 11; 28; 73
str. of the body, 12;
strychnine, 4527
stubborn, 1634
stuff, 5398
stuffing, 3996
stumble, v., 2372
stupefactive, 4206
stupefy, v., 4205
stupid, 2617
stupidity, 2619
stupor, 2459-60, 2582; 2584-5
stye (on the eyelid), 3442
stylet, 3666
subcutaneous, 3736
subcutaneous injection (a), 4236
subject, v., 5367
subjected, to be —, 5368
subligation, 3613
subsequent disease, 1523
subside, v., 3115; 5298
subsist, v., 5031; 5392
substance, 80; 202
subsultory, 2001
succumb, v., 2447
suck, v., 751
suck one's mother, 1294
suck, to give —, 1295-7
suckle, v., 1295
sudden 1724

sudorific, 4227
sudriparous canal, 939
suffer, v., 5028
s. pain, 1548
suffering, 1549; 1551
sufficient, 5218; 5499
sufficient, to be —, 5361
sufficiently, 5189
suffocation, 1766; 2023
suffused with blood, 3338
sugar, 2294
s. of milk, 4462
suggest (a thought), to —, 4937
suggillation (a), 3338
suit, v., 4862; 5449
suitable, 5452; 5515
sulphate, 2290
sulphur, 2289
sulphurated water, 4314
sulphuric ether, 4291
summer rash, 3207
sun, 909
sun-stroke, 2766
sup, v., 5260
superficial, 722
s. wound, 3347
superficies, 1763
superfluity, 2275
superimpose (one's arms), 4778-9
supinator, 576-8
supper, 3927-9
supply the place of one, 5585
support, 5500
suppose, v., 5315
suppository, 4260
suppurate, v., 2409
suppuration, 2410; 5030
spurt, 5309
surface, 1763
s. covered with cartilage, 182
surgeon, 3820-1
surgeon general, 3819
surgeon's forceps, 3708
surgeon's scissors, 3644
surprise, v., 3788
suspenders, 4707
suspension of consciousness, 2676
suspensory, 3603
sustain, v., 5028; 5202
suture, 108
s. (of the skull), 109
suture needle, 3662
swaddling, 5520
**swallow**, v., 3880; 5116
s. eagerly, 3879
s. down, 4783
s. up, 3886-7

swallow the wrong way, 3888
swamp, 5375
swampy, 5376
swathing band, 5204
**sweat,** *v.,* 1963
sweat, *n.,* 937
s. of agony, 1978
sweating, 1964
s., excessive —, 1973
s., general —, 1974
sweaty, 1979-80
sweet, 4455
s. flag, 4402
s. -meats, 4063
sweeten, *v.,* 4438
swell, *v.,* 2398-9; 2402; 2405; 5336
swelled, to be —, 2406
**swelling,** 1761; 1766; 2400; 5062
s. badly, 2045
s., hard —, 1962
swim, *v.,* 5447
swing, *v.,* 4552
swollen, 1870
swoon, *v.,* 2599
swoon, *n.,* 2598
swoon, to be in a —, 2600
sycosis, 3192
sympathetic, 1618; 1657; 2962; 4138
s. nervous system, 904
sympathy, 836
synarthrosis, 173
synchondrosis, 166
syncope, 2598
synovia, 216
synovial capsule, 199
s. fluid, 215
s. membrane, 217-18
syphilis, 3236
syphilitic, 3005; 5064
s. ophthalmia, 3245
s. orchitis, 3246
syringe, *v.,* 4244
syringe, *n.,* 3731-3; 3742
s. of hard india-rubber, 3734
syringe, ear —, 3748-9
syringing, 4245-7
syrup, 4261
system, 78
s. of bones, 74; 79
s. of the heart, 747

T

table cloth, 4667
t. linen, 4666
t. spoon, 4671

tail, 470
**take,** *v.,* 2390
t. (food or drink), 3895; 4879
t. in addition, 2391
t. away, 3390; 5446
t. care, 4814; 5148; (of), 5136
t. care of one's health, 4887
t. courage, 4826
t. heed, to —, 4814
t. hold of, 5510
t. medicine, 4807
t. no notice of, 5268
t. place, 5027
t. up, 5060
t. upon one's self, 5301
taking, 5229
t. of food, 5047
tallow, 935
tansy, 4476
tape worms, 3541
tar, 4312
tar water, 4313
tarantula, 4505
tarry appearance, of —, 2264
tarsus, 502
tart, 4057
tartar, 5572
tartaric acid, 4303
taste, *v.,* 1433
t. (food or drink), 4879
**taste,** *n.,* 1434
t., sense of —, 1435
t., to have good —, 1497
t., to have lost one's —, 1498
t., be out of —, 1498
**tea,** 4079
t., black —, 4082
t., camomile —, 4346
t., elder —, 4348
t., fennel —, 4347
t., flax-seed —, 4351
t., green —, 4081
t., lungwort —, 4350
t. of medicinal herbs, 4345
t., pectoral —, 4344
t., peppermint —, 4349
t., primrose —, 4350
t. spoon, 4672-3
teach, *v.,* 5316
tear, *n.,* 270; 3342
tear, *v.,* 3376; 3341
t. up, to —, 3381
tearing, 1701
teat, 377; 1232
**teeth.** caries of the —, 3491-2

teeth, cut one's —, 321
t., fungus of the —, 5567
t., grinding of —, 2481
t., looseness of the —, 2895
t., rottenness of the —, 2894
t., row of —, 5544
t., set of —, 292
t., set of false —, 293
t., tartar of the —, 2899
teething, 327
t. convulsions, 2903
tell, *v.,* 4837; 4866
temper, ill —, 2626
temperament, 33-4
temperature, 2102
t., high —, 2105
t. to be lowered, 2106
temple, 233
temporal fossa, 339
tench, 4001
tender, 385; 1859
tender (of sick persons), 3846-8
tenderness in the pit of the stomach, 1822
tendinous, 584
tendinous membrane, 542
tendon, 208; 541
tendon of Achilles, 620-1
tenesmus, 2247; 5281
tense, 1992; 1863
tension, 1833
tepid, 5246
terrified, 2507
terror, 5477
testicle, 1191
tetanus, 2746
tetter, 1945; 3201
tetters, 1932
texture, 85
therapeutics, 3803
therefore, 5596
thermal waters, 3766
thick, 484
thigh, 478; 485
th., lower part of the —, 497
th., upper part of the —, 483
thigh-bone, 479
thin, 1867; 2431
thin, to make —, 5363
thing, 4829
think, *v.,* 2646; 5559
thirst, 2171
th., deficient sensation of —, 2179
th., feeling of —, 2173
th., morbid —, 2178

thirsty, 2172
thirsty, to get —, 4876
thoracic cavity, narrowing of —, 1002
thorax, 365 ; 374
thorn-apple, 4516
thoroughly, 5571
thought, 2664
thread, 3545
thread-worms, 3546
thready, 2007
throat, 310 ; 1009
th., sore —, 4974
th. trouble, 2781
th., obstruction in the —, 1795
throes, 1286
thrombus, 3358 9
throng, v., 2246
through, 5119
throughout, 5119
throw, v., 2344
throw back, 5534
thrush, 2891-2
thrust, v., 736; 2144
thrust out, 2484
thumb, 442
th. lancet, 3649
tibia, 495-6
tic douloureux, 2780
tickle, v., 1228
tickler, 1229
tickling, 1954 ; 5103
tie, v., 4755 ; 4760
t. round, 3611
t. together, to —, 4761
t. under, 3360
t. up, 4762-3
tight, 219; 1863; 4178
till, not —, 5099
till now, 4960
tilt over, to —, 3470
time, 1266
t., at a —, 5264
t., at the —, 5494
t., at the same —, 5008; 5249
t., before the proper —, 1272
t., for the first —, 4932
t., from — to —, 2680 ; 5574
t., since that —, 5079
t., this —, 4926
t., wrong —, 1267
timed, wrong-, 1268
times, at —, 2680
tincture, 4262
t. of camphor, 4434-5
t. of soap, 4436
tinged with yellow, 2252
tingle, v., 2563
t. my ears, 1484
tinkle, v., 2563
tinnitus aurium, 2565-9
tipple, v., 3902
tippler, 2708
tired, 2443
t., to grow —, 2441 ; 2444
t., to make —, 2441
tissue, 85
t., adipose —, 929
t., bone —, 86
t., connective —, 540
t., investing cellular —, 914
t., muscular —, 518
t., vascular —, 623
toad, 4506
toast (bread), v., 3940
toast-bread, 3944
tobacco fumes, 5164
t. leaves, 4343
toe, 510
t., big —, 511
t., great —, 511
t., little —, 512
tænia, 3541
together with, 5488
toilet, 4680
Tokay, 4092
tolerate, v., 2164
to-morrow, 4874
tone, 2055
tongs (pair of), 3706
tongue, 328
t., back of the —, 333
t., base of the —, 331
t., bite one's —, 1491
t., border of the —, 1438
t., end of the —, 332
t., fur of the —, 2143
t., ligament of the —, 329
t., papilla of the —, 1440
t., raspberry —, 2147
t. scraper, 3720
t., strawberry —, 2147
tonic, 2744
tonsillitis, 2921-2
tonsils, 1455
tooth, 294
t., back —, 306
t. -brush, 4592
t., canine —, 297
t., corner —, 297
t., crown of the —, 300
t., decayed —, 2898
t., deciduous —, 301
t. -drawer, 3685
t. -forceps, 3116
t., front —, 305
t. -gap, 314
t., incisive —, 299
t., jaw —, 295
t., milk —, 301
tooth, molar —, 295
t. powder, 5577
t., socket of a —, 307
t., wisdom —, 304
toothache, 1791-2
torment, 1604 ; 1606
torment, v., 1603 ; 1605 ; 5232
torments, full of —, 5248
torpid, 2756
torpidity, 2457-8
torpor, 2457-8 ; 2660
torticollis, 3081
tortoise, 4012
torture, 1604 ; 1608
torture, v., 1607
torturing, 1704-5 ; 1716
toss, v., 2724
toss about, to —, 553
total, 2602
totter, v., 2366 ; 2369
tottering, 2455
touch, 1500
touch, v., 2123 ; 2572 ; 2765
t. upon, to —, 5408
tough, 2087
toward the front, 2358
towel, 4596
toxicology, 4496-7
trachea, 1010
t., bifurcation of the —, 1022
trachelagra, 1796
trade, 4908
tragacanth, 4472
trance, 2767-8
transfer, v., 5550
transferable, 1645
transient, 2683 ; 4990
traumatic, 1667 ; 4229
treat, v., 3807
treatment, 3808
t., to be under —, 4115
tree, 4488
trembling, 2371 ; 2479
tremor, 2479
t., muscular —, 2474
trepan-key, 3727
trephine, 3726 ; 3728
trepidatio cordis, 2852
trichina, 3555
trickle, v., 1921 ; 1981 ; 5437
tricocephalus dispar, 3553-4
trim a candle, to —, 4615
trivial, 5042
trochanter, 364
t., great —, 489
t., lesser —, 490
trochlearis, 1317

334

trouble, 1289; 1547; 2375; 4952
trouble, v., 2155
troublesome, 5068
trout, 4005
true, 155
trumpet, 1221
trunk, 340
truss, 3603; 3607; 3619-21
t. (abdominal) for hernia, 3609
t. for umbilical hernia, 3600
truss-maker, 3840-1
try, v., 4800
tube, 96; 622; 3740-1
tubercle, 1907; 3389
tuberculous disease, 3017
t. d. of the bladder, 3045
tubular, 97
tug, v., 3697
tumefied, 1870
tumor, 2400; 3215
t. of the brain, 2778
t. in the throat, 2789
t. of the tongue, 2906
tune, v., 2623
t., to be in —, 2623
tunic, 530; 601
tunica uvea, 1367
turbid, 2253
turgescence, 2407-8
turgid, 2403
turkey, 3989
turn, v., 196; 5525
t. about, 363
t. away, 5525
t. in, 5459
t. inwards, 5432; 5459
t. over, 3470
t. the stomach, 2946
turn-key, 3689
turnips, 4037
turpentine, 4396
t. liniment, 4468
turtle, 4012
tweezers, 3707; 3712
twilight, 5424
twin, 3293
twine, v., 681
twinging, 1719
twirl, v., 363
twist, v., 681; 846; 2978
t. out of shape, 3125
typhus, hunger —, 3123-4
tympanites, 3011
t., chronic —, 3012
tympanum, 1394-6

U

ulcer, 2414; 3233

ulcer of the bladder, 3042
u. of the heart, 2878
u. of the tongue, 2907
ulcerate, v., 2412-13
ulcerated, 2141
u. gums, 2885-6
ulceration, 5086
ulcerous, 2411
umbilical cord, 1292
u. region, 3016
unable, 2355
unadulterated, 5595
unaltered, 2500
unbind, v., 4765
unbutton, v., 4758
unceasing, 2175
uncertain, 2365; 5000
unclean, 4806
uncommon, 4793
unconcealed, 4839
unconfined, 2343
unconscious of, 5385
unconsciousness, 2595; 2597
uncover, v., 5512
under, 395
under-bed, 4566
under-dress, 4720
undergarment, 4720
undergo, v., 5301
understanding, 2662
undertake, v., 4896; 5458
undress, v., 4744; 4748-50; 4751
undress one's self, 4745
undulating, 2010
uneasiness, 1538; 1602
u. after meals, 2210
u. of the mind, 2626
uneasy, 1601
u. feeling in the abdomen, 2228
u., to feel very —, 1539
u., to make —, 4844
u. sensation, a mere —, 1752
unendurable, 1703
unequal, 1996
uneven, 1996
unfermented, 5331
unhealthy, 1533
uniform, 5006; 5171
unilateral, 2737
unimportant, 5042
uninterrupted, 5314
union, 138
unite, v., 137; 3353
United States, 4906
unlace, v., 4764
unless, 5360
unmarried, 4910
unmeaning, 2521
unmixed, 5325

unnoticed, 5267
unpleasant, 1601
unquenchable, 5230
unravel, v., 3697
unreasonable, 5046
unripe, 1168
unsound, 1533
unsteady, 2365
untie, v., 4754; 4765; 4772; 4773
untimely, 1268
untouched, 5211
unwell, 1534
u., feel one's self —, 1528
upper, 394
upright, 2340
uræmia, 3031
urates, 2286
urea, 1144
ureter, 1152
urethra, 1160
u., structure of the —, 3055
urethralgia, 1841
urethral ulcers, 3080
urethritis, 3052-4
uretic phosphorus, 2288
uric acid, 1145
urine, v., 1138-40
urine, 1136-7; 2279
u., constituents of the —, 2283
u., to discharge —, 2276
u., excessive flow of —, 2333
u., incontinence of —, 3061
u., inspection of —, 2305-6
u., involuntary discharge of —, 2332
u., retention of —, 2326-8
u. sand, 2300-1
u., spontaneous discharge of —, 2332
u., suppression of —, 3056
u. of vesical catarrh, 2280
urinary apparatus, 1141
u. calculi, 3039
u. passages, 1153-4
u. salts, 2285
u. secretion, 1143
u. vessels, 1142
urinate, v., 2276; 2278
uriniferous canaliculés, 1155
urinous, 2281
uroscopy, 2305-6
urtica, 1908
urticaria, 3168-9
use, v., 5034
use, 5187
use (of food, drink, etc.), 5047

use of, to make —, 4113; 4961; 5172
use of (certain) remedies, make —, 4114
use of mineral waters, 4119-20
use up, 2449; 5543
used, 5034
useful, 5286
u. ligament, 1216
u. spasm, 3276
u. syringe, 3750
uterus, 1212; 1220
u., cavity of the —, 1214
u., neck of the —, 1213
utter abruptly, to —, 2484
uvula, 5091

## V

vaccina, 3146-7
vaccinate, v., 3149 ]
vaccinating needle, 3676
vaccination, 3150-2
vaccine matter, 3153
vagina, 546; 1207
vaginitis, 3278
valerianic acid, 4304
valve, auricular —, 635
v., auriculo-ventricular —, 636
valvular affection, 2883
vanilla, 4446
vapor, 4581
vapors of mercury, 5088
varicella, 3154-5
variola, 3135-6
varioloid, 3137; 3156
variolous, 3157
v. virus, 3499
various, 5085
varnish, 316
varying in intensity, 1738
vascular membrane, 1343-6
v. texture, 623
vasomotor system, 903
vault, v., 2428
vaulting, 2429
veal, 3957
veal-cutlet, 3975
vegetable, 3519
vegetables, 4027
v. (cooked for eating), 2170
vein, 662; 692-3
velum palatinum, 1452
vena cava, 774
venereal, 3237
v. crown, 3248
v. disease, 3236; 3238

venison, 3960
venom, 3494
v. of serpents, 3500
venomous animals, 4498
v. herbs, 4508
v. plants, 4507
v. serpents, 4499
venous, 711
vent-door, 4623
ventilate, v., 5124
ventricle, 629-30
v., dilatation of —, 2867
verdigris, 4526
verily, 5489
vermiform process, 1122
vermin, 3530
vermis, 3535
verruca carnea, 3373
vertebra, 341; 346; 349-50
v., abdominal —, 380
v., coccygeal —, 477
v., dorsal —, 349-50
v., lumbar —, 381
v., sacral —, 469
v., thoracic —, 371
vertebrae, cervical —, 354
v. of the neck, 354
vertebral articulation, 342
v. canal, 877
v. column, 344; 348
vertex, 225-6
vertigo, 4798
vesicatory, 4426
vesicle, 1218; 1909; 1924
vesper time, 5179
vessel, 622
vestibule, 631-2; 1400
v. window, 1402
veterinary science, 3834
v. surgeon, 3832-3
viands, 3917
vibrating, 2009
victuals, 3919
view, v., 2303-4
vigor, 17
vigorous, 1554; 2352
villous, 957
vinegar, 4066
violence, 5147
violent, 1626; 1628
violet, 1891
viper, 4504
virgin, 1225
virility, 1171
virus, 4525; 4532-4
v., a rabid —, 3495
visage, 241
viscera, 919
viscid, 2087
viscous, 2087
vision, double —, 2560

visit, v., 4851
visual organ, 1308-9
vital, 45
v. strength, 47
vitreous body, 1377
v. humor, 1378-9
v. membrane, 1380
Vitus dance, St., 2774
vocal cord, 5071
v. organs, 1457
voice, 1458
voice (organs), v., 2623
voluntary, 552
voluntas, 2655
voluptuous, 1248
voluptuousness, 1247
vomer, 289
vomit, v., 2181-3
v., make an effort to —, 2190
v., provocation or tendency to —, 2189
vomiting, bilious —, 2216
v. of blood, 5138
v. of stercoraceous matter, 2222-3
vulnerary balsam, 4421
vulva, 1208

## W

wadding, 4182
wagon, 3567
waistcoat, 4683
wakefulness, 2589-90
wake up, to —, 5152
walk, 2361
walk, v., 2350
walk along, to —, 2351
walking in sleep, 2610-12
wall, 127
walnut, 4053
wan, 1591
wandering, 2520
want, 4951
want, v., 4788; 5357
want of, for —, 5462
wanting, to be —, 1552; 4852; 4976
ward off, to —, 4208
wardrobe, 4638
warm, 51
warm, to make —, 4353
warming-pan, 4582-4
warmth, 52; 2206
warmth, vital —, 53
wash, 3800; 4266
wash, v., 5532
wash-basin, 4589
washings of meat, resembling the —, 2266

washstand, 4587-8
wash out, to —, 5532
waste, v., 2896
waste, n., 2436
waste away under sickness, 1569
watch, v., 5538
water, 705
w. blebs, 3183
w. closet, 4597-8
w. -porridge, 4070
w. -proof, 4687; 5397
water, to make —, 2277
watering pot, 1467
watery, 2533
waver, v., 2369
wax, 4478
waxlike, 2494
wax-stand, 4620
wax taper, 4620
waxy, 1882-3
weak, 1540; 2949; 5423
weak from old age, 1587
weaken, v., 1543
weakly, 1544
weakness, 1542
w., condition of —, 1542
w. of old age, 1588
wean, v., 1305
weaning, 1306
weariness, 2445
wearing apparel, articles of —, 5134
weary, 2443
weary, v., 2444
weave, v., 84
weaving, 85
wedge, 144
wedge in, to —, 145
week, 1274
weeks, for —, 5245
weight, 1832; 2321
well, 5121
well, to be —, 1515
well-fed, 1127
wet, v., 3171; 5056
wet all over, 501
wet-nurse, 1299
wet the swaddling, to —, 5521
what, about —, 4929
what, of —, 4929
wheat, 4481
wheat flour, 4483
wheel-chair, invalid —, 3586
wheeze, v., 2041; 2071
when? 4925
whereas, 5403
whereby, 4934

wherefore, 4930
where from, 4903
wherein, 5007
whey, 4123
whim, 2631
whimsical, 1256; 2632
whistle, v., 2040
white, 1880
white, almost —, 2262
whites, the —, 3272
whiting, 4002
whitish, 2127
whit-low, 3227-8
whiz, v., 2038
whole days, 5560
wick, 4613
wicked, 2630
wide, 999; 3243
widen, v., 1001
widow, 4912
widower, 4911
will, 2655
willingly, 4885
wind, 2195; 2197
wind, v., 846
windings, 847
window, 1401
window-shutter, 4632
wine, 4083
w., Hungary —, 4088
w., red —, 4085
w., Rhenish —, 4086
w., white —, 4084
wing, 1027; 1208
wipe out, to —, 5308
wiry, 2008
wise, a., 302
wise, n., 4921
wisdom, 303
witchcraft, 4141
withdraw, to —, 5403
within, 4978
woe, 1285
wolf, 3216
woman lying in, 1278-80
womb, 1212; 1220
w., mouth of the —, 1215
wonder, 4143
wont, 5034
wood-charcoal, 4321
woodland air, 5169
wool, 4487
work, v., 567
worm, 3535
worms, 3536
w., flat —, 3540
w. -seed, 4440-1
wormwood, 4450
worn out, 2450
worse, to make —, 1758

wound, v., 3303
w. from a chisel, 3324
w. powder, 3702
w., contused —, 3329
w., cross —, 3347
w., edge of a —, 3349
w., fatal —, 3310
w., gun-shot —, 3326
w., furrowed gun-shot —, 3331
w., incised —, 3320
w., lacerated —, 3343
w., longitudinal —, 3346
w., margin of a —, 3349
w., oblique —, 3349
w., penetrating —, 3348
w., punctured —, 3322
w., slight —, 1928
w., suppurating —, 3310
w., transverse —, 3347
w., ulcerating —, 3310
wounding, 3304
wrap, v., 3610
wrap up, 528; 913
wrap up in, 4180
wrapper, 4723
wrapping up, 4181
wrath, 1123
wrenching, 3428
wring out, to —, 5193
wrinkle, 1899
wrinkle, v., 1319
wrinkled, 1864
wrist, 432-3
write, v., 4900
writing-desk, 4644 5
wrong, 3295
wry-neck, 3081

Y

yellow, 1894
yellow, bright —, 2308
yellowish-brown, 2260
yellowish-green, 1895
yield, v., 5107
yielding, 1993
yoke, 266
yolk, 4020

Z

zona, 3179-81
zone, 3178
zonula ciliaris, 1371
zoo-sperm, 1196
zygoma, 266

www.ingramcontent.com/pod-product-compliance
Lightning Source LLC
Chambersburg PA
CBHW031848220426
43663CB00006B/537